FRACTALS AND MULTIFRACTALS IN ECOLOGY AND AQUATIC SCIENCE

FRACTALS AND MULTIFRACTALS IN ECOLOGY AND AQUATIC SCIENCE

LAURENT SEURONT

CRC Press
Taylor & Francis Group
Boca Raton London New York

CRC Press is an imprint of the
Taylor & Francis Group, an **informa** business

CRC Press
Taylor & Francis Group
6000 Broken Sound Parkway NW, Suite 300
Boca Raton, FL 33487-2742

First issued in paperback 2017

ISBN 13: 978-1-138-11639-9 (pbk)
ISBN 13: 978-0-8493-2782-7 (hbk)

Library of Congress Cataloging-in-Publication Data

Seuront, Laurent.
    Fractals and multifractals in ecology and aquatic science / author, Laurent Seuront.
      p. cm.
    "A CRC title."
    Includes bibliographical references and index.
    ISBN 978-0-8493-2782-7 (hardcover : alk. paper)
    . Biomathematics. 2. Mathematics in nature. 3. Ecology--Mathematics. 4. Aquatic sciences--Mathematics. 5. Fractals. 6. Multifractals. I. Title.

QH323.5.S458 2010
570.15'1--dc22

2009025699

Visit the Taylor & Francis Web site at
http://www.taylorandfrancis.com

and the CRC Press Web site at
http://www.crcpress.com

Dedication

To my sweetheart, for the past, present, and future.

Contents

Preface

This book originally owed its existence to chance, although it also represents a fair amount of work. Apart from a brief encounter with fractals as an undergraduate in marine biology in 1992, through the reading of a paper entitled "Applications of Fractal Theory to Ecology" by S. Frontier picked up by pure chance on the dusty shelf of a university library, I had never heard the word *fractal* and was struggling to find something really appealing in my studies. The seed was planted, though, and my main stimulus came a few months later from my very first encounter with Professor Serge Frontier. He was to become the mentor—without him, I would probably never have been offered the chance to try to apply fractals and multifractals to the characterization of plankton patchiness in turbulent flows for my Ph.D. thesis. His legacy is everywhere in this volume, and I wish to thank him for his inspirational guidance that has lasted over the years.

I also greatly benefited from passionate discussions on fractals shared with Christophe Luczak during the stammering of our scientific careers in the mid-1990s. I am indebted to Jim Mitchell who persuaded me to turn a rough manuscript into the present book. His friendship and support extend well beyond the piece of work involved in this volume. A few significant encounters contributed to my personal and professional developments, and as such I wish to thank here, Mark Doubell, Li Hua, Sophie Leterme, Justin Seymour, Yugi Tanaka, Raechel Waters, Dr. Fabian "95%" Wolk, and Hidekatsu Yamazaki.

I wish to thank John Sulzycki from CRC Press / Taylor & Francis Group for his patience and trust.

Last, but not least, I thank my parents for their indefectible support despite their ongoing struggle to understand my passion for science.

About the Author

Laurent Seuront, Ph.D., is a professor in biological oceanography at Flinders University (Adelaide, Australia) and a senior research scientist at the South Australian Research and Development Institute (West Beach, Australia). His education includes a B.S. in population biology and ecology from the Université des Sciences et Technologies de Lille (1992); an M.S. in marine ecology, data analysis, and modeling from the Université Pierre et Marie Curie, Paris (1995); and a Ph.D. in biological oceanography from the Université des Sciences et Technologies de Lille (1999). Prior to his present position, Dr. Seuront was a research fellow of the Japanese Society for the Promotion of Science at the Tokyo University of Fisheries (1999–2000) and a research scientist at the Centre National de la Recherche Scientifique (CNRS) in France (2001–2008).

Dr. Seuront's current research concerns biological–physical coupling in aquatic and marine system environments, with a focus on the effect of microscale (submeter) patterns and processes on large-scale processes. Aspects of his work combine field, laboratory, and numerical experiments to study the centimeter-scale distribution of biological (nutrient, bacteria, phytoplankton, microphytobenthos, and microzoobenthos) and physical (temperature, salinity, light, turbulence) parameters, as well as the motile behavior of individual organisms in response to different biophysical forcings. His work to date has been the subject of over 100 publications in international journals and contributed books, more than 80 presentations at international conferences, and invited seminars at over 50 locations throughout the world, including at prestigious institutions such as Cambridge University and the Massachusetts Institute of Technology (MIT). Among multiple awards, Dr. Seuront recently received the CNRS Bronze Medal in France (2007) in recognition of his early career achievements, and a prestigious Australian Professorial Fellowship from the Australian Research Council.

1 Introduction

As suggested by the titles of several seminal earlier works, such as *The Fractal Geometry of Nature* (Mandelbrot 1983) and *Fractals Everywhere* (Barnsley 1993), one could expect to find fractal and multifractal properties everywhere. This might indeed be the case, as the Science Citation Index returned more than 12,500 and 1,600 articles respectively containing *fractal* and *multifractal* in their title between 1987 and 2008.

The fractal character of a system has many implications for its properties. In particular, a fractal set tends to fill the whole space in which it is embedded and has a highly irregular structure, while it possesses a certain degree of self-similarity; that is, when viewed at increasing levels of magnification, a fractal set appears to be the union of many ever smaller copies of itself (Figure 1.1). This character is captured by the so-called fractal dimension, D_F, of the set and can be regarded as one measure of the complexity of the system and as the degree at which a set fills the Euclidean space in which it is embedded. The compelling reasons for the emerging fractal theory in many scientific fields are based on the hope that complex systems could be explained using a relatively low number of parameters, say, the fractal dimension D_F.

One of the most illustrative processes where the concept of fractals applies relates to atmospheric and oceanic turbulence. Originally, the word *turbulence* referred to the random motion of a crowd, *turba* being the Latin for crowd. The complex and self-similar nature of flows is present in the early work of Leonardo da Vinci (1442–1519). In his study of flowing and running water (1508–1510), he captured the transition from ordered to chaotic fluid motions (Figure 1.1A) and the complexity and self-similar nature of turbulent flows (Figure 1.1B). Later pictorial work by Katsushika Hokusai (1760–1849) (Figure 1.2A), Utagawa Hiroshige (1797–1858) (Figure 1.2B,C), and Vincent van Gogh (1853–1890) (Figure 1.2D) clearly expressed the multiscale nature of the surface of the turbulent ocean (Figure 1.2A,B,C) and atmospheric flows (Figure 1.2D). The self-similarity evidenced by Jonathan Swift in his satiric verse, "So, Nat'ralists observe, a Flea Hath smaller Fleas that on him prey,

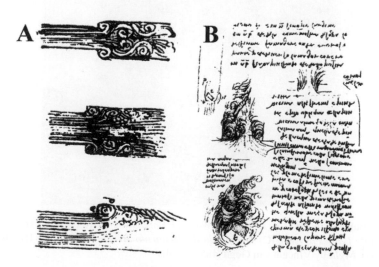

FIGURE 1.1 Flowing and running water (1508–1510) by Leonardo da Vinci (1442–1519).

FIGURE 1.2 Turbulence in art. (A) *The Great Wave of Kanagawa* by Katsushika Hokusai (1760–1849); (B) *Naruto Straight Eddies at Awa*; (C) *Vortices in the Konaruto Stream* by Utagawa Hirshige (1797–1858); and (D) *The Starry Night* by Vincent van Gogh (1853–1890).

And these have smaller yet to bite 'em, And so proceed ad infinitum" (Figure 1.3A), seemingly led to the crude, but picturesque, seminal description of turbulence by L. F. Richardson (1922), "Big whirls have little whirls that feed on their velocity, and little whirls have lesser whirls and so on to viscosity" (Figure 1.3B). This qualitative description of fully developed turbulence was later formalized by the Kolmogorov turbulent cascade, which describes how turbulent kinetic energy generated at large scales L, cascade through a hierarchy of eddies of decreasing size down to the viscous scale l_k, where energy is dissipated into heat (Figure 1.3C). This self-similarity of turbulent flows is clearly visible from natural and simulated turbulent flows (Figure 1.4), as well as from large-scale patterns such as von Karman vortex streets and cloud cover (Figure 1.5).

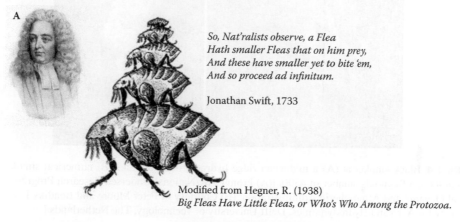

So, Nat'ralists observe, a Flea
Hath smaller Fleas that on him prey,
And these have smaller yet to bite 'em,
And so proceed ad infinitum.

Jonathan Swift, 1733

Modified from Hegner, R. (1938)
Big Fleas Have Little Fleas, or Who's Who Among the Protozoa.

Williams and Wilkins, Baltimore.

Big whirls have little whirls that feed on their velocity,
and little whirls have lesser whirls and so on to viscosity

Lewis Fry Richardson, 1922

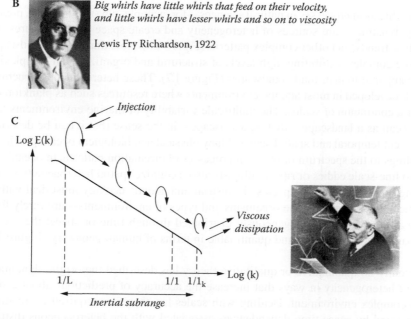

FIGURE 1.3 The verse of the Irish satirist Jonathan Swift on fleas, originally meant as a swipe at lesser poets (A); parodied by the meteorologist L. F. Richardson to describe the multiscale nature of turbulence (B); which was later formalized by the Kolmorogov turbulent cascade (C).

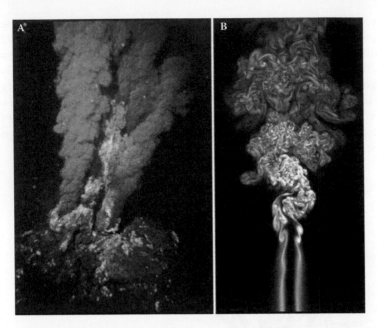

FIGURE 1.4 Black smoker at (A) a midocean ridge hydrothermal vent and (B) numerical simulation of a turbulent jet at a Reynolds number of 4500. [(A) From OAR/National Undersea Research Program (NURP), National Oceanic and Atmospheric Administration. (B) Courtesy of Peter Moore and Bendiks Jan Boersma, Laboratory for Aero and Hydrodynamics, Delft University of Technology, The Netherlands.]

More specifically, most processes in natural sciences—for example, physical forcings, population and community dynamics—are sources of heterogeneity and create space-time structures such as gradients, patches, trends, and other complex patterns (Figure 1.6). Note that natural landscapes are consistently more complex, exhibiting high levels of structural and organizational complexity that can barely be compared to man-made landscapes (Figure 1.7). These heterogeneous structures are particularly well developed in most aquatic environments where resources such as plankton exhibit patchiness over a continuum of scales. The multiscale variability of marine environments leads to a view of the ocean as a landscape—that is, a seascape—in the sense that it can be described by patterns of different temporal and spatial scales. Many physical and biological oceanographers then relate their findings to the spectrum of physical processes of circulation patterns in oceanic basins or large gyres to fine-scale eddies or rips. Ecologists also recognize spatial heterogeneity as a major factor regulating the distribution of species. Terrestrial and aquatic ecology must deal with scale, because the objects it focuses on—the organisms and types of environments—are rarely found to have regular shapes and to be homogeneously distributed through time or space; the "geometry of Nature" is barely understandable and quantifiable in terms of human geometry (Figure 1.8 and Figure 1.9).

Yet until recently no quantitative or qualitative theory has described the origin, dynamics, and consequences of heterogeneity in ways that increase the accuracy of predictions about ecological processes in a complex environment. Dealing with scales has therefore required overcoming the difficulties generated by space-time dependencies associated with the heterogeneous distribution of ecological variables. Classical statistical theory works well to predict changes in variance due to different sizes of sampling units or different grains of sampling strategies when the sampling units are independent. The basic independence of replicates assumption, however, is rarely verified in natural science, and therefore the use of classical theory is questionable. Moreover, the more traditional, widely used mathematical descriptors have little meaning in a multiscale spatial context.

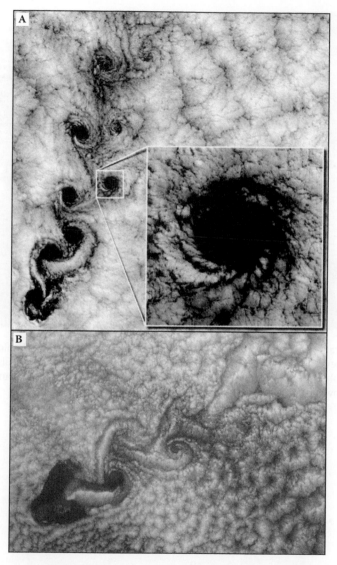

FIGURE 1.5 Self-similarity in the atmospheric von Karman vortex streets observed off the Mexican coast near Guadalupe Island on June 11, 2000 (A) and off the Chilean coast near the Juan Fernandes Islands on September 15, 1999 (B). Also note the self-similarity in cloud patterns. [(A) From Robert F. Cahalan, NASA/GSFC (see Cahalan et al. 2001, http://climate.gsfc.nasa.gov/viewPaperAbstract.php?id=69). (B) From NASA/GSFC/JPL.]

Scale is undoubtedly one of the central themes of landscape ecology (see, for example, Peterson and Parker 1998; Wiens 1989, 2001). Most, if not all, of the landscape properties playing a role in the biology and ecology of populations (such as gradients, patch quality, boundaries, connectivity, and organism response) change with changes in scale. The notion of scale (*sensu lato*) is quite broad and involves a wide range of terms and concepts that can be clustered under key categories such as heterogeneity, hierarchy, and size. The first one (heterogeneity) includes spatial patchiness and temporal variability, and has been acknowledged as an essential property of nature (Kolasa and Pickett 1991). The second one (hierarchy) is an intrinsic property of ecosystems, which are always hierarchically organized (O'Neill et al. 1986; Kolasa 1989), and this implies the consideration of an organizational scale. Finally, the evident, thought widely neglected, size-dependence of species'

FIGURE 1.6 Natural landscapes and seascapes. (A) Vertical vegetation zones in Honfleur Harbor, France; (B) a "snail's-eye-view" of an intertidal rocky shore, Lincoln National Park, South Australia; (C) reefs built by the colonial tube-polychaete *Ficopomatus enigmatus* in the Coorong, South Australia; and (D) local alternation between meadows and thickets, La Cauchie, France.

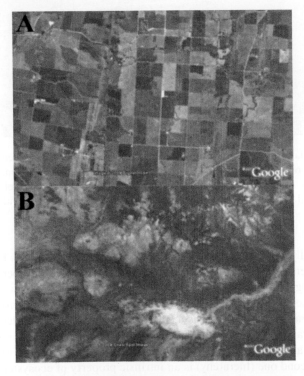

FIGURE 1.7 South Australian landscapes, with (A) and without (B) anthropogenic influences. Both pictures were taken from an altitude of 20 km. (**See color insert following page 80.**)

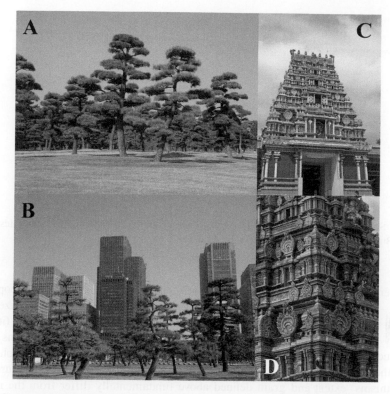

FIGURE 1.8 Contrast between the complexity of the geometry of nature (A) and the Euclidean geometry of human architecture (B) in Tokyo (Japan), and an example of self-similarity in Hindu architecture, the Sri Siva Subramaniya Temple (Nadi, Fiji) in an ensemble (C) and detail view (D).

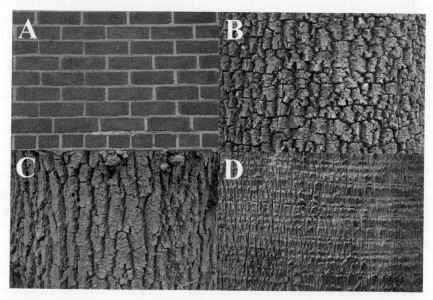

FIGURE 1.9 Contrast existing between the geometry of a man-made surface, a brick wall (A) and (B) the bark patterns of white fig (*Ficus virens*), (C) the English oak (*Quercus robur*), and (D) the cotton palm (*Washingtonia filifera*). **(See color insert.)**

FIGURE 1.10 Self-similar rabbits, originally used to illustrate concepts in population dynamics. (Courtesy of Professor M. Bull, Flinders University, Adelaide, Australia.)

features (Peters 1983) is critical to our understanding of how organisms—hence, populations and communities—respond to the abiotic and biotic properties of their environment. The size of the "window" through which an organism views or responds to the structure of its landscape (its extent), for example, may differ for organisms of different body sizes or mobility, and organisms may discern the patch structure of the landscape within the window with different levels of resolution (grain). As a result, the organism-defined "landscape" is intrinsically scale dependent (Figure 1.10). Note that the organisms' extent and grain defined above fundamentally differ from the measurement scales imposed by the observer.

In heterogeneous data sets, where estimates of quantities such as biomass vary precisely with the scale at which measurements are made, fractal dimension then appears to be a useful measure of space-time complexity and provides several advantages over other descriptive indices of ecological patchiness. However, despite some insightful description of various possible applications of fractals in ecology (Frontier 1987; Sugihara and May 1990a; Hastings and Sugihara 1993) and successful applications in landscape ecology, entomology, and behavioral ecology, they are still hardly ever used. The situation is even more dramatic for multifractals, where use is often restricted to the fields of nonlinear dynamical systems, fully developed turbulence, rainfall modeling, spatial distribution of earthquakes, financial time-series modeling, and Internet traffic modeling. Applications of multi-fractals in ecology appear limited to a few papers published over the past 10 years dealing with the characterization of the dynamics of forested systems (Scheuring and Riedi 1994; Solé and Manrubia 1995, 1996; Manrubia and Solé 1996; Drake and Weishampel 2000, 2001), patchiness of marine systems (Pascual et al. 1995; Seuront et al. 1996a, 1996b, 1999, 2001, 2002; Seuront and Schmitt 2005a, 2005b; Lovejoy et al. 2001; Seuront and Spilmont 2002), and species-area relationships (Borda-de-Água et al. 2002).

Two main reasons are suggested for the still very limited applications of fractals and multifractals in ecology and aquatic sciences. First, the fractal and multifractal formalisms, mainly developed and used in the fields of nonlinear dynamical systems and physical sciences, might be impenetrable, at least for ecologists without a reasonable mathematical and statistical background or for those who do not have the time to devote to such studies. Second, unlike most of the numerical techniques used to analyze spatial data sets and time series, no software is commercially available for fractals and multifractals. As a consequence, the main aim of this book is to bridge the gap between the potentially obscure fractal and multifractal concepts and tools and the end-user ecologists. Fractals and multifractals have thus been theoretically, mathematically, and practically treated at a level that is reasonably accessible to the ecologists willing to fully understand and use them. Detailed considerations on the construction and properties of theoretical fractals, such as the Cantor set, Sierpinski gasket and carpet, Pascal triangle, and Koch curve or the Mandelbrot and Julia sets widely investigated elsewhere (such as Peitgen et al. 1992; Schroeder 1991; Barnsley 1993, 2000; Falconer 1985, 1993) were intentionally omitted to put more focus on real case studies.

The book naturally starts with basic definitions and illustrations related to Euclidean and fractal geometries and dimensions. In particular, a special effort was made to define the too-seldom-used concepts of fractal codimension and sampling dimension. In Chapters 3 and 4, the concepts of self-similar and self-affine fractals are introduced and the fundamental differences existing between them are discussed, as well as the concepts of statistical self-similarity and statistical self-affinity. Chapter 5 introduces a family of fractal dimensions derived from frequency distributions. Chapter 6 has subsequently been devoted to clarify the relationship between fractal theory and concepts such as chaos theory, strange attractors, self-organization, and self-organized criticality. In Chapter 7, the intrinsic limitations of fractal analysis are addressed in detail, and some criteria and easy-to-handle procedures to ensure the relevance of fractal analysis are provided. In Chapter 8, the concept of a multifractal is defined, the different multifractal analysis techniques available are reviewed and exemplified, and a very intuitive, "without the math" multifractal technique is introduced and illustrated using a step-by-step procedure applied to a real case study. The seldom-used joint multifractal framework is also introduced, defined, and illustrated.

It is finally stressed that the motivation to write the present book stems from a report that non-mathematically acquainted ecologists might not be able to appreciate the strength of fractals and multifractals in analyzing their data sets because of the lack of nontechnical—hence, accessible—books on the subjects. As such, the present work has been thought, designed, and written with ecologists in mind. It has been written in a "handbook fashion" to promote the understanding and the use of fractals and multifractals in ecological sciences. More technical sections are nevertheless provided throughout the text for readers interested in getting into the (more mathematical) details of fractal and multifractal techniques. As a consequence, it is, of course, statistically and mathematically colored. As such, the readers willing to get the details behind what could be referred to as the "fractal/multifractal black box" can understand where a given equation comes from. However, what ecologists do care about is ecology! Most of the techniques presented and discussed here have then been illustrated with concrete examples from recent works but mostly using original data sets to allow the readers to understand what they could get out of fractal and multifractal analysis without the hassle of going through the math, or at least before eventually feeling the need to go through the math. The less-mathematical readers will hopefully find the hooks they need to appreciate the strength and usefulness of fractals and multifractals in the field of ecological sciences. Each example has been treated as a short paper, including a description of the species and the system considered, and the experimental procedures used to get the data, before presenting their results and discussing them in an ecological context. More generally, the relevance of fractals and multifractals to describe branched patterns and growth processes, habitat complexity, organism distribution, behavioral processes, predator–prey and population dynamics, turbulent processes, and species diversity and evolution are reviewed, exemplified, and discussed.

2 About Geometries and Dimensions

2.1 FROM EUCLIDEAN TO FRACTAL GEOMETRY

The geometries of shores, rocks, plants, waves, hydrodynamic flow, organism trajectories, and many other natural phenomena are important in different scientific disciplines, and each field tends to adapt specific concepts to describe the complexity of Nature. Ecological models often approach natural shapes as simple geometrical approximations. Lakes are approximated as circles, particles as spheres, patches as squares and rectangles, and trees as cones (Figure 2.1). Many patterns and shapes in Nature, however, are so irregular and fragmented that they present not simply a higher degree but an altogether different level of complexity, as compared with Euclidean approximations. Curves, surfaces, and volumes in Nature can thus be so complex that ordinary measurements become meaningless. Mandelbrot (1977, 1983) coined the term *fractal geometry*, introducing a new concept that has rapidly provided a unifying and cross-disciplinary basis to the description of Nature's complexity. Many natural phenomena have a nested irregularity and may look similarly complex under different resolutions (for example, turbulent water flow or clouds) (Figures 2.2 and 2.3). Although this nested structure, referred to as *scale invariant*, could be thought of as an additional source of complexity, it becomes a source of simplicity in fractal geometry.

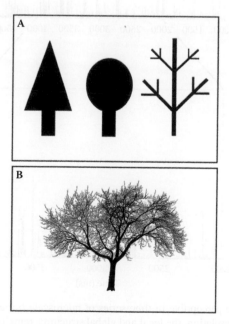

FIGURE 2.1 Illustration of the fundamental differences between human schematic depictions (A) of natural forms such as trees (B).

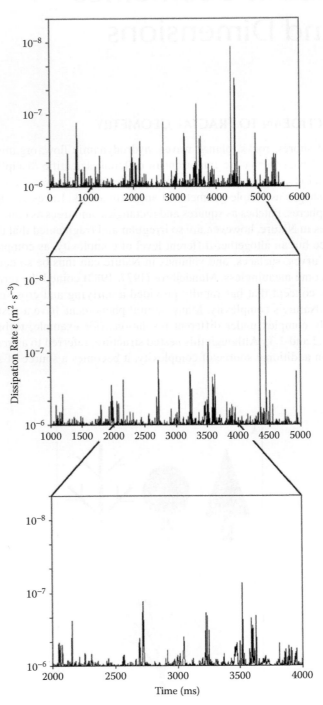

FIGURE 2.2 Nested structure perceptible in time series of microscale turbulent kinetic energy dissipation rates ($m^2 \cdot s^{-3}$). At increasing resolution, the local and global structures remain very similar.

FIGURE 2.3 Nested structure perceptible in the geometry of clouds. At increasing resolution, the local and global structures remain very similar. **(See color insert following page 80.)**

Scale invariance means that the observed structure remains unchanged under magnification or contraction. A scale-invariant pattern is thus *scale dependent* and cannot be characterized by a single scale. Fundamental to most definitions of fractals is then the idea of "measurement at scale δ." For each δ, we measure a set in a way that ignores irregularities of size less than δ, and we see how these measurements behave as $\delta \to 0$. For instance, if one considers a plane curve C, then the measurements, $M(\delta)$, might be the number of steps required by a pair of dividers set at length δ to traverse C. In case of a fractal, the relationship between the measurements $M(\delta)$ and the scale δ must obey a power-law form:

$$M(\delta) = k\delta^{-\phi} \tag{2.1}$$

where k and ϕ are empirical constants, the constant ϕ being referred to as the *scaling exponent*. Taking logarithms, Equation (2.1) can be written as:

$$\log M(\delta) = \log k - \phi \log \delta \tag{2.2}$$

These relationships are appealing for computational and experimental purposes, since ϕ can be estimated as the slope of a log-log graph plotted over a suitable range of δ, and k is the intercept (see Figure 2.4A). Over a wide range of scales—typically many orders of magnitude—the same relationships among critical structural and functional variables are maintained. One may nevertheless note that for real phenomena, we can only work with a finite range of δ. In the ideal case, theory and experiment diverge before an atomic scale is reached, but practically, there may be several scaling regions, separated by breakpoints, that are fractal within each region but failing when a breakpoint is crossed (Figure 2.4B). This question will be studied more thoroughly in Chapter 7. Such similarity is said to be fractal, and the relationship among variables can be described by a fractal dimension or a power law. Although the concept of fractals is fairly new (Mandelbrot 1983), the use of power functions to characterize scaling laws has a venerable history.

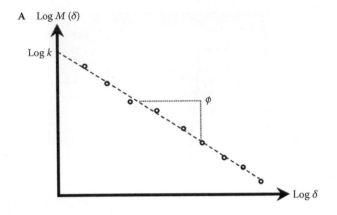

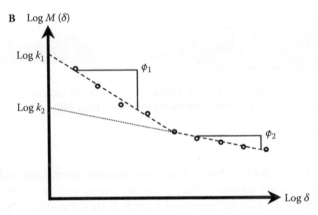

FIGURE 2.4 Schematic illustration of the expected behavior of $M(\delta)$ vs. δ in a log-log plot in case of a single scaling regime (A) and multiple scaling (B). The slope of the linear parts of the graph provides an estimate of the scaling exponent ϕ.

In particular, the use of power laws is so well established that they are called *allometric equations* (McMahon and Bonner 1983; Schmidt-Nielsen 1984). They are of the form:

$$y = ax^b \tag{2.3}$$

where y is some dependent variable, a is a normalization constant, x is some independent variable (typically a body mass), and b is referred to as the scaling exponent introduced above.

Biological scaling relationships are called *allometric* because the exponent, b, typically differs from unity. If $b = 1$, the relationship is called *isometric*, and it plots as a curve on linear axes. When $b \neq 1$, the relationship is called allometric, and it plots as a curve on linear axes. However, power functions have the nice property that they are linear when plotted on logarithmic axes. This is readily seen by taking the logarithms of both sides of Equation (2.3)

$$\log y = \log a - b \log x \tag{2.4}$$

which is conceptually equivalent to Equation (2.2). This is equivalent to the equation for a straight line, where the dependent variable, $\log y$, is equal to an intercept, $\log a$, plus the product of the slope, b, times the independent variable, $\log x$. As stated above for Equation (2.2), the scaling exponent of the power function is the slope of the linear plot on logarithmic axes. The mathematical equivalence of Equations (2.3) and (2.4) means that it is fairly straightforward to derive empirical allometric relationships by using a least-squares regression technique to fit a linear regression to log-transformed data. We will, nevertheless, see in Chapter 7 that great care must be exercised to choose the appropriate regression procedure.

The most familiar example of allometry is simple geometric scaling. If we have spheres or any objects of self-similar shapes, one can describe changes in surface area, A, or volume, V, as a function of a linear dimension, the radius, r, as follows:

$$A = \pi r^2 \tag{2.5}$$

and

$$V = \frac{4}{3}\pi r^3 \tag{2.6}$$

in which the scaling exponents are the dimensions of the objects. In addition, if the objects maintain a constant density as they vary in size, then the mass, M, is proportional to the volume, V (that is, $M \propto V$), and we can express their linear dimension, L, or surface areas, A, as functions of their mass M as:

$$L = c_1 M^{1/3} \tag{2.7}$$

and

$$A = c_2 M^{2/3} \tag{2.8}$$

where the values of the normalization constants, c_1 and c_2, depend on the units of measurements. Since the same equations apply to any shape, if living organisms preserve self-similar shapes as they vary in size (Peters 1983; Schmidt-Nielsen 1984), then their linear dimensions and their surface areas should vary as one-third and two-thirds of their body mass respectively, but with some intrinsic restrictions.

Although the power laws shown in Equations (2.7) and (2.8) provide sharp limits on the form and metabolic requirements of many families of living organisms, organisms do not usually exhibit such simple geometric scaling as expected, for example, in the ideal case of size-nested painted wooden dolls from Russia. This is because there are powerful constraints on structure and function that do not allow organisms to maintain the same geometric relationships among their components as size changes over several orders of magnitude. This was pointed out by Galileo, who noticed that some laws of physics and biology are not necessarily unchanged under changes of scale. Referring to the strength of bones, he argued that an animal twice as long, wide, and tall will weight eight times more. He nevertheless pointed out that bones that are twice as wide have only four times the cross-section and can only support four times the weight. Thus, to support the full weight, bone width must be scaled by a factor greater than 2. This deviation from simple similarity introduces a natural scale in the design of organisms, both animal and vegetal, land bound and aquatic. At some roughly predictable size, the bones become larger than the rest of the animal, and scaling breaks down; see Haldane (1928). For instance, as trees increase in size, the cross-sectional areas of their trunks and the total surface areas of their leaves increase more rapidly than expected from purely geometric considerations, as $M^{3/4}$ rather than $M^{2/3}$. The differential increase in trunk area provides for mechanical resistance to buckling due to gravity and wind, while the scaling of leaf area allows for increased gas exchange to support the increased phytomass. Similarly, as mammals increase in size, there is a differential increase in the thickness of their bones to provide mechanical support and in the surface area of their lungs to provide gas exchange for metabolism. Another well-known instance of scaling in biology is the energy dissipation of homeothermic animals as a function of their weight or mass. A "naïve" approach would expect the energy dissipation, E, to be proportional to the animal's surface area, which for similar animals is proportional to the two-thirds power of its volume or mass M (see Equation 2.8), leading to $E = C_3 M^{2/3}$ where C_3 is a constant. However, it appears that larger animals dissipate more energy than the relation $E = C_3 M^{2/3}$ would predict. The data for a wide variety of species, ranging from unicellular organisms to whales, are much better fitted by an exponent of 3/4 (Peters 1983; Schmidt-Nielsen 1984), suggesting that larger animals are less energy efficient.

Although the origin and the phenomenological relevance of the previous allometric (scaling) exponents have been widely discussed elsewhere (Brown and West 2000), one must note here from the comparison of Equations (2.5) and (2.6) with Equations (2.1) and (2.3) that the scaling exponents ϕ and b are conceptually similar to a *fractal dimension*. Before going further into the refinements of fractal geometry, I will discuss extensively the meaning of what we usually call a "dimension"— with regard to "dimensions" in the fractal framework specifically—and some related concepts.

2.2 DIMENSIONS

2.2.1 EUCLIDEAN, TOPOLOGICAL, AND EMBEDDING DIMENSIONS

2.2.1.1 Euclidean Dimension

We learn from an early age that lines and curves are one-dimensional, planes and surfaces are two-dimensional, and solids such as a cube are three-dimensional. The concepts refer to the traditional Euclidean geometry and *Euclidean dimension*. More generally, any space that can be conceived of has a characteristic number associated with it called a dimension. But what is a dimension? Surprisingly, despite the *a priori* naïve character of the question, it is far from being easy to provide a complete definition of dimension.

A definition of dimension could be the number of real-number parameters needed to uniquely describe all the points in a space. Thus, the real-number line is one-dimensional as it only takes one parameter to describe each point. Dimension is invariant so that a plane, for example, requires two parameters in rectangular (x, y) or polar (r, θ) coordinates. Other suitable examples come to mind. The set of lines in a plane is two-dimensional, as describing any one of them uniquely requires two parameters: the slope and y intercept or the x and y intercepts, for example. The set of all circles in a plane is three-dimensional (two for the coordinates of the center and one for the radius), and the set of all conic sections in a plane is five-dimensional. More formally, we say a set is d-dimensional if we need d independent variables to describe a neighborhood of any point.

Another way to think of dimension is as the degree of freedom available within the space (see, for example, Grassberger 1983; Hentschel and Procaccia 1983). Physical (Euclidean) space is three-dimensional because there are three independent directions that objects within the space can move (up/down, left/right, and forward/backward). The surface of the Earth, on the other hand, is two-dimensional, as we are only free to move in one of two directions (left/right and forward/backward). Any vertical motion is the result of moving in the other two directions. Under these constraints, a countable set of points is now zero-dimensional as we have zero degrees of freedom. It is not possible to move through such a space from one point to another without leaving the space.

2.2.1.2 Topological Dimension

From earlier work in topology, the dimension of any set can be defined as one greater than the dimension of the object that could be used to completely separate any part of the first set from the rest. A line has thus a dimension 1 since it can be separated by a point $(0 + 1 = 1)$, a plane has dimension

2 since it can be separated by a line ($1 + 1 = 2$), and a volume has dimension 3 since it can be separated by a plane ($2 + 1 = 3$). This notion of dimension is called the topological dimension D_T of a set (Hurewicz and Wallman 1941; Dugundji 1966). Strictly speaking, the topological dimension of any set is defined as one greater than the dimension of the object that could be used to completely separate any part of the first space from the rest. However, when referring to composite sets such as an x-shaped set (×) or the union of a point and a filled circle (· ●), the above definition seems, however, incomplete. Indeed, locally the former set is one-dimensional except at the intersection of the two segments where it becomes zero-dimensional (that is, a single point), and thus is obviously one-dimensional. The latter is a bit more challenging, as it is a union of completely separated components, where the point component (·) is zero-dimensional while the circle component (●) is two-dimensional.

Introducing the concepts of *local dimension* and *global dimension*, one can thus characterize the composite set (· ●) via the local dimensions of its components. To get the global dimension of the set, the above definition needs to be slightly modified. The dimension of any set should be the maximum of its local dimensions where the local dimension is defined as one more than the dimension of the lowest-dimensional objects needed to separate any neighborhood of the space into two parts. According to this definition, the composite set (· ●) is indeed two-dimensional.

More practically, the dimension of the union of finitely many sets is the largest dimension of any one of them, so if we "grow grass" on a plane, the result is still a two-dimensional set. We should nevertheless note here that if we take the union of an infinite collection of sets, the dimension can grow. For example, a line, which is one-dimensional, is the union of an infinite number of points, each of which is a zero-dimensional object.

2.2.1.3 Embedding Dimension

There can nevertheless occasionally be a little confusion about the dimension of an object. Sometimes people call a sphere a three-dimensional object because it can only exist in space, not in the plane. However, a sphere is two-dimensional. Any little piece of it looks like a piece of the plane, and in such a small piece, you only need two coordinates to describe the location of a point. More formally speaking, this is only a different measure of dimension, called the embedding dimension D_E: A set has embedding dimension D_E if D_E is the smallest integer for which it can be embedded into D_E without intersecting itself. Thus, the embedding dimension of a plane is 2 and the embedding dimension of a sphere is 3, even though they both have (topological) dimension 2.

A topological property of an entity is one that remains invariant under continuous, one-to-one transformations or homeomorphisms. A homeomorphism can best be envisioned as the smooth deformation of one space into another without tearing, puncturing, or welding it. Throughout such processes, the topological dimension does not change. A sphere is topologically equivalent to a cube since one can be deformed into the other in such a manner. Similarly, a line segment can be pinched and stretched repeatedly until it has lost all its straightness, but it will still have a topological dimension of 1.

The meaning of dimension can be questioned, however, when dealing with geometric constructs initially referred to as "mathematical monsters." For the sake of illustration, consider two case-study mathematical constructs (Figure 2.5). First, we consider the Koch curve, or Koch snowflake (Koch 1904, 1906). To build the Koch curve (Figure 2.5A), consider a triangle. First, take each line segment and divide it into thirds. Second, place the vertex of an equilateral triangle in the middle third, copy the whole curve, and reduce it to 1/3 its original size. Place these reduced curves in place of the sides of the previous curve. This procedure is subsequently iterated n times. With each iteration, the curve length increases by a factor of 4/3. An infinite repeat of this procedure would send the length off to infinity. Such a geometric construct is unusual but not disturbing regarding the above definition of dimension.

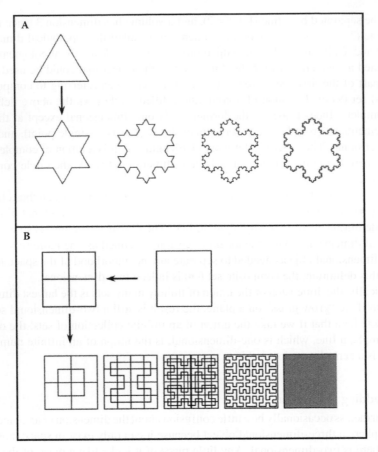

FIGURE 2.5 Building process of two theoretical fractals: the Koch snowflake (A) and the Peano curve (B).

This is not the case, however, with the second construct investigated here, the Peano curve (Schroeder 1991) (Figure 2.5B). First, take a square and divide it into four identical copies of the original. Second, draw a line starting in one square so that it passes through the center of every other square until it returns to the starting position. Iterating this procedure n times leads to a curve so twisted that it has infinite length. The resulting object is specifically referred to as a Peano monster curve (Mandelbrot 1983), so called because of its monstrous or pathological nature; note the reference to the demiurgical nature of such objects. More remarkable is that it will ultimately visit every point in the initial square. This construct thus generates a one-to-one mapping from the points in the unit interval to the points in the unit plane. An object with topological dimension 1 can then be transformed into an object with topological dimension 2. This iteration procedure could also be implemented in a cube and would ultimately lead to a space-filling curve (Gilbert 1984). Simple bending and stretching should leave the topological dimension unchanged, however. These apparently paradoxical results thus raise questions about the meaning of dimension, especially when one knows that the Koch and Peano curves are both regarded as basic examples of geometrical fractals.

2.2.2 FRACTAL DIMENSION

What about the dimension of the so-called fractal objects? For example, what is the dimension of the Koch snowflake (Figure 2.5A)? It has topological dimension 1, but it is by no means a curve; the length of the "curve" between any two points on it is infinite. No small piece of it is linelike, but neither is it like a piece of the plane or any other d. In some sense, we could say that it is too big to

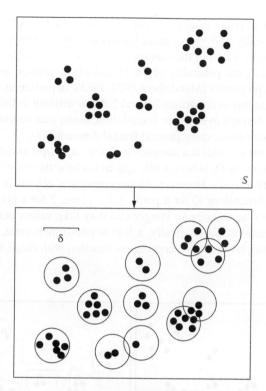

FIGURE 2.6 Hausdorff dimension of a set S. The Hausdorff dimension D_H is estimated from counting the number of open circles of radius δ needed to cover the set S, where $\delta \to 0$.

be thought of as a one-dimensional object but too small to be a two-dimensional object. Maybe its dimension should be a number *between* one and two. In order to make this kind of thinking more precise, one must look at the dimension of familiar objects another way.

Strictly speaking, a mathematical fractal is defined as any patterns for which the dimension exceeds the discrete topological dimension (Mandelbrot 1977, 1983). Formally, the concept of scaling exponent defined above can be extended and generalized through the concept of *Hausdorff dimension* (Carathéodory 1914; Hausdorff 1919), which can be regarded as the "core" of a whole family of fractal dimensions. The Hausdorff dimension D_H of a subset S embedded in an Euclidean space of dimension D_E (that is, $S \in D_E$) arises from asking "What is the size of S?" Note that this question is fairly general and can be applied to a wide variety of sets as "How long is S?" referring to the length of a coastline or the Koch curve, "How large is S?" referring to the surface of an island or a vegetation patch, or "How big is S?" referring to the volume of a cloud or a sponge. The answer comes from counting the number of open balls needed to cover the set S (Figure 2.6). For each $\delta > 0$, consider $N(\delta)$ the smallest number of open balls of radius δ needed to cover S. One can then show that the limit:

$$D_H = \lim_{\delta \to 0} \left(-\log N(\delta) / \log \delta \right) \tag{2.9}$$

exists (see, for example, Mandelbrot 1977). D_H is the Hausdorff dimension of the set S. Equation (2.9) is equivalent to the approximate power law

$$N(\delta) \approx k\delta^{-D_H} \tag{2.10}$$

where k is a constant and $\approx$ refers to an asymptotic behavior. Equation (2.10) must then be written as "$N(\delta)$ scales asymptotically as δ^{-D_H}," or more loosely as "$N(\delta)$ scales as δ^{-D_H}." Although several equivalent definitions can be found in the literature (see, for example, Rogers 1970; Federer 1969; Falconer 1985) (Figure 2.7), the generality of the Hausdorff dimension makes it difficult to compute and to determine its properties (Mandelbrot 1977, 1983). In particular, the intrinsic asymptotic condition, $\delta \to 0$ (compare this to Equations 2.9 and 2.10), is difficult to fulfill in most applications. More practical methods devoted to estimate fractal dimensions can be found in Chapters 3 and 4. Hereafter, we will thus refer to the more general fractal dimension D_F.

It can be understood that a fractal is a complex geometrical shape, constructed of smaller copies of itself. The fractal dimension D_F subsequently quantifies how the "size" of a fractal set changes with decreasing observation scales. However, while geometrical objects in Euclidean geometry are described using integer dimensions (0 for a point, 1 for a line, 2 for a plane, and 3 for a volume), fractal dimensions are not necessarily an integer and may take values between the boundaries of integer topological dimensions. Topologically, a line is one-dimensional. The dimension D_F of a fractal pattern on the plane, however, is a continuous function with range $1 \le D_F \le 2$. A completely

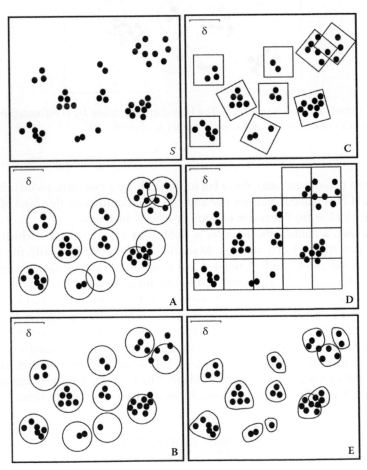

FIGURE 2.7 Different ways of estimating the Hausdorff dimension D_H of a set S. The number $N(\delta)$ (see Equation (2.10), is taken as (A) the least number of circles of radius δ that cover S; (B) the greatest number of disjointed circles of radius δ with centers in S; (C) the least number of boxes of radius δ that cover S; (D) the number of boxes of size δ that intersect S; and (E) the least number of sets of diameter at most δ needed to cover S. In all cases, the procedure is iterated until $\delta \to 0$.

differentiable series has a fractal dimension $D_F = 1$ (the same as the topological dimension), while a Brownian motion completely occupies two-dimensional topological space and therefore has a fractal dimension $D_F = 2$. Fractal dimensions $1 \leq D_F \leq 2$ quantify the degree to which a pattern fills the plane. In the same way, a planar curved surface is topologically two-dimensional, while a fractal surface has dimension $2 \leq D_F \leq 3$.

Consider now a measure unit defined as δ^{D_E}; for $D_E = 1$, 2, and 3, one refers to a length, a surface, and a volume, respectively expressed in m, m², and m³. Following Equation (2.10), a given set S thus measures $N(\delta)\delta^{D_E} \approx \delta^{D_E - D_F}$ "meters at the power $(D_E - D_F)$." A set S characterized by a fractal dimension $D_F = 2$ will then have a finite surface ($D_E = 2$), while its length ($D_E = 1$) will be infinite, and its volume ($D_E = 3$) nil. When D_F is a noninteger, length, surface, and volume become useless to characterize S because these metrics are nil or infinite.

So what is the dimension of the Koch snowflake? For such a self-similar mathematical fractal (we will see in Chapter 3 that the fractal concept can nevertheless be significantly complex) that can be divided into N similar parts, each of which is a copy of the whole reduced k times, the fractal dimension D_F can simply be written as

$$D_F = \frac{\log N}{\log k} \tag{2.11}$$

In the case of the Koch snowflake (Figure 2.5A), each part of the "curve" can be decomposed into four rescaled copies of itself, contracted by a linear factor of 3. Equation (2.11) thus leads to a fractal dimension $D_F = 1.262$ for the Koch snowflake. Consider now two other basic geometrical fractal objects: the Sierpinski carpet and gasket (Figure 2.8). Each component of the Sierpinski carpet and gasket can be decomposed into eight and nine copies of itself, contracted by linear factors of 3 and 4, respectively. The related fractal dimensions are then $D_F = 1.893$ and $D_F = 1.585$ for the Sierpinski carpet and gasket, respectively. In other words, the Sierpinski carpet covers space more intensively than the Sierpinski gasket and the Koch curve. One must finally note that any set exhibiting integer fractal dimensions can simply be thought as a specific case of fractal patterns. The fractal dimension can thus be thought of as a *measure of sparseness* of any set embedded in a Euclidean space. Consider a set embedded in a two-dimensional space. Homogeneous, regularly spaced sets will then be characterized by a higher fractal dimension than more sparse sets (Figure 2.9). At the limit, a plane-filling set has a fractal dimension $D_F = 2$, while the fractal dimension of a set so sparse that it is reduced to a single point is nil, $D_F = 0$ (Figure 2.9). These different statements can nevertheless be refined with the introduction of the concepts of *fractal codimension* and *sampling dimension*.

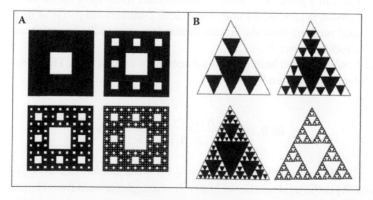

FIGURE 2.8 Building process of two theoretical fractals: the Sierpinski carpet (A) and the Sierpinski gasket (B).

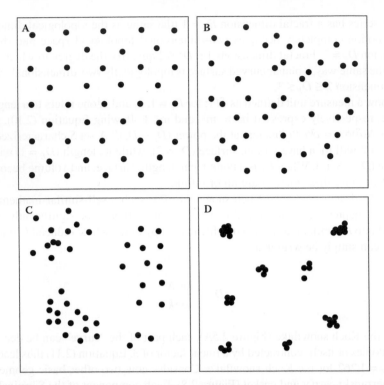

FIGURE 2.9 Fractal dimensions and codimensions of different point patterns. (A) Regular point pattern, $D_F = 2$ and $c_F = 0$; (B) random point pattern, $D_F = 1.8$ and $c_F = 0.2$; (C) random clumped point pattern, $D_F = 1.4$ and $c_F = 0.6$; and (D) aggregated clumped point pattern, $D_F = 1.1$ and $c_F = 0.9$.

2.2.2.1 Fractal Codimension

Consider a set S characterized by a fractal dimension D_F embedded in a space of topological dimension D_T. The *fractal codimension*, c_F, of the set S is given by:

$$c_F = D_T - D_F \tag{2.12}$$

The fractal codimension thus appears as a measure of the *relative sparseness* of the set S, while the fractal dimension is a measure of its *absolute sparseness*. As a consequence, Koch and Sierpinski constructs considered in a plane or in a three-dimensional space have the same fractal dimension, while their codimension is increased a unit. The fractal codimension can thus be regarded as being a more fundamental measure than the fractal dimension, especially in a probabilistic framework where it can be introduced directly.

Consider the number of open balls of radius δ needed to cover a set S. The probability $\Pr(B_\delta \cap S)$ for a ball B_δ to intersect S is given by:

$$\Pr(B_\delta \cap S) \approx \frac{N(B_\delta \cap S)}{N(B_\delta)} \approx \frac{\delta^{-D_F}}{\delta^{-D_T}} \approx \delta^{c_F} \tag{2.13}$$

where $N(B_\delta \cap S)$ [$N(B_\delta \cap S) \approx \delta^{-D_F}$ (see Equation 2.10) is the number of balls B_δ intersecting S, and $N(B_\delta)$ [$N(B_\delta) \approx \delta^{-D_T}$] is the total number of boxes. It is straightforward from Equation (2.13) that the most infrequent events are characterized by the highest fractal codimensions and thus the lowest fractal dimensions (see Figure 2.9). One may note here that Equations (2.12) and (2.13) are equivalent when $c_F \leq D_T$, or equivalently $D_F \geq 0$. Equation (2.13), however, does not imply any constraint on the fractal codimension c_F. When $c_F > D_T$, Equations (2.12) and (2.13) thus lead to $D_F < 0$. This statement is totally inconsistent with the essence of a fractal dimension, defined as a strictly positive measure (Mandelbrot 1977, 1983) (see, for example, Equations (2.1), (2.2), (2.9), (2.10), and (2.11). A purely geometrical definition is no longer satisfactory in a probabilistic framework where the effective dimension of the probability space is a function of the sampling effort.

2.2.2.2 Sampling Dimension

Mainly for practical reasons present in most scientific areas, statistics implicitly deal with samples of finite size. The dimension of the probability space can nevertheless increase with the number of independent samples considered (Figure 2.10). Considering N_S independent samples of dimension D_T, the related quantity of information can be expressed as:

$$N \times N_S = \delta^{-(D_T + D_S)} \tag{2.14}$$

where D_S is the *sampling dimension* (Seuront 1998) defined as:

$$D_S \approx (-\log N_S / \log \delta) \tag{2.15}$$

In particular, Equations (2.14) and (2.15) show that the dimension of the probability space can be increased above D_T (a single sample) and to overcome the *a priori* paradoxical limitation related to the occurrence of negative fractal dimensions. Thus, considering a rare event S such as $c_F > D_T$, Equation (2.12) can be rewritten as:

$$D_{FS} = D_T + D_S - c_F \tag{2.16}$$

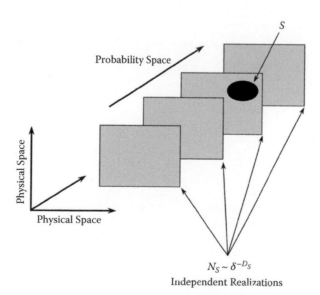

FIGURE 2.10 Sampling fractal dimension. Considering an increasing number of independent realizations (N_S) increases the effective dimension of the probability space. The probability of finding a set S embedded in a D_T-embedding space increases with N_S. One may also note that as $N_S \rightarrow +\infty$, the entire probability space is explored.

where the *sampling fractal dimension* D_{FS} verifies $D_{FS} > 0$ for $D_S > c_F - D_T$. The extreme case $D_{FS} = 0$ corresponds to single points isolated in the sample; when $D_S > c_F$ and $D_S > c_F$, S is present and absent in the sample, respectively.

The so-called fractal dimension and codimension referred to in the previous section are commonly estimated from the regression slope of a log-log power-law plot; see, for example, Equations (2.1), (2.3), (2.10), (2.14), and (2.15). However, this procedure is not necessarily as straightforward as it may appear at first glance and relies on many successful consecutive steps, the minimum prerequisite being to choose the appropriate analysis techniques (that is, an appropriate power law). To achieve this goal, one needs first to know the difference between self-similar and self-affine fractals as well as to identify the limits of fractal analysis, such as those related to both anisotropy and nonstationarity conditions often encountered in aquatic ecology. Objective criteria are also needed to select the appropriate range of scales to include in the regression analysis. Then comes the question of distinguishing scaling from multiple scaling behaviors. The final question that needs to be addressed is to know whether fractal concepts can be powerful enough to measure the extreme complexity emerging from the highly intermittent patterns encountered in both terrestrial and aquatic ecosystems.

3 Self-Similar Fractals

As stated in Chapter 2, fractals are defined to be scale-invariant geometric objects. However, scale invariance can be dichotomized into self-similar and self-affine fractals. Strictly speaking, an object is called self-similar if it may be written as a union of rescaled copies of itself with the rescaling isotropic (that is, uniform in all directions). Regular fractals such as the Koch snowflake (Figure 2.5a) and the Sierpinski carpet and gasket (Figure 2.8) display exact self-similarity. Random fractals such as the random Koch snowflake (Figure 3.1) display a weaker, statistical version of self-similarity or, more generally, statistical self-similarity. More formally, a geometric object is called self-affine if it may be written as a union of rescaled copies of itself, where the rescaling is anisotropic (that is, dependent on the direction). Thus the trace of particulate Brownian motion in two-dimensional space is self-similar, whereas a plot of the x coordinate of the particle as a function of time is self-affine (Figure 3.2).

3.1 SELF-SIMILARITY, POWER LAWS, AND THE FRACTAL DIMENSION

Mathematical fractals exhibit exact self-similarity across all spatial or temporal scales, such that successive magnifications reveal an identical structure. A self-similar object is composed of N copies of itself (with possible translations and rotations), each of which is scaled down by a scale ratio δ in all directions of the D_E dimensional available space. More formally, consider a set S of points at positions $\vec{x} = (x_1, x_2, \ldots, x_{D_E})$ in Euclidean space of dimension D_E. Under a similarity transform with a scale ratio δ ($0 < \delta < 1$), the set S becomes δS with points at positions $\delta \vec{x} = (\delta x_1, \delta x_2, \ldots, \delta x_{D_E})$. A bounded set S is self-similar when S is the union of N nonoverlapping subsets, each of which is identical (under translations and rotations) to δS. A basic example of a self-similar fractal is the Cantor set (Cantor 1883). Consider a line segment ([0, 1]), divide it into thirds, and remove the central part. Repeat the procedure on the two remaining thirds, and after an infinite number of iterations, one converges to a set of points or Cantor set, also referred to as Cantor dust (Figure 3.3).

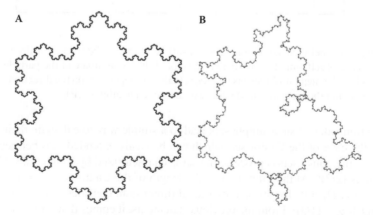

FIGURE 3.1 Difference between self-similar and statistical self-similar fractals. The Koch snowflake (A) displays exact self-similarity, while the random Koch snowflake (B) displays a weaker, statistical version of self-similarity, referred to as self-affinity.

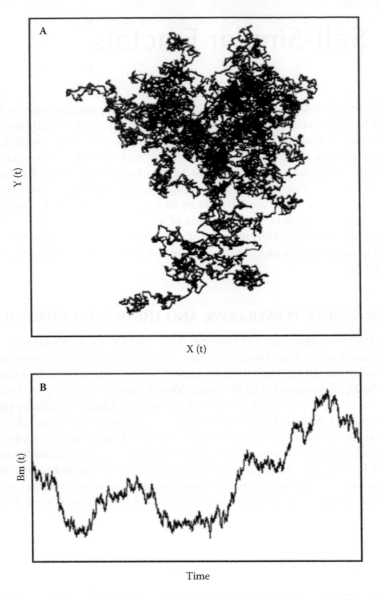

FIGURE 3.2 Difference between self-similar and self-affine fractals. The trace of a Brownian motion in a two-dimensional space is self-similar (A), whereas the plot of the x coordinate of the particle as a function of time is self-affine (B). The major difference is that the rescaling is dependent on the direction in the latter case; that is, the horizontal and the vertical axes do not have the same meaning in (B).

The construction rules of such simple sets lead to a simple way to calculate their related fractal dimensions. In the case of the Cantor set (which can be easily extended to other geometrical fractals), at stage n of the construction process, the set is characterized by 2^n intervals of length 3^{-n}; its total length is thus $(2/3)^n$. At the limit $n \to \infty$, the length of the Cantor set is then nil, and its topological dimension is $D_T = 0$. To estimate its fractal dimension, consider a cover of the sets by line segments of length $\delta_n = (1/3)^n$. From the previous statements, it comes that only $N(\delta_n) = 2^n$ segments cover a part of the Cantor set. The length of the Cantor set can thus be expressed as $L_{n+1} = (2/3) L_n$, whose solution is of the form:

$$L(\delta_n) = \delta_n^{1-D_F} \tag{3.1}$$

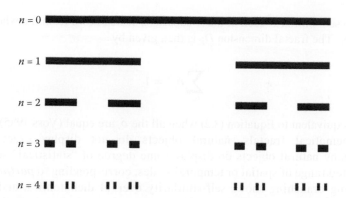

FIGURE 3.3 Construction of the Cantor set, by repeated removal of the middle third of each interval; at each step, there are two elements that are three times smaller, leading to a fractal dimension $D_F = \log 2/\log 3 = 0.631$.

The number of length elements required to cover the total length is given by $L(\delta_n)/\delta_n$ and equal to:

$$N(\delta_n) = \delta_n^{-D_F} \tag{3.2}$$

The fractal dimension D_F can finally be generally expressed as:

$$D_F = \log N(\delta_n)/\log(1/\delta_n). \tag{3.3}$$

or equivalently:

$$D_F = \log N(\delta_n)/\log \lambda^n \tag{3.4}$$

where the scale-ratio λ is defined as:

$$\lambda = \delta_{n+1}/\delta_n \tag{3.5}$$

where δ_{n+1} and δ_n are respectively the length elements required to cover a piece of the Cantor set at steps $n + 1$ and n of the construction process. For the Cantor set, $\lambda = 3$, and the fractal dimension $D_F = \log 2/\log 3 = 0.631$. A generalization for the areas and volumes of fractal surfaces and volumes can be easily derived from Equation (3.1) as:

$$S(\delta_n) = \delta_n^{2-D_F} \tag{3.6}$$

and

$$V(\delta_n) = \delta_n^{3-D_F} \tag{3.7}$$

It is stressed here that the fractal dimensions D_F introduced in Equations (3.2), (3.6), and (3.7) can be referred to as "fractal line dimension," "fractal surface dimension," and "fractal volume dimension," respectively. The exponents in Equations (3.1), (3.6), and (3.7) also directly refer to the codimension concept introduced in Section 2.2.2.1, Equation (2.12).

A set S is also self-similar if each of the N subsets is scaled down from the whole by a different similarity ratio δ_i. The fractal dimension D_F is then given by

$$\sum_{i=1}^{N} \delta_i^{D_F} = 1 \tag{3.8}$$

which is strictly equivalent to Equation (3.2) when all the δ_i are equal (Voss 1985).

Unlike mathematical fractals, natural objects do not display exact self-similarity. Nevertheless, many natural objects do display some degree of "statistical" self-similarity, at least over a limited range of spatial or temporal scales, corresponding to *partial self-similarity*. For example, lung branching shows self-similarity over 14 dichotomies, and tree branching over 8 dichotomies (Schroeder 1991). In addition, the existence of stepwise behavior (that is, changes in fractal dimension when shifting between scales) implies that in place of *true self-similarity*, we observe only *partial self-similarity* over limited ranges of scales separated by transition zones, where the environmental properties or constraints acting upon organisms are probably changing rapidly (Frontier 1987; Seuront and Lagadeuc 1998; Seuront et al. 1999). This is of prime interest in aquatic ecology since the extent of a given power law (referred to as the *scaling range* hereafter) allows the identification of the characteristic scales of organization of any pattern or process. Several objective procedures devoted to the identification of scaling ranges (that is, a key step for the result of fractal analysis to be meaningful) will thus be introduced in Chapter 7.

Statistical self-similarity refers to scale-related repetitions of overall complexity but not of the exact pattern. Specifically, details at a given scale are similar, though not identical, to those seen at coarser or finer scales. A set S is statistically self-similar if it is composed of N distinct subsets, each of which is scaled down by a ratio δ from the original and is identical in all statistical respects to δS. The related fractal dimension is still given by Equation (3.1) and Equation (3.2). In that way, the prevalence of power law with respect to the scale of observation in a certain range is commonly used to discern a fractal, especially when the scale of similarity is statistical and cannot be identified by sequential enlargement of segments of the fractal object. A collection of methods devoted to the characterization of different kinds of self-similar natural objects (for example, branching processes such as vascularization, lung systems, or stream orders and discrete patterns such as islands or pancreatic islets) are provided in Section 3.2.

It should be emphasized that self-similarity is not a prerequisite to applying fractal theory. Self-similar or statistically self-similar patterns are characterized by fractal dimensions that remain constant for each subpart of the whole (Mandelbrot 1983; Tricot 1995). Geographic lines, such as coastlines, are nevertheless very complex curves, whose local dimensions (see Section 2.2.1) are not the same everywhere. Such curves are not self-similar, not even statistically (Normant and Tricot 1993). As a consequence, we stress that (strictly speaking) fractal does not imply self-similar, and thus coastlines are not self-similar but fractal. Self-similarity is thus a restrictive point of view. This has also briefly been addressed by Voss (1985), who stressed that "in practice it is impossible to verify that all moments of the distributions are identical, and claims of statistical self-similarity are usually based on only a few moments." This specific point is nevertheless beyond the scope of this section and will be discussed more extensively hereafter with the introduction of the concept of multifractals (see Chapter 8).

3.2 METHODS FOR SELF-SIMILAR FRACTALS

There is no unique definition of self-similar fractal dimension; rather, there are a variety of methods used to measure it. Unfortunately, this statement also implies that the fractal dimensions are not all the same. As a consequence, in order for comparisons between fractal dimensions to be

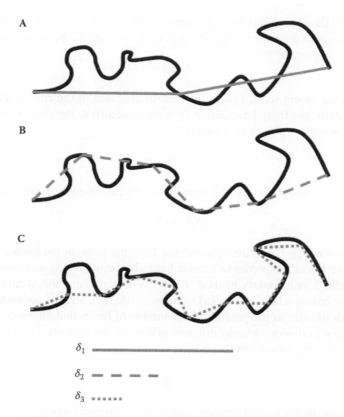

FIGURE 3.4 Schematic illustration of three successive steps of the divider procedure using dividers δ_1 (A), δ_2 (B), and δ_3 (C) of decreasing lengths.

meaningful, I strongly recommend the reader to be aware of the methods used in the studies he or she refers to.

The choice of a method is usually a matter of convenience, as different methods are tailored to different types of data sets. This section includes methods suitable for measuring the dimension of a set of zero- and one-dimensional objects lying in a plane (such as organisms, trajectories, and coastlines) and of sets of two-dimensional objects (such as islands, patches, and mountains) lying in a plane.

Loosely speaking, the methods described are quite similar. In all cases, one measures some characteristic of the data set that should be related through a power law to a length scale. The results are plotted in log-log space and, if the set is fractal, they should follow a straight line. The fractal dimension is a simple function of the exponent of the power law, that is, of the slope of the straight line in log-log space. The slope is estimated by fitting a line using the objective methods described and discussed in Chapter 7.

3.2.1 DIVIDER DIMENSION, D_d

3.2.1.1 Theory

Consider the problem of estimating the fractal dimension of a convoluted line, such as a coastline, vegetation patch edge, or movement pathway. Using the dividers procedure, the fractal dimension D_d is estimated by measuring the length of a curve at various scale values δ (Figure 3.4). The procedure is analogous to moving a set of dividers (like a drawing compass) of fixed length δ along the curve. The estimated length of the curve is the product of $N(\delta)$ (the number of compass dividers required to "cover" the object) and the scale factor δ. The number of dividers necessary to cover the object increases with decreasing measurement scale, leading to the power-law relationship:

$$L(\delta) = k\delta^m \tag{3.9}$$

where δ is the measurement scale, $L(\delta)$ is the measured length of the curve, $L(\delta) = N\delta$, and k is a constant. Practically, the fractal dimension D_d is estimated from the slope m of the log-log plot of $L(\delta)$ versus δ for various values of δ where:

$$D_d = 1 - m \tag{3.10}$$

One must note that, because $L(\delta) = N(\delta)\delta$, Equation (3.10) can be equivalently written as:

$$N(\delta) = k\delta^{-D_d} \tag{3.11}$$

The fractal dimension D_d is then directly estimated from the slope of the log-log plot of $N(\delta)$ versus δ. Note that the fractal properties of a curve have also been investigated using the mosaic tile amalgamation method, or boundary method (Kaye 1989). In this method, a curve is covered by a grid with each box having a length δ, called the mosaic tile size, and the perimeter of the curve is calculated for each tile size as the product of the number of boxes that intersect the curve and the size of the box. By superimposing many different grid sizes, the perimeter of a curve is then given by the scaling relationship (Kaye 1989):

$$P \propto \delta^{-D_{1b}} \tag{3.12}$$

where D_{1b} is the boundary fractal dimension, and $\propto$ means proportionality.

The fractal dimension D_d is bounded between 1 and 2 when the curve is Euclidean and space filling, respectively. In the former case, the length of the curve is a constant independent of the length scale δ. This lower bound is nevertheless proven to be true only when δ is sufficiently small when compared to the characteristic external scale of a given object. For instance, the perimeter of a circle of radius r is constant when $\delta = r$. When $D_d = 2$, the curve is so convoluted that it fills the whole available (two-dimensional) space; the length of the curve is linearly related to δ.

Even though coastlines, patch boundaries, and movement pathways are curves, their wiggliness is so extreme that it is practically infinite. For example, it is not useful to assume that they have either well-defined tangents (Perrin 1913) or a well-defined finite length (Mandelbrot 1967). Specific measures of the length depend on the method of measurement and have no intrinsic meaning. For example, let a pair of dividers "walk" along the coast; as the step length δ is decreased, the number $N(\delta)$ of steps necessary to cover the coast increases faster than $1/\delta$. Hence, the total distance covered, $L(\delta) = N(\delta)\delta$, increases without bound.

3.2.1.1.1 *Divider Dimensions and the Length of Coastlines*

The dividers method was originally used to describe the fractal nature of coastlines. From Mandelbrot's (1967) seminal paper "How Long Is the Coast of Britain? Statistical Self-Similarity and Fractional Dimension," chapters in standard monographs and textbooks on fractals (for example, Feder 1988; Turcotte 1992) discussed the fractal nature of coastlines at length. Mandelbrot (1983) examined a small set of coastlines and found their fractal dimension to be in the range of 1.2 to 1.3. Later work investigated the fractal dimension D_d for a limited sample of other coastlines (Dietler and Zhang 1992; Feder 1988; Carr and Benzer 1991; Pennycuick and Kline 1986; Korin 1992; Paar et al. 1997; Jiang and Plotnick 1998; Zhu et al. 2000, 2004) (see Table 3.1). For instance, Jiang and Plotnick (1998) showed that the Atlantic coast of the United States was much more complex

TABLE 3.1

Fractal Divider Dimension D_d of a Range of Coastlines

Coastline	D_d	Sources
Adak Island, Alaska (USA)	1.20	Pennycuick & Kline (1986)
Amchitka Island, Alaska (USA)	1.66	Pennycuick & Kline (1986)
West Coast (Great Britain)	1.27	Carr & Benzer (1991)
North Coast (Australia)	1.19	Carr & Benzer (1991)
South Coast (Australia)	1.13	Carr & Benzer (1991)
West Shore, Puget Sound (USA)	1.19	Carr & Benzer (1991)
East Shore, Puget Sound (USA)	1.15	Carr & Benzer (1991)
West Shore, Gulf of California (USA)	1.03	Carr & Benzer (1991)
East Shore, Gulf of California (USA)	1.02	Carr & Benzer (1991)
Ria Coast from Kamaishi (Japan)	1.21–1.37	Korin (1992)
Louisiana (USA)	1.20	Lam & DeCola (1993)
Pacific Coast (USA)	1.00–1.27	Jiang & Plotnick (1998)
Pacific Coast (USA)	1.00–1.70	Jiang & Plotnick (1998)
Jiangsu Province (China)	1.07	Zhu et al. (2000)
Coastline of China	1.04–1.24	Zhu et al. (2004)
Van Koch Curve	1.21	Carr & Benzer (1991)
Fractal Brownian Coastline	1.30	Mandelbrot (1977)

than the Pacific coast. This is consistent with previous qualitative statements describing the Pacific coast as being unique among the coastlines of the world in that the evenness of its contour is almost unbroken by embayments (Keen 1971). In contrast, the Atlantic coast is extensively embayed, with numerous fjords and river systems. In addition, no latitudinal gradient was detected in the fractal dimensions estimated over 1° latitudinal ranges, although they significantly increased from North to South on the Atlantic coast. Hypothesizing that the spatial complexity of a coastline can be regarded as a proxy of the bathymetric complexity of the related ocean floor, Jiang and Plotnick (1998) suggested that the bathymetric complexity of the Atlantic Ocean floor adjacent to the United States is far more complex than that of the corresponding Pacific Ocean floor. In an attempt to assess the potential impact of the lithologic properties of faults on the development of the east coast of Britain, Philip (1994) found that in some areas the coastlines were parallel to the faults. This was later confirmed by a fractal analysis of the coastline of China showing both direct and indirect effects of faults on coastline complexity (Zhu et al. 2004). Although the general trends of coastlines are forced by the geometry of the faults, the more active the faults are, the smaller the fractal dimension of the coastline is (Zhu et al. 2004). The potential implications of the fractal nature of coastlines (and ultimately of any potential habitat) on species diversity and species extinction are discussed hereafter.

3.2.1.1.2 Coastline Complexity and Marine Species Diversity

Many explanations for diversity patterns have been proposed, and there have been several early reviews of the subject (for example, Pianka 1966, 1974; Pielou 1975). High diversity has been attributed both to intense competition, which forces niche restriction (MacArthur and Wilson 1967; Hutchinson and MacArthur 1959; MacArthur 1965), and reduced competition resulting from predation (Risch and Carroll 1986). However, along with a variety of ecological and evolutionary processes, historical events, and geographical circumstances, habitat complexity on a wide range of scales plays an important role in community structure (Schluter and Ricklefs 1993; Rahbek and

Graves 2001). Environmental heterogeneity is critical for species coexistence, with structurally complex habitats offering a great variety of different microhabitats and niches, thereby allowing species to coexist and contributing to within-habitat diversity (Pianka 1988). Habitat heterogeneity provides a diversity of resources that can lead to coexistence of competitors, which would not be possible in homogeneous environments (Levin 1992) and is, *de facto*, a critical mechanism in the maintenance of biological diversity (Levin 1981). Although theoretical support for the importance of habitat heterogeneity is overwhelming, empirical evidence is not always clear and can be confounding (Kareiva 1990). Many studies conducted in a variety of ecosystems thus support a positive relationship between habitat complexity and species diversity (Petren and Case 1998; Kerr et al. 2001; Rahbek and Graves 2001), although evidence exists for diversity decreasing with or being independent of habitat heterogeneity (Eadie and Keast 1984; Kelaher 2003; Taniguchi et al. 2003). According to the above-mentioned statements, the increase in complexity of the Atlantic coast of the United States identified by Jiang and Plotnick (1998) should favor a higher species diversity when compared to the Pacific coast. Valentine (1989) identified 468 shallow-water gastropod species from the Californian faunal province of the Pacific coast, while for similar latitudes in the Atlantic, Allmon et al. (1993) found 778 gastropod species from the east coast of Florida. These results support the long-standing hypothesis of a positive relationship between habitat complexity and species diversity, and the potential for fractal analysis to be related with more traditional biological and ecological approaches in entangling the complex relationship between habitat heterogeneity and species diversity.

3.2.1.1.3 Coastline Complexity and Species Extinction

A mass extinction of coastal marine mollusks along the Atlantic coast of the United States in the late Pliocene has been well documented (Schopf 1970; Stanley 1981; Allmon et al. 1993); only 22% of Early Pliocene bivalve species have survived (Stanley 1986). In contrast, 80% and 75% of the Pacific coast Pleistocene fossil species of bivalves and gastropods, respectively, are still living (Valentine 1989). Stanley (1986) has suggested that the western Atlantic extinction was produced by cooling associated with the onset of glaciation. The cooling of the Pacific was proposed to be much weaker, so that no extinction resulted. However, an alternative explanation may lie in the topographic changes related to sea-level drop and climate change (Jiang and Plotnick 1998). As Valentine (1989) suggested, the early Pliocene coastline of the U.S. Atlantic coast was less complex than that of today, so an increase in coastline complexity may have increased speciation rates, resulting in the observed increase in diversity. Considering the observed higher complexity of the coastlines of the Atlantic coast, it is likely that the sea-level changes related to climate forcing induced sharper changes in the local properties of habitat complexity, which might have also contributed to species extinction.

3.2.1.2 Case Study: Movement Patterns of the Ocean Sunfish, *Mola Mola*

3.2.1.2.1 Study Organism

The ocean sunfish, *Mola mola*, inhabit tropical and temperate regions of the Mediterranean Sea and the Atlantic, Indian, and Pacific oceans (Fraser-Brunner 1951; Wheeler 1969; Miller and Lea 1972). Key aspects of their biology and behavior, such as annual movement and the mode and location of breeding, are still largely unknown (Fraser-Brunner 1951; Reiger 1983). It has been suggested that the main part of their life is spent in deep water (Fraser-Brunner 1951; Lee 1986); however, ocean sunfish are frequently observed during daylight hours at the sea surface (Fraser-Brunner 1951; McCann 1961; Schwartz and Lindquist 1987; Sims and Southall 2002; Seuront et al. 2003). In a recent study of their fine-scale movements, ocean sunfish were found to exhibit nocturnal vertical movements within the surface mixed layer and thermocline, while diurnal vertical movements were characterized by repeated dives below the thermocline (Cartamil and Lowe 2004). Although ocean sunfish have been previously regarded as planktonic fish, primarily passively transported by oceanic currents (McCann 1961; Lee 1986), Cartamil and Lowe (2004) showed that ocean sunfish are active swimmers not significantly affected by the velocity or the direction of the currents, and with cruising speeds similar to those found for yellowfin tuna *Thunnus albacares* (Block et al. 1992).

3.2.1.2.2 Experimental Procedures and Data Analysis

The data used here to investigate the diel variability of the fine-scale properties of ocean sunfish movement patterns were originally described in a previous study (Cartamil and Lowe 2004). As a consequence, we only provide hereafter the basics of capturing, tagging, and tracking procedures (Figure 3.5) and refer the reader to Cartamil and Lowe (2004) for further details. Eight ocean sunfish were captured by dipnetting while basking at the surface or found in association with kelp patties, and measured and tagged with a temperature and depth-sensing acoustic transmitter (Vemco, Model V22TP, 22 mm diameter × 100 mm length, frequencies 34 to 40 kHz). The acoustic output of the transmitters was detected using a fixed directional hydrophone mounted through the vessel's hull, and decoded by a receiver unit (Vemco, Model VR-60) mounted above the boat console (see Cartamil and Lowe 2004 for more details). Depth, temperature, and GPS-derived location of the vessel were recorded every 3 minutes.

Although the fractal nature of fish movement patterns have seldom been studied (Coughlin et al. 1992; Dowling et al. 2000; Faure et al. 2003), it may nevertheless be thought of as a unifying framework to model and compare movement patterns of other groups of organisms (Turchin 1998; Faure et al. 2003). Although movement patterns of ocean sunfish have been limited to scale-dependent measurements, such as rate and directionality of movement, the same data used to determine these metrics can be used to calculate and compare the fractal dimension of these movements (Cartamil and Lowe 2004). The use of the dividers dimension D_d is illustrated on the basis of acoustic telemetry tracking data for *Mola mola* off the southern California coast (Figure 3.6A) and demonstrates the fractal nature of diurnal and nocturnal movement patterns (Figure 3.6B).

3.2.1.2.3 Results

For the eight trajectories considered, log-log plots of $L(\delta)$ versus δ (Equation 3.9) exhibit very strong linear behaviors over the whole range of considered scales, with coefficients of determination ranging from 0.98 to 0.99 (Figure 3.6C). This unambiguously demonstrates the scale-dependent (fractal)

FIGURE 3.5 An ocean sunfish, *Mola mola*, released immediately after tagging with a temperature and depth-sensing acoustic transmitter. (Courtesy of Dr. C. G. Lowe, California State University, Long Beach, California.)

nature of sunfish movement patterns. Although no differences were "visually" perceptible between diurnal and nocturnal movement patterns, the resulting fractal dimensions, plotted as a function of diurnal and nocturnal movement patterns for each individual's swimming path, ranging from 1.05 to 1.31, show a high individual variability (Figure 3.6D). Except for the paths 5, 6, and 8, nocturnal fractal dimensions were significantly higher than diurnal ones (analysis of covariance and subsequent Tukey test, $p < 0.01$, Figure 3.6D). A significant negative correlation ($p < 0.05$) was found between sunfish size and both diurnal and nocturnal fractal dimensions. A significant ($p < 0.05$) positive correlation was found between nocturnal temperature and fractal dimensions. This relation was nonsignificant ($p > 0.05$) during daylight hours.

3.2.1.2.4 Ecological Interpretation

The difference between diurnal and nocturnal fractal dimensions suggests two specific swimming strategies (Figure 3.6D). The relative linearity associated to the low fractal dimension during daylight hours suggests individuals are moving in a direct manner. In contrast, the more complex, convoluted

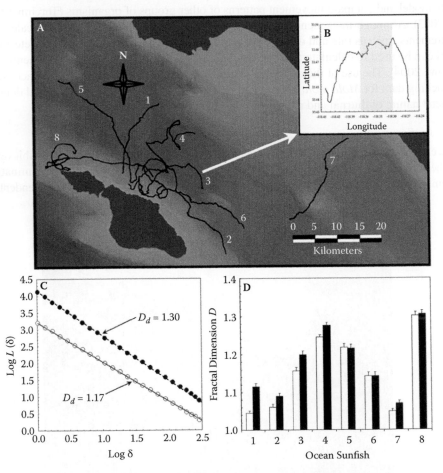

FIGURE 3.6 (A) Movements of eight ocean sunfish, *Mola mola,* tracked by acoustic telemetry, near Santa Catalina Island, California. (B) Details of diurnal and nocturnal (shaded area) movement patterns for the individuals. (C) Illustrations of the related diurnal (open symbols) and nocturnal (black symbols) scaling behaviors of the log-log plot of $L(\delta)$ versus δ (see Equation 3.9). (D) Fractal dimensions D_c are shown for the diurnal (white bars) and nocturnal (black bars) movement patterns of the eight individuals.

paths observed during the nighttime period suggest that sunfish are interacting with environmental heterogeneity on a finer scale (Wiens et al. 1995). As shown for a variety of organisms ranging from minute invertebrates (Crist et al. 1992; With 1994; Hoddle 2003; Seuront 2006) to large mammals (Bascompte and Vilà 1997; Ferguson et al. 1998a, 1998b; Mouillot and Viale 2001; Mårell et al. 2002; Laidre et al. 2004), an increase in the complexity of spatial movements may also indicate an increase in foraging or searching efforts in a localized area. Therefore, this may suggest that sunfish were searching for more clumped resources at night. Ocean sunfish vertical dive profiles observed during daylight hours may enable them to feed on vertically migrating gelatinous zooplankton at their diurnal depths below the thermocline (Cartamil and Lowe 2004). Here, the higher fractal dimension observed at night for five of the eight ocean sunfish investigated might be related to an increased foraging activity on gelatinous zooplankton occurring near the surface nocturnally. This result does not contradict previous work suggesting that sunfish are feeding primarily during the day. Instead, this may further substantiate the hypothesis that the movement patterns of ocean sunfish could have evolved as a means of foraging on vertically migrating organisms, and previous observations of nocturnal vertical movements were confined to the surface layer and thermocline (Cartamil and Lowe 2004). Ocean sunfish may also be feeding during both day and night but use different movement patterns to fully exploit prey that vertically migrate. The potential link between environmental heterogeneity and sunfish behavior was previously investigated by looking at sunfish movements relative to sea-surface temperature fronts (Cartamil 2003). However, the cloud cover that limited sea-surface temperature image quality and availability over the study area hampered the identification of any relationship (Cartamil 2003). The observed changes in fractal dimensions of swimming trajectories may then provide an efficient, alternative tool to infer changes in environmental properties.

The significant negative correlation between ocean sunfish size and the fractal dimension ($p < 0.05$) of their swimming paths suggests that larger individuals interact with their environment at a finer scale resolution than do smaller ones. Assuming that larger sunfish have increased remote sensing ability and motility, they may achieve more convoluted (that is, high D value) swimming paths, likely to increase their encounter rates with their intrinsically patchy zooplankton prey. A more convoluted swimming strategy also increases predation risk (see, for example, Tiselius et al. 1997), and ocean sunfish are typically hunted by fast-swimming predators such as large sharks (Fergusson et al. 2000) and California sea lions (Cartamil and Lowe 2004). However, since predation risk is less likely at large sizes, their size may enable them to use this type of movement relatively safely. In contrast, the more linear swimming behavior of the smallest and slowest sunfish (that is, $D = 1.14 \pm 0.01$; $\bar{x} \pm SD$) could then be thought as an antipredator strategy.

The significant positive correlation between temperature and the tortuosity of nighttime movement patterns suggests that the foraging activity of ocean sunfish may be temperature dependent. This may be directly related to an increased motility in warmer waters but also to different foraging strategies. This is consistent with the high individual variability observed in the movement patterns (Figure 3.6D) as different individuals moved over similar areas (see Figure 3.6A) but at different times. These areas likely differed in their biotic and abiotic properties during each tracking period. In particular, studies have documented the hierarchical nature of marine species' responses to food patch structure (Fauchald et al. 2000), indicating that organisms may be specifically responding to different prey distribution or density (Seuront et al. 2001).

3.2.1.3 Methodological Considerations

3.2.1.3.1 How to Start the Analysis?

As illustrated in Figure 3.4, the implicit easiest way to conduct the analysis is to use the first point of the object under consideration as a starting point for the divider algorithm. However, the values $L(\delta) = N\delta$ may vary depending on the starting position along the curve, especially at large scales (Sugihara and May 1990a; With 1994; Nams 1996). This issue can nevertheless be circumvented by starting the dividers procedure at different, randomly chosen positions, walking forwards and backwards, and using the distribution of the resulting divider dimensions (also referred to as "compass dimension" in the literature) as an estimate of D_d. This procedure has been applied on nine three-dimensional swimming paths of the freshwater crustacean, the water flea *Daphnia pulex* (see Figure 3.12). Taking 10 random starting positions for each of the swimming paths available, Seuront et al. (2004a) did not identify any spurious effect related to the first point considered in the analysis, as all the divider dimensions returned were not found significantly different (analysis of covariance, $p > 0.05$). It is strongly advised, however, to systematically infer this potential bias to ensure the relevance of the divider dimension returned by the analysis.

3.2.1.3.2   On the Relevance of Fractal Analysis in Movement Ecology

3.2.1.3.2.1   Fractal Analysis in Movement Ecology

Since its introduction to characterize tortuosity of animal trails (Dicke and Burrough 1988), fractal analysis has been widely used to describe movement pathways of a wide variety of organisms ranging from invertebrates to large vertebrates (Table 3.2). The fractal analysis of animals' paths appeared to be an alternative and promising method for measuring tortuosity of foraging behavior. Although D_d is primarily a measure of the extent to which a given (fractal) object fills the space in which it is embedded, the logic of the use of fractals to characterize motion behavior lies behind

TABLE 3.2
Nonexhaustive Literature Survey of Movement Pathways Analyzed Using Fractal Dimensions

Organism	Terrestrial	Aquatic	Sources
Vertebrate	+		Cody (1971)
Invertebrate*	+		Shlesinger (1986)
Invertebrate	+		Dicke & Burrough (1988)
Invertebrate*		+	Levandowski et al. (1988)
Invertebrate	+		Wiens & Milne (1989)
Invertebrate	+		Turchin et al. (1991)
Invertebrate	+		Crist et al. (1992)
Invertebrate	+		Johnson et al. (1992)
Invertebrate		+	Coughlin et al. (1992)
Invertebrate		+	Bundy et al. (1993)
Invertebrate	+		Gautestad & Mysterud (1993)
Invertebrate	+		With (1994)
Invertebrate		+	Erlandson & Kostylev (1995)
Invertebrate	+		Wiens et al. (1995)
Invertebrate	+		Cole (1995)
Invertebrate		+	Brewer (1996)
Vertebrate*		+	Viswanathan et al. (1996)
Invertebrate*		+	Schuster & Levandowsky (1996)
Vertebrate	+		Nams (1996)
Vertebrate*	+		Focardi et al. (1996)
Invertebrate	+		Turchin (1996)
Invertebrate		+	Jonsson & Johansson (1997)
Vertebrate*	+		Kafetsopoulos et al. (1997)
Vertebrate	+		Bascompte & Vila (1997)
Invertebrate	+		Turchin (1998)
Vertebrate		+	Ferguson et al. (1998a)
Vertebrate		+	Ferguson et al. (1998b)
Vertebrate	+		Etzenhouser et al. (1998)
Vertebrate		+	Dowling et al. (2000)

Invertebrate		+	Schmitt & Seuront (2001)
Vertebrate		+	Mouillot & Viale (2001)
Vertebrate*	+		Marell et al. (2002)
Vertebrate*	+		Atkinson et al. (2002)
Vertebrate		+	Fritz et al. (2003)
Invertebrate	+		Hoddle (2003)
Invertebrate*		+	Bartumeus et al. (2003)
Vertebrate		+	Laidre et al. (2004)
Vertebrate		+	Faure et al. (2003)
Invertebrate	+		Martin (2004)
Vertebrate*		+	Austin et al. (2004)
Invertebrate		+	Seuront et al. (2004a)
Invertebrate		+	Seuront et al. (2004b)
Invertebrate		+	Seuront et al. (2004c)
Vertebrate*	+		Ramos-Fernández et al. (2004)
Invertebrate	+		Biesinger & Haefner (2005)
Vertebrate	+		Nams (2005)
Vertebrate	+		Garcia et al. (2005)
Invertebrate		+	Uttieri et al. (2005)
Invertebrate		+	Uttieri et al. (2007)
Invertebrate		+	Seuront (2006)
Invertebrate*	+		Reynolds (2006)
Invertebrate*	+		Reynolds & Frye (2007)
Invertebrate*	+		Reynolds et al. (2007)
Invertebrate*		+	Seuront et al. (2007)
Vertebrate**		+	Bertrand et al. (2007)
Vertebrate**	+		Brown et al. (2007)
Vert/Invert*	+	+	Edwards et al. (2007)
Vertebrate*		+	Sims et al. (2008)

*Movement pathways analyzed indirectly as Lévy flights using cumulative frequency distributions, see Chapter 5.
**Humans.

the two extreme representations of organism movement. The first extreme is an organism moving along a perfectly linear path. As previously described, since the path is a line, its length $L(\delta)$ will always be the same whatever the measurement scale δ (that is, the size of the divider) and its fractal dimension is unity. The second extreme is the Brownian motion, which results in a fractal dimension $D_d = 2$, suggesting that all points in the available space have been visited, leaving no area unfilled. The fractal dimension is then expected to provide a measure of the path sinuosity or tortuosity. However, the critique made by Turchin (1996), and to a lesser extent Benhamou (2004), might throw some doubt on the reliability of the fractal dimension for measuring organisms' path tortuosity. The advantages of using fractal analysis instead of more traditional behavioral metrics and the rationale for criticism against the fractal approach in behavioral ecology are addressed in details hereafter.

3.2.1.3.2.2 Scale Independence or Scale Dependence of Fractal Dimension of Movement Pathways?

The key assumption of fractal analysis is that the fractal dimension is scale independent; that is, the fractal dimension estimated from a movement path that is tens of meters long is the same for paths measured at the scale of meters to kilometers. This is not the case, however, for traditional behavioral metrics (Box 3.1).

BOX 3.1   ON THE SCALE DEPENDENCE OF
TRADITIONAL BEHAVIORAL METRICS

Movement pathways have been characterized by a variety of measures, including path length (the total distance traveled, or gross displacement), move length (the distance traveled between consecutive points in time), move duration (time interval between successive pauses, as well as between successive spatial points), speed (move length divided by move duration), turning angle (the difference in direction between two successive moves), turning rate (turning angle divided by move duration), net displacement (the linear distance between starting and ending points, often used as a metric when making comparisons with diffusion or correlated random walk models; see, for example, Kareiva and Shigesada 1983; McCulloch and Cain 1989; Turchin 1991; Johnson et al. 1992), and NGDR (net to gross displacement ratio; Wilson and Greaves 1979). As discussed by Seuront et al. (2004a), the values of all the metrics are implicitly a function of their measurement scale. The scale dependence of these ratio metrics—that is, the path length and the turning angle (Figure 3.B1.1)—implies that there is no single scale at which movement paths can be unambiguously described.

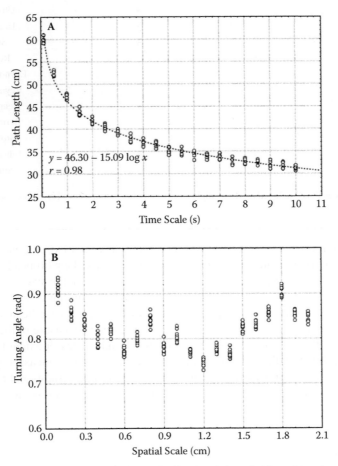

FIGURE 3.B1.1 Illustration of the scale dependence of path length (A) and turning angle (B) of the swimming trajectories of the water flea, *Daphnia pulex,* investigated at different temporal resolution and spatial scale, respectively. (Modified from Seuront et al., 2004a.)

Scale measurement is the central tenet of one of the main issues faced by landscape ecologists: understanding how to meaningfully extrapolate ecological information across spatial scales (Gardner et al. 1989; Turner and Gardner 1991). The scale independence of the fractal dimension is mainly driven by the belief that the same process is controlling the observed patterns from the smallest to the largest available scales. Strictly speaking, fractals are mathematical constructs characterized by a never-ending cascade of similar structural details that are revealed upon magnification on all scales. A fractal object, considered in its mathematical sense, then requires "many" orders of magnitude of power-law scaling, and subsequent interpretation of empirical data as indicating fractality must require "many" orders of magnitude (Avnir et al. 1997). While the concept of "many orders of magnitude" might be thought of as fairly vague in itself, it refers to several key processes involving equilibrium-critical phenomena (such as magnets, liquids, percolations, and phase transitions) and some nonequilibrium growth models (such as diffusion-limited aggregation, stopovers of Lévy flights) that are backed by intrinsically scale-free theories and lead therefore to power-law scaling behavior on all scales. It is then legitimate to expect constant fractal dimensions over up to five decades (Avnir et al. 1997). This is, however, very unlikely to happen in ecology due to the heterogeneous and hierarchical organization of landscapes and ecosystems.

Wiens et al. (1993) concluded their investigation of insect movement in microlandscape mosaics, stating that "because the fractal dimension has the desirable feature of being constant over a finite range of measurement scales, it is useful in comparing movements of insects that may respond to the patch structure of the environment at different absolute scales." A fractal motion behavior is also implicitly limited by the sizes of the organisms and their home range, or more prosaically by the size of the enclosure in which their behavior has been investigated. These potential changes in fractal dimensions then define transitions between different spatial domains for which the fractal dimension is scale independent (Wiens 1989). As a consequence, the fractal dimensions of organisms' movements are expected to apply to a limited range of scales or to change with scales (Box 3.2).

BOX 3.2   ON THE SCALE INDEPENDENCE OF FRACTAL DIMENSIONS IN ANIMAL BEHAVIOR

The traditional dividers method (Section 3.2.1.1) can be adapted to measure D_d over different ranges of scales (Krummel et al. 1987; Sugihara and May 1990a; Nams 1996). First, consider a regression window varying over a range of divider lengths δ (where a minimum of five values is expected to ensure the statistical relevance of a regression analysis; Seuront et al. 2004a), and estimate the slope of the log-log plot of $L(\delta)$ vs. δ over that range, thus the fractal dimension D_d. Then shift the range of the δ value along the x axis, and estimate D_d again. This procedure is illustrated using swimming paths of adult males and females of the subtropical copepod, *Oncaea venusta* (Figure 3.B2.1), which have been investigated using the dividers method (Section 3.2.1.1) and the sliding regression-window method described above (Figure 3.B2.2).

The very strong linear behavior of log-log plot of $L(\delta)$ vs. δ for the male trajectory over more than two decades (Figure 3.B2.2A) and the lack of variability in the fractal dimension D_d obtained from the sliding regression-window procedure (Figure 3.B2.2C) suggest the fractal character of the motion behavior of *O. venusta*. Similarly, the female swimming path clearly exhibits two distinct scaling behaviors below and above a critical scale of 10 mm (Figure 3.B2.2B), which have been confirmed by the sliding window

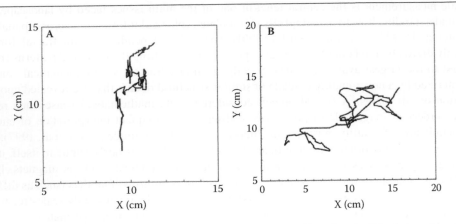

FIGURE 3.B2.1 Examples of swimming paths of adult female (A) and male (B) of the subtropical copepod, *Oncaea venusta*.

regression procedure (Figure 3.B2.2D). As the identified behaviors significantly differed from a correlated random walk model ($p < 0.05$; see the procedure described in Box 3.3), it is legitimate to conclude the fractal character of the movement pathways of both male and female *O. venusta*.

Because different scales are often associated with different driving processes, the fractal dimension may have the desirable feature of only being constant over a finite, instead of an infinite, range of measurement scales. It is then useful for (1) identifying characteristic scales of variability, and (2) comparing movements of organisms that may respond, for instance, to the patchy structure of their environment at different absolute scales. Changes in the value of the fractal dimension with scale may indicate that a new set of environmental or behavioral processes is controlling movement behavior (for example, decreased influence of patch barriers or the effect of home-range behavior). Thus, the scale dependence of the fractal dimension over finite ranges of scales may carry more information, both in terms of driving processes and sampling limitation, than its scale independence over a hypothetical infinite range of scales. Alternatively, although the point of slope change may indicate the operational scale of different generative processes, it may simply reflect the limited spatial resolution of the data being analyzed (Kenkel and Walker 1993; Gautestad and Mysterud 1993). However, discussed elsewhere (Seuront et al. 2004a), the effect of spatial resolution in the data will manifest itself as a gradual change of the fractal dimensions toward $D_d \rightarrow 1$ or $D_d \rightarrow 2$, and cannot be confused with a transition zone between two different scaling regions as illustrated here.

In order to measure if—and then eventually how and why—organisms use habitats at different scales, it is critical not only to measure the overall fractal dimension but also to measure how the fractal dimension changes with scale. The signatures of the log-log plot of $L(\delta)$ vs. δ returned by

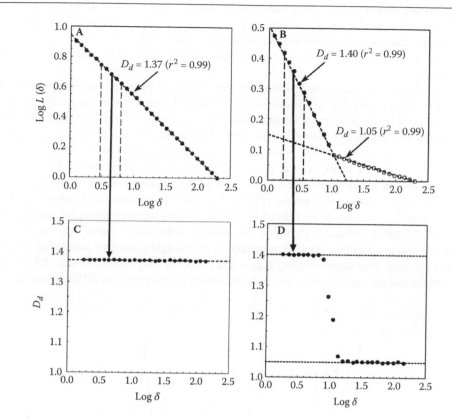

FIGURE 3.B2.2 Illustration of the "traditional" log-log plots of $L(\delta)$ vs. δ showing (A) the unique scaling behavior of the swimming path of an adult male for scales ranging from 1.2 to 200 mm ($D_d = 1.37$; A), and (B) the two distinct scaling behaviors of the swimming path of an adult female for scales ranging from 1.2 to 10 mm ($D_d = 1.40$) and from 10 to 200 mm ($D_d = 1.05$; B). The fractal dimension D_d calculated from the plots (A) and (B) using a sliding window of five points along the x axis clearly confirms the results of the divider method. The dashed gray lines in (A) and (B) identify the limits of the regression window, and the gray arrow indicates the related fractal dimension D_d estimate.

the swimming paths of ocean sunfish (Figure 3.6) and the subtropical copepod, *Oncaea venusta* (Figure 3.B2.2A), typically spanning more than two decades and satisfying objective optimization criteria (see Chapter 7), can then reliably be considered as fractal and exhibit the same movement pattern over the whole range of available scales. In contrast, the two distinct movement patterns observed in the female of the copepod, *Oncaea venusta,* above and below 10 mm (Box 3.2) reveal two distinct foraging strategies over two distinct ranges of scales. Similar results have been reported for the American marten (*Martes americana*), which displayed different responses to their environment at scales smaller and greater than 3.5 m (Nams and Bourgeois 2004). At scales smaller than 3.5 m, marten moved in a more direct way (that is, at a lower fractal dimension D_d) than at larger scales, suggesting different habitat use. This is consistent with earlier work by Benhamou (1990), who showed that at a smaller scale wood mice (*Apodemus sylvaticus*) travel in a directed path toward individual bushes, but at a larger scale they move from bush to bush randomly. At the smaller scale, the directed path would show a constant D_d, but at the larger scale the correlated random walk would show an increasing D_d. What is critical for a proper interpretation of fractal dimensions is then the identification of the range of scales over which a fractal dimension is invariant, as shown here for the ocean sunfish (*Mola mola*) and the copepod (*Oncaea venusta*).

However, changes in fractal dimensions are not always as clear as the one observed for *O. venusta* (Box 3.2) (Figure 3.B2.2D). Instead, the fractal dimension can smoothly change with scales as observed for the wandering albatrosses (Fritz et al. 2003) (Figure 3.7A) and the American marten (Nams and Bourgeois 2004) (Figure 3.7B). The limits of the transition scales have been defined as the first significant break in the slope of the relationship between the path length and the divider length (Nams 1996; Fritz et al. 2003; Nams and Bourgeois 2004). Transition scales are then centered on a minimum or maximum fractal dimension, with the surrounding dimensions gradually increasing or decreasing (Figure 3.7). The above-mentioned definition of transition scales should consequently be refined as "the transition between different spatial domains for which the fractal dimension is changing continuously with scale." Using this concept allowed Fritz et al. (2003) to identify a consistent pattern in the foraging paths of wandering albatrosses (*Diomedea exulans*) over scales ranging across five orders of magnitude (10 m to 1000 km). The 11 birds considered thus consistently adjust the tortuosity of their paths to different environmental

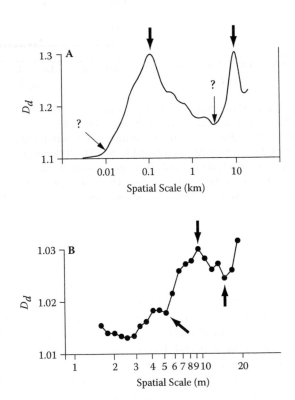

FIGURE 3.7 Patterns of the fractal dimension D_d estimated at different spatial scales (that is, divider length δ) using the method described in Box 3.2, and used to identify transition scales in the movement pathways of (A) wandering albatrosses (*Diomedea exulans*) (modified from Fritz et al. 2003); and (B) the American marten (*Martes americana*) (modified from Nams and Bourgeois 2004). The transition scales originally chosen by the authors are indicated by the thick arrows, and the question marks and thin arrows indicate additional transition scales that might also have been chosen given the shape of the relationship between D_d and the spatial scale and the definition found in Nams (1996), Fritz et al. (2003), and Nams and Bourgeois (2004).

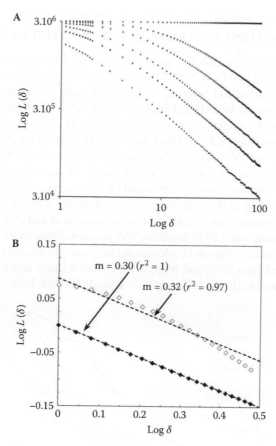

FIGURE 3.8 Influence of the mean cosine of turning angles c ($c = 0.0, 0.4, 0.6, 0.8, 0.9,$ and 1.0 from bottom to top) (modified from Benhamou, 2004) of a simulated correlated random walk on the log-log plot of $L(\delta)$ vs. δ (A); and (B) illustration of the r^2 values returned by the power-law $L(\delta) \propto \delta^{-0.3}$ (black symbols) and a surrogate slightly nonlinear curve (open symbols).

and behavioral constraints over three distinct scale-dependent nested domains. At small scales, they exhibit a zigzag movement to adjust for optimal use of wind; at intermediate scales, the movement shows changes in tortuosity consistent with food-searching behavior; and at a large scale, the movement relates to commuting between patches and to large-scale weather systems. The absence of such transitions in sunfish trajectories implies that they were using the same foraging strategies over the whole range of scales considered. It is nevertheless acknowledged that the choice of transition scales can sometimes seem fairly arbitrary (Figure 3.8) (see also Fritz et al. 2003, Figure 2A), stressing the need to use objective, statistically sound procedures to ensure the relevance of the measured fractal dimensions and transitions scales. The uses, misuses, and abuses of fractal analysis that may lead to spurious results and conclusions are addressed in detail in Chapter 6.

3.2.1.3.2.3 *Movement Pathways, Correlated Random Walks, and Fractal Analysis*
An alternative hypothesis is that "the fractal dimension is not constant but changes continuously with scales" (Turchin 1996). Turchin (1996) suggested that if an organism moves according to a correlated random walk (Box 3.3), then fractal analysis of that movement is not justified.

BOX 3.3   MOVEMENT PATTERNS AS CORRELATED RANDOM WALKS

The correlated random walk (CRW) model has proved its worth in many applications (Kareiva and Shigesada 1983; Cain 1990, 1994; Turchin 1991). Here, the applicability of CRW to describe organism movement behavior is illustrated using the movement patterns of the ocean sunfish, *Mola mola,* investigated in Section 3.2.1.2. CRW formulation assumes that move lengths and turning angles are not correlated serially. This was tested by calculating the autocorrelation function (ACF) and the Ljung–Box Q-statistic for all lags up to six moves for move length (Turchin 1998). Angular correlation was determined by defining sequential turns as left or right and performing a run test to check for nonrandomness (Turchin 1998). For each individual *Mola mola*, move lengths and turning angles were pooled together within two groups corresponding to the trajectories recorded during daylight hours and during the night to calculate an average expected net squared displacement and 95% confidence intervals using a bootstrap simulation of 1000 iterations. Net squared displacements were used because of the prohibitive difficulty inherent in calculations necessary to predict nonsquared displacement (McCulloch and Cain 1989) and because they show a linear relationship with time and thus are directly related to the rate of population spread (Turchin 1998).

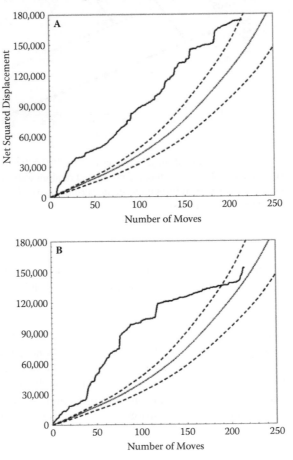

FIGURE 3.B3.1 Examples of observed (continuous thick line) and expected (dotted line) net squared displacements in case of a correlated random walk (CRW) model for diurnal (A) and nocturnal (B) movements of individual eight (see Figure 3.6). The dashed lines are the upper and lower limits of the 95% confidence interval.

We calculated the expected net squared displacement R_n^2 following Kareiva and Shigesada (1983):

$$R_n^2 = nl^2 + 2l^2 \frac{c}{1-c}\left(n - \frac{1-c^n}{1-c}\right) \tag{3.B3.1}$$

where R_n^2 (km²) is the squared displacement from the first location, n the number of moves from the first location, l the mean move length (km), and c the mean of the cosines of the turning angles. The total number of moves per path simulated by the model varied among *Mola mola* tracks and was equal to the largest number of moves taken by an *M. mola* in a given track. The observed net squared displacements were then compared to the value predicted by the CRW model (Figure 3.B3.1).

It is usually unambiguous if the *M. mola* fit the model or not (Figure 3.B3.1A), but in some cases the track crossed the 95% confidence interval for some portion of the displacement (Figure 3.B3.1B). To determine if *M. mola* fit the model, we used the *I* statistics as an objective index of the proportion of observations that lie outside of the 95% confidence intervals of the model prediction (Austin et al. 2004):

$$I = \sum_{i=1}^{p}\left[j \times \frac{\left(E_u(i) - O(i)\right)}{E_u(i) - E_m(i)} + k \times \frac{\left(E_l(i) - O(i)\right)}{E_l(i) - E_m(i)}\right] \tag{3.B3.2}$$

where $i = 1, \ldots, n$ are the successive moves; $j = 1$ if $O(i) > E_u(i)$ and $j = 0$ if $O(i) < E_u(i)$; $k = 1$ if $O(i) > E_l(i)$ and $k = 0$ if $O(i) < E_l(i)$; u and l are the upper and lower limits of the 95% confidence intervals; $O(i)$ the observed R_n^2, and $E_l(i)$, $E_u(i)$, and $E_m(i)$ the expected lower, upper, and mean R_n^2 values. The 95th percentile of the expected R_n^2 is used as the critical value and compared to the observed trajectories. Paths with an *I* value greater than the critical value were considered to significantly differ from the CRW model. The movements of the eight *M. mola* were systematically underpredicted by the CRW model (Figure 3.B3.1) for daytime and nighttime movements. Sunfish then exhibited greater directionality of movement and longer move lengths than expected under the CRW assumption.

Benhamou (2004) also stated that "animals' random search paths, at least when they are modeled as correlated random walks, are not fractal." Although this is common sense since correlated random walks are classical (nonfractal) models, they may erroneously return a fractal signature (that is, a linear behavior of the log-log plot of $L(\delta)$ vs. δ), especially when the range of scales available in the analysis (that is, the number of data points) is limited (Turchin 1996; Benhamou 2004), which is a recurring issue in ecological studies and may lead to what is called *apparent fractality* (Hamburger et al. 1996; Halley et al. 2004). A typical example of a mishandling of both correlated random walks and fractal analysis is provided by, for example, Uttieri et al. (2005). Despite the nonfractal character of correlated random walks (Turchin 1996; Benhamou 2004); Uttieri et al. (2005) analyzed the fractal properties of simulated three-dimensional correlated random walks of increasing length (5,000, 10,000, and 50,000 points) using a three-dimensional box-counting method (see Section 3.2.2). Although they did not provide any figure to support their findings, they claim that "the regression lines of the log-log plots fit the points with good accuracy" (i.e. $r^2 \approx 0.99$) and their fractal analysis returned decreasing fractal dimensions (1.69, 1.48, and 1.41), interpreted as the expression of "a decrease in morphological complexity of the trajectories." These results are irrelevant given the nonfractal nature of correlated random walks. They may nevertheless be explained by the shape of the log-log plots of $L(\delta)$ vs. δ returned by the fractal analysis of a correlated random walk (Figure 3.8A). These plots can indeed appear relatively, even very, linear depending on the parameters used to simulate the correlated random walk (Figure 3.8A). As a consequence, in some instances, linear fits can return very high values of r^2 and hence lead to identification as linear a nonlinear plot of log $L(\delta)$ vs. log δ (Figure 3.8B). It is also likely from their subsequent analysis of the three-dimensional swimming path of the water flea *Daphnia pulex* that Uttieri et al. (2005) did not use any objective criteria to choose the range of scales used to estimate fractal dimensions (see their Figure 4C).

To avoid erroneously considering a correlated random walk as returning a fractal signature prior to fractal analysis, it is then necessary to test, as a null hypothesis, whether a correlated random walk model adequately describes the path properties (Box 3.3). If the null hypothesis is to be rejected, it is still necessary to assess objectively the nature of the signature of the log-log plot of $L(\delta)$ vs. δ (see Box 3.2 and Chapter 6). This is critical to ensure the relevance of fractal analysis, as a number of previous works on fractal analysis have implicitly made the assumption that the slopes of log-log plots were linear without preliminary critical assessment (for example, Crist et al. 1992; Crist and Wiens 1994; With 1994; Laidre et al. 2004; Uttieri et al. 2005, 2007). It is stressed that the "fractal signature" of correlated random walks can only be erroneously considered as the expression of a fractal behavior if the scaling range is narrow, that is, typically smaller than one decade (see Figure 3.8A).

3.2.2 Box Dimension, D_b

3.2.2.1 Theory

This procedure, like the divider method, can be used to measure the fractal dimension of a curve (Longley and Batty 1989). In addition, it can be applied to overlapping curves (Peitgen et al. 1992) and structures lacking strict self-similar properties such as vegetation (Morse et al. 1985). Formally, the method finds the "δ cover" of the object—that is, the number of boxes of length δ (or circles of radius δ) required to cover the object (Voss 1988). A more practical alternative is to superimpose a regular grid of pixels of length δ on the object and count the number of "occupied" pixels. This procedure is repeated using different values of δ. The volume occupied by a curve is then estimated with a series of counting boxes spanning a range of volumes down to some small fraction of the entire volume. The number of occupied boxes increases with decreasing box size, leading to the following power-law relationship (Loehle 1990):

$$N(\delta) = k\delta^{-D_b} \tag{3.13}$$

where δ is the box size, $N(\delta)$ is the number of boxes occupied by the curve, k is a constant, and D_b is the box fractal dimension. D_b is estimated from the slope of the linear trend of the log-log plot of $N(\delta)$ versus δ. Note that Equation (3.13) can indifferently be used with one-, two- or three-dimensional objects.

It is stressed that both interior and border boxes contribute to the total number of boxes $N(\delta)$ intersected by the set. Interior boxes are fully contained within the fractal set; that is, they only contain a part of the set. In contrast, border boxes contain at least one white pixel and contain or adjoin at least one black pixel. Thus,

$$N(\delta) = N_b(\delta) + N_i(\delta) \tag{3.14}$$

The border-box fractal dimension, D_{b_b}, is then estimated as the linear trend of the log-log plot of $N_b(\delta)$ versus δ. The interior-box fractal dimension, D_{b_i}, is similarly estimated as the linear trend of a log-log plot of $(N(\delta) - 0.5N_b(\delta))$ against δ. The substraction term is necessary to avoid an overestimate of the area of the structure at large box sizes since the border boxes are not entirely filled by the set (Kaye 1989). Note that the border box dimension D_{b_b} is strictly equivalent to the boundary fractal dimension D_{1b} (see Equation 3.12, Section 3.2.1.1).

In the case of real data sets, one must always work with a finite set S (which may or may not be interpreted as a sample of the points from some infinite set). The above limit of a finite set is thus always zero ($D_b = 0$) because eventually δ will be so small that there will only be one point in each occupied cell. Once δ becomes sufficiently small, $N(\delta)$ becomes equal to the number of points in the set, and the limit above this is (see Equation 3.9):

$$D_b = \lim_{\delta \to 0} \frac{\log N(\delta)}{\log(1/\delta)} \tag{3.15}$$

where $D_b = 0$, because $\log N(\delta)$ is a constant and $\log(1/\delta) \to +\infty$ as $\delta \to 0$. This is a consequence of all finite sets being zero-dimensional; they have the same dimension as a single point (see Section 2.2.1 for further details). In practice, the dimension D_b will thus be estimated using a range of δ values that are always greater than the resolution of the studied set, which can be significantly greater than a pixel in the case of digitized objects. This potential limitation has to be taken into account when writing (or, more prosaically, using) a computer program for automatic estimation of D_b. In addition, this method does not take into account the frequency with which the set in question might visit the covering cells, and thus local properties of the set (that is, properties pertaining to neighborhoods of individual points) are not distinguishable. It will nevertheless be shown below with the introduction of the cluster dimension, D_c, and the family of dimensions related to frequency distributions that this difficulty can easily be overcome.

Finally, for Euclidean objects, Equation (3.13) directly defines their dimensions. A number of boxes proportional to δ^{-1} are needed to cover a smooth line, and proportional to δ^{-2} and to δ^{-3} to cover a curve convoluted on a plane and in a volume, respectively.

3.2.2.2 Case Study: Burrow Morphology of the Grapsid Crab, *Helograpsus Haswellianus*

3.2.2.2.1 Study Organism

The Australian grapsid crab or mud shore crab, *Helograpsus haswellianus* (Figure 3.9A) is common in sheltered bays and estuaries along the eastern coastline of Australia from Queensland south to Tasmania. *H. haswellianus* is a nocturnal species often found well above high-tide level in areas of mud, and forages widely on the shore between tides (Breitfuss 1982). These crabs can be especially abundant on salt-marsh flats (Figure 3.9B), and some are found well upriver in fairly low salinity areas (Marsh 1982). They are also found among mangrove roots, especially those of *Avicenna marina*, often in association with the red-fingered marsh crab, *Sesarma erythrodactyla*, and the semaphore crab, *Heloecius cordiformis* (Campbell and Griffen 1966).

H. haswellianus burrow in a variety of soft sediments, ranging from dirty sand to moist clay, and shelter under debris or rocks. Such burrowing may create quite distinct systems of interconnecting burrows in muddy estuaries. Burrows increase the surface area available for tidal infiltration of seawater (Smith et al. 1991), thus maintaining a critical chemical pathway between anoxic sediments and seawater (Nomann and Pennings 1998), and provide crabs daytime protection from desiccation and predation as well as being used for courting, breeding, and molting (Morrisey et al. 1999).

Typically, studies of burrow shape have examined metrics such as burrow system shape, burrow system area, number of segments, linearity, turn angle, number of branches, segment length, and branch length; see Romaña et al. (2005) for a detailed explanation of these terms. However, because the interactions between these variables are not clear (Le Comber et al. 2006), recent studies have begun to use fractal dimension to provide a single measure of shape that has the desirable advantage of being independent of burrow length (Puche and Su 2001; Sumbera et al. 2003; Romaña and Le Comber 2004).

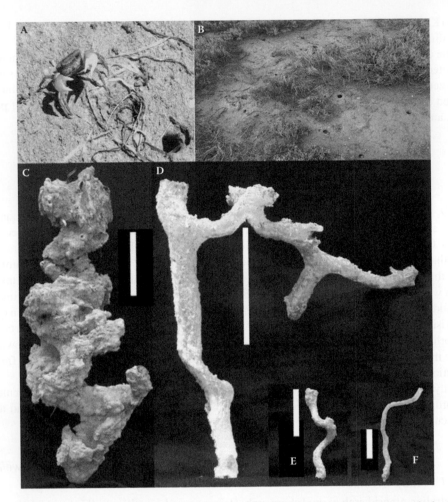

FIGURE 3.9 (A) Grapsid crab, *Helograpsus haswellianus*, shown together with (B) its typical salt-marsh environments, and illustrations of the resin casts sampled at (C) Goolwa, (D) Torrens Island, (E) Middle Beach, and (F) Port Noarlunga. The scale bars represent 10 cm. (Courtesy of G. Katrak, Flinders University, Australia.)

3.2.2.2.2 Experimental Procedures and Data Analysis

The morphology of *H. haswellianus* burrows was investigated by pouring a liquid epoxy resin into openings, leaving it to harden, and digging the cast out after 36 hours. The three-dimensional complexity of the burrows (Figure 3.9C,D,E,F) was investigated using Equation (3.13), a procedure previously successfully applied to other branched structure–like vegetation (Morse et al. 1985; Critten 1993; Zeide and Gresham 1991; Zeide 1998; Eshel 1998; Alados et al. 1998, 1999; Forountan-pour et al. 1999, 2000, 2001). Seaweeds (Corbit and Garbary 1995; Kübler and Dugeon 1996; Davenport et al. 1996), corals (Basillais 1997), sponges (Kaandorp 1991; Kaandorp and de Kluijver 1992; Abraham 2001) and gorgonians (Burlando et al. 1991; Mistri and Ceccherelli 1993) provide instances of marine organisms that have been shown to have fractal properties using the box-counting method (Figure 3.10A,B,C). However, because the fractal characterization of three-dimensional (3D) structures would have nontrivially requested a rebuild of the digitized version of the original burrow, the problem has been simplified by obtaining two orthogonal photographic images of the burrow and subsequently estimating their two-dimensional (2D) fractal dimensions.

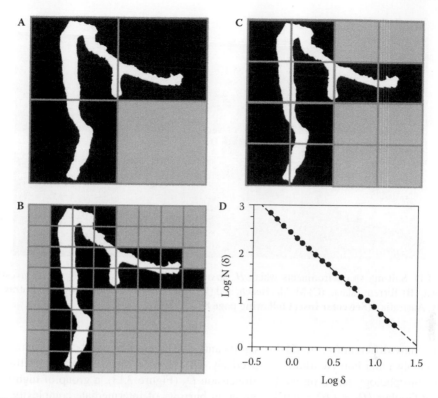

FIGURE 3.10 Schematic illustration of the box-counting method used to describe the complexity of burrow architecture with fractal dimension. Three steps are shown, using three different characteristic scales δ_1 (A), δ_2 (B), and δ_3 (C) defined as $\delta_1 = 2\delta_2 = 4\delta_3$. The gray areas are the squares of size that do not include a part of the burrow. The scaling behavior of the log-log plot of $N(\delta)$ vs. δ for the burrow shown in (A, B, C) is shown in (D).

H. haswellianus burrow morphology was investigated from four distinct South Australian salt marshes located at Goolwa (Figure 3.11A; $N = 2$), Torrens Island (Figure 3.11B; $N = 7$), Middle Beach (Figure 3.11C; $N = 13$), and Port Noarlunga (Figure 3.11D; $N = 14$). The vegetation of the four sites is dominated by *Sarcocornia quinqueflora* and characterized by the presence of *Halosarcia haswellianus*.

3.2.2.2.3 Results

All the burrows investigated exhibited very strong scaling behavior for Equation (3.13) (Figure 3.10D) with the coefficient of determination r^2 ranging from 0.96 to 0.99 over the whole range of available scales, that is, between 0.5 and 15 cm, and 0.5 and 50 cm for the smallest and largest burrows, respectively. This is illustrated by the log-log plot of $N(\delta)$ versus δ (Figure 3.10D) estimated for the burrows shown in Figure 3.10A,B,C. The two-dimensional fractal dimensions $D_b(0°)$ and $D_b(90°)$ were never significantly different (covariance analysis, $p > 0.05$; Zar 1996) for the burrows investigated at Goolwa and Port Noarlunga, and only one burrow each at Torrens Island and Middle Beach exhibited significant differences between $D_b(0°)$ and $D_b(90°)$ (Figure 3.12). The resulting mean two-dimensional fractal dimensions ranged between 1.59 and 1.62 at Goolwa ($D_b = 1.60 \pm 0.02$; $\bar{x} \pm$ SD), 1.34 and 1.67 at Middle Beach ($D_b = 1.56 \pm 0.05$), 1.39 and 1.54 at Port Noarlunga ($D_b = 1.49 \pm 0.03$), and 1.50 and 1.62 at Torrens Island ($D_b = 1.55 \pm 0.04$). Significant differences were found between the fractal dimensions obtained from the burrow resin casts sampled at Goolwa, Middle Beach, Port Noarlunga, and Torrens Island (Kruskal-Wallis test, $p < 0.01$). A subsequent multiple-comparison procedure showed that the fractal dimensions estimated at Middle Beach and Torrens Island were

FIGURE 3.11 Salt-marsh environments where *H. haswellianus* burrow morphology was investigated in (A) Goolwa, (B) Torrens Island, (C) Middle Beach, and (D) Port Noarlunga. (Courtesy of G. Katrak, Flinders University, Australia.) **(See color insert following page 80.)**

not significantly different ($p > 0.05$), but significantly lower and higher ($p < 0.05$) than those estimated at Goolwa and Port Noarlunga, respectively. This leads to the identification of three groups of burrow morphology based on the box dimension D_b (Figure 3.13): a group of highly complex burrows at Goolwa ($D_b = 1.62 \pm 0.02$), a group of burrows of intermediate complexity at Middle Beach and Torrens Island ($D_b = 1.56 \pm 0.05$), and a group of less complex burrows at Port Noarlunga ($D_b = 1.49 \pm 0.03$).

3.2.2.2.4 Ecological Interpretation

No significant differences were found between the fractal dimensions of the burrows investigated at Port Noarlunga whether the dominant vegetation was *Sarcocornia quinqueflora* (20 to 30%)

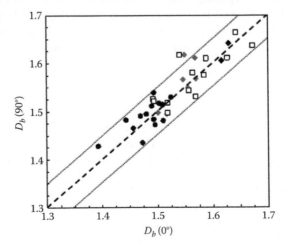

FIGURE 3.12 Comparisons of the two-dimensional fractal dimensions estimated from two orthogonal projections of the original burrow resin casts sampled at Goolwa (black diamonds), Middle Beach (open squares), Port Noarlunga (black dots), and Torrens Island (gray diamonds). The dashed line is the first bisectrix $D_b(0°) = D_b(90°)$, and the dotted lines indicate the corresponding 5% confidence intervals.

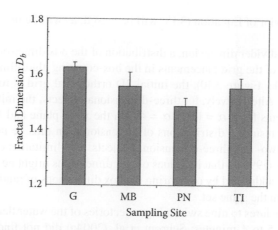

FIGURE 3.13 Fractal dimensions of burrows architecture, estimated as $D_2 = [D_2(0°) + D_2(90°)]/2$ at Goolwa (G), Middle Beach (MB), Port Noarlunga (PN), and Torrens Island (TI). $D_2(0°)$ and $D_2(90°)$ are the fractal dimensions of two random, orthogonal views of the burrows resin casts.

or *Suaeda australis* (20 to 35%). This suggests that the qualitative nature of the vegetation cover does not influence the complexity of burrows. In turn, the quantitative nature of the vegetation cover might influence the complexity of *H. haswellianus* burrows. The fractal dimensions of the burrows investigated at Goolwa, Middle Beach, Torrens Island, and Port Noarlunga on substrates respectively covered at 100%, 85 to 90%, 85%, and 20 to 30% are significantly decreasing with the vegetation cover (Figure 3.13). Finally, the nonsignificant differences found between the burrows investigated at Middle Beach on bare substrate and on a substrate covered by *S. quinqueflora* might also suggest that within a given site, the percentage of vegetation cover has a rather limited effect on burrow morphology. The nonsignificant differences found between the two 2D fractal dimension estimates $D_2(0°)$ and $D_2(90°)$ (Figure 3.12) indicate that the morphology of *H. haswellianus* burrows is mainly isotropic, suggesting a fully three-dimensional burrowing behavior.

As burrowing might be thought as a way of foraging underground, the use of fractals to quantify burrow architecture is conceptually equivalent to the use of fractals to characterize the complexity of three-dimensional trajectories of swimming organisms (Seuront et al. 2004a, 2004b; Uttieri et al. 2005). As most behavioral metrics are also scale dependent (see Seuront et al. 2004a, 2004b for a review), there is no single scale at which swimming paths can be unambiguously described. As a consequence, when using standard metrics, there is no single scale at which swimming paths (and burrow morphology) can be compared without leading to potentially spurious conclusions. Fractal dimension is a natural choice for a measure of burrow shape complexity, as it is essentially a measure of the extent to which a one-dimensional structure fills a plane, with low fractal dimension ($D_b \approx 1$) describing a burrow that explores relatively little of the surrounding area, and high fractal dimension ($D_b \approx 2$) describing a burrow that explores the surrounding area more thoroughly. Fractal dimension seems thus particularly well suited as a measure of burrow shape when burrows are used for foraging.

3.2.2.3 Methodological Considerations

The procedure described above can be used to estimate the box dimension of two- and three-dimensional objects through the superposition of squares (or circles) and boxes (or spheres) of different side lengths and radii to the object of interest. Two potential limitations intrinsically related to the method have nevertheless been identified in both cases: (1) slight reorientation of the overlying grid can produce different values of $N(\delta)$ (Equation 3.13; Appleby 1996), and (2) the values of box dimensions may be positively correlated to the object length (Erlandson and Kostylev 1995).

Consequently, the behavior of Equation (3.13) will be biased, as will be the subsequent box dimension estimates.

As described for the divider dimension, a distribution of the box dimension, D_b, can be obtained from random replicates of the grid placements in the box-counting algorithm. For two-dimensional objects (see, for example, Figure 3.10), the initial 2D orthogonal grid is rotated in 5° increments from $\alpha = 0°$ to $\alpha = 45°$. Alternatively, for three-dimensional objects, the initial 3D orthogonal grid is rotated in 5° increments from $\alpha = 0°$ to $\alpha = 45°$ in the $x - y$ plane and from $\beta = 0°$ to $\beta = 45°$ in the $x - z$ plane. The resulting distributions of dimensions can thus be used as estimates of the box dimensions of the two- and three-dimensional objects. The limitation of the method raised by Erlandson and Kostylev (1995)—that is, values of box dimensions might be positively correlated to a path's length—can be addressed by comparing the box dimensions of randomly chosen subsets of decreasing length within the same set.

Applying these procedures to nine swimming trajectories of the water flea, *Daphnia pulex*, ranging in duration from 1.5 to 4 minutes; Seuront et al. (2004a) did not find any effects related to the orientation of the three-dimensional grid or to the length of the trajectories. It is nevertheless advised that this potential bias should be thoroughly investigated to ensure the robustness of any box dimension estimated through Equation (3.13).

3.2.2.4 Theoretical Considerations

3.2.2.4.1 Two-Dimensional versus Three-Dimensional Fractal Dimensions

The ability to characterize 3D paths based on 2D projections of these paths is an attractive proposition, as the reduction in complexity of both the data-gathering equipment and the analysis procedures is significant. However, the reliability of conclusions based on such a procedure is not clear. The consequences of extrapolating fractal dimensions estimated in a 2D framework to three dimensions have been assessed by the following:

- Testing the validity of the extrapolation procedures proposed in the literature (Morse et al. 1985; Shorrocks et al. 1991; Gunnarsson 1992).
- Investigating the potential disparity among the fractal dimensions estimated from the three orthogonal two-dimensional projections of three-dimensional objects.
- Demonstrating the necessity of three-dimensional isotropy of a given object as a prerequisite for extrapolating two-dimensional fractal information into a three-dimensional space.

The philosophy behind the extrapolation of two-dimensional fractal estimates to three dimensions is simple. Morse et al. (1985) described a box-counting method for estimating the fractal dimension of vegetation habitats ($2 \leq D_3 \leq 3$, where the subscript 3 indicates a fractal object embedded in a three-dimensional space). Consider now the problem of estimating the fractal dimension of a tree. In theory, a three-dimensional grid system could be superimposed on the tree and the size of "counting-cubes" varied. Such a procedure would nevertheless require a digital reconstruction of the tree from photographs, which is still extremely challenging (see, for example, Shlyakhter et al. 2001). Morse et al. (1985) simplified the problem by obtaining a 2D photographic image of the habitat, the fractal dimension of which was determined using the box-counting method ($1 \leq D_2 \leq 2$, where the subscript 2 indicates a fractal object embedded in a two-dimensional space). Following Mandelbrot (1983), they determined heuristic lower ($D_{3min} = D_2 + 1$) and upper ($D_{3max} = 2D_2$) limits of the "habitat" fractal dimension under the assumption that the photograph is a randomly placed orthogonal plane. This procedure has subsequently been used to estimate the fractal dimensions of various habitats (for example, Shorrocks et al. 1991; Gunnarsson 1992). However, it is argued here, on the basis of both simple theoretical and empirical arguments, that the use of this procedure to characterize three-dimensional movement pathways is highly questionable. This issue is illustrated using records of the mud shore crab burrow morphology (Section 3.2.2.2) and the three-dimensional motion behavior of the water flea, *Daphnia pulex* (Box 3.4, Figure 3.14).

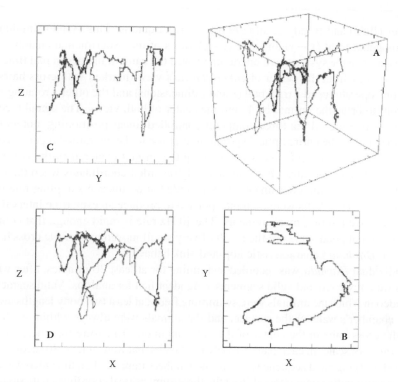

FIGURE 3.14 Illustration of the three-dimensional swimming path of the water flea, *Daphnia pulex* (A), and the corresponding two-dimensional projections on the planes $x - y$ (B), $x - z$ (C), and $y - z$ (D).

BOX 3.4   THREE-DIMENSIONAL ANALYSIS OF THE WATER FLEA, *DAPHNIA PULEX*, SWIMMING PATH

A clone of *Daphnia pulex* was cultured in aged tap water under cool white fluorescent bulbs, in a 16–8 light–dark cycle. The cultures were maintained at the experimental temperature (20°C) and fed every day with a 1:1 mixture of the green algae *Ankistrodesmus* sp. and *Scenedesmus* sp. at a final concentration of about 5×10^5 cells/ml^{-1}. Algae were grown in multiple 250 ml batch cultures under cool white fluorescent bulbs, in an 18–6 light–dark cycle, at 20°C, in Bold's basal medium.

All paths analyzed here are the movements of solitary *D. pulex* swimming in the 5-liter (18 × 18 × 15.5 cm high) Plexiglas recording vessel of the CritterSpy, a high-resolution 3D recording system. All recordings were made with animals swimming in an algal concentration of 5×10^4 cells/ml^{-1}, which is an intermediate food concentration, well below *D. pulex*'s incipient limiting concentration (Lampert 1987). The test chamber was illuminated with a diffused, fiber-optic light placed 0.5 meter directly overhead that resulted in an illumination of about 12 μEm^{-2}s^{-1} in the vessel, approximately equal to full daylight. At least 1 hour prior to experiments, adult, gravid females (2.1 ± 0.2 mm) were transferred from their culturing vessels and acclimated to experimental light and food conditions in holding vessels. A single animal was then transferred from its holding vessel to the recording chamber with a large-bore pipette and allowed to acclimate for at least 10 minutes before recording began.

The CritterSpy uses a schlieren optical system consisting of a collimated red laser beam (λ = 623 nm) that serves as the light source for two orthogonally mounted video cameras, two frame-number generators, two 20" video monitors, and two VHS videocassette recorders;

see Strickler (1985) and Bundy et al. (1993) for further details. This system simultaneously records orthogonal front $(x - y)$ and side $(x - z)$ views of the experimental chamber as dark field images. To run the system, two operators viewed the camera images in real time. As the animal swam away from the center of either camera's view (marked with cross hairs on the monitors), one operator used a trackball (x and z dimensions) and the other a rotating cylinder (y dimension) to bring the animal back into the center of both views. The actual recentering of the image was achieved via three computer-controlled linear positioning motors (one for each axis) that moved the entire optical system in response to the operators' input. A computer recorded the motor movements necessary to keep the animal centered in the two views as x, y, and z coordinates. Because the computer only recorded coordinates when the trackball or cylinder was moved, the coordinates were recorded at an uneven sampling rate (ranging from about 5 to 15 Hz). Paths were then interpolated to produce an even time interval (10 Hz) between successive position measurements. The 10 Hz rate is rapid enough that coordinates recorded at that temporal scale are the result of very small movements of the cross hairs corresponding to *Daphnia*'s characteristic hop-and-sink behavior.

Each individual *Daphnia* was recorded swimming for at least 30 minutes, after which the videotapes were reviewed and valid segments were identified for analysis. Valid segments consisted of video in which the animals were swimming freely, at least two body lengths away from any of the chamber's walls or the surface, and the animals were always within one half-body length of the cross hairs in the center of the video monitors. To ensure that there would be a significant range of scales in each path, we only used paths that were at least 30 seconds in duration. After identifying valid sequences, the frame numbers imprinted on the video were used to isolate the corresponding time interval from the three-dimensional coordinate data stored on the computer. These time series of coordinates formed the 3D trajectories used in the analysis.

First, from a purely theoretical perspective, it appears that the limits of the extrapolated three-dimensional fractal dimension D_3 are not constant (Table 3.3). Instead they increase with increasing values of the two-dimensional fractal dimension D_2. The disparity between the upper and lower limits range from 4.8 to 31.0% for values of the two-dimensional fractal dimension, D_2, bounded between 1.10 and 1.90, respectively (Table 3.3). Moreover, for values of D_2 greater than 1.5, the upper limit of the extrapolated fractal dimension D_3 is beyond the maximum space-filling limit $D_3 = 3$. Consider now the extreme case of an organism moving according to a Brownian motion model, that is, $D_2 \rightarrow 2$. The resulting D_3 values would always be unrealistically found beyond the space-filling limit, $D_3 = 3$, that is, $D_{3min} \rightarrow 3$ and $D_{3max} \rightarrow 4$. This is illustrated using the fractal dimensions estimated from the two-dimensional swimming paths of barramundi fish larvae, which drop from 1.8 prior to metamorphosis to 1.1 following metamorphosis (Dowling et al. 2000). Extrapolating the fishes' behavior to three dimensions will lead to reasonable values of D_3 for fish before metamorphosis, $D_3 \in [2.10 - 2.20]$. However, for postmetamorphosis fishes, 75% of the D_3 values are beyond the space-filling limit, $D_3 \in [2.80 - 3.60]$, and cannot therefore be considered legitimate. Similar conclusions can be reached from the three-dimensional extrapolation of the mud shore crab two-dimensional burrow morphology (Section 3.2.2.2b). Although the lower three-dimensional fractal dimensions ranged from 2.49 to 2.60, their upper limits are consistently higher than the space-filling limit $D_3 = 3$, with $D_3 \in [2.98 - 3.20]$ (Table 3.3). The validity of this extrapolating procedure is then highly questionable. In addition, to be meaningful such extrapolation procedures should, *de facto*, only be applied to perfectly isotropic three-dimensional objects, a property impossible to assess objectively *a priori*.

The similarity or the difference between the fractal dimensions estimated from orthogonal two-dimensional projections of three-dimensional objects on the $x - y$, $x - z$, and $y - z$ planes may instead be much more informative than trying to extrapolate two-dimensional fractal estimates to

TABLE 3.3

Evaluation of the Procedures Used to Extrapolate the Two-Dimensional Fractal Dimension, D_2, of a Three-Dimensional Object to the Corresponding Three-Dimensional Fractal Dimension, D_3

D_2	$D_3 = D_2 + 1$	$D_3 = 2D_2$	$\%D_3$	$\%D_3 > 3$
1.10	2.10	2.20	4.76	-
1.20	2.20	2.40	9.09	-
1.30	2.30	2.60	13.04	-
1.40	2.40	2.80	16.67	-
1.50	2.50	3.00	20.00	-
1.60	2.60	3.20	23.08	33.33
1.70	2.70	3.40	25.93	57.14
1.80	2.80	3.60	28.57	75.00
1.90	2.90	3.80	31.03	88.89

Note: $D_3 = D_2 + 1$ and $D_3 = 2D_2$ are respectively the lower and upper limits of the extrapolated 3D fractal dimensions, $\%D_3$ is the percentage of increase between $D_3 = D_2 + 1$ and $D_3 = 2D_2$ for a given D_2, and $\%D_3 > 3$ is the percentage of extrapolated 3D fractal dimensions that falls outside the space-filling limit, $D_3 = 3$.

three dimensions. This has been illustrated above for the morphology of the grapsid crab *Helograpsus haswellianus* burrow morphology (Section 3.2.2.2). The nonsignificant differences between the fractal dimensions of the two orthogonal projections of the actual burrow observed in most cases (see Figure 3.12) thus illustrate three-dimensional isotropy in the whole burrow. This is, however, not always the case, as illustrated by the three-dimensional swimming behavior of the water flea, *Daphnia pulex* (Figure 3.14A). The box dimensions estimated from the $x - y$, $x - z$, and $y - z$ projections of the same three-dimensional swimming path (D_{2xy}, D_{2xz}, and D_{2yz}, respectively) are always significantly different (Kruskal-Wallis test, $p < 0.05$). More specifically, the dimensions D_{2xz} and D_{2yz} (side views; Figure 3.14C,D) cannot be distinguished (Jonckheere test, $p > 0.05$; Figure 3.15) and are both significantly higher than the dimension D_{2xy} (top view; Figure 3.14B) (Jonckheere test, $p < 0.05$; Figure 3.13). The complexity of the vertical components of the *D. pulex* swimming path is then higher than that of its horizontal components, suggesting that the vertical swimming behavior of *D. pulex* is more complex than the horizontal ones. On the other hand, the average of D_{2xy}, D_{2yz}, and D_{2yz} is not significantly

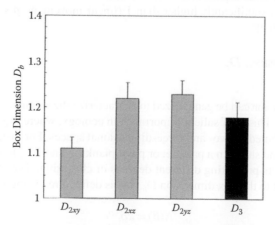

FIGURE 3.15 Box fractal dimension D_b of the two-dimensional projections of the actual three-dimensional path on the planes $x - y$ (D_{2xy}), $x - z$ (D_{2xz}), and $y - z$ (D_{2yz}), compared to the box dimension of the three-dimensional path (D_3).

different from $D_3(p > 0.05)$ due to the intrinsic three-dimensional integrative properties of Equation (3.13). Finally, as expected following the results presented in the previous subsection, the three-dimensional extrapolations of the two-dimensional fractal dimensions D_{2xy}, D_{2yz}, and D_{2yz} are always significantly higher than the actual three-dimensional fractal dimensions. This has been systematically verified for both divider and box dimensions, estimated in two and three dimensions. Consequently, it appears that a two-dimensional fractal dimension is not sufficient to characterize three-dimensional swimming behavior if the swimming path is not isotropic.

3.2.2.4.2   Three-Dimensionally Branched Processes versus Three-Dimensionally Convoluted Processes

The seemingly paradoxical result that fractal dimensions estimated from three-dimensional paths are always significantly smaller than 2, the expected lower bound of values for objects embedded in a three-dimensional space, is detailed hereafter. Following basic fractal theory, an object embedded in a d-dimensional space should have a fractal dimension bounded between $d - 1$ and d (see Section 2.2). A linear succession of spaced dust particles will thus have a dimension bounded between 0 and 1 as they occupy a fraction of the available space greater than a single point (dimension 0) and lower than a line (dimension 1). Similarly, a convoluted curve—a coastline, for instance—will occupy a fraction of space between a line (dimension 1) and a surface (dimension 2), while the dimension of a tree will be bounded between 2 (a surface) and 3 (a volume). Now consider again the case of movement paths. The path of an ant foraging on a flat surface occupies a fraction of a two-dimensional space (see, for example, Figure 3.14B,C,D). Its dimension is then bounded between 1 (a perfectly linear path) and 2 (a plane-filling path). Similarly, the swimming path of *Daphnia pulex* is obviously embedded in a three-dimensional space, the volume of water (Figure 3.14A). However, it does not present a three-dimensional branching structure as does a tree, and each change of direction occurs within a two-dimensional space. Therefore, even in three-dimensional space, a zooplankton swimming path or the flying path of a foraging bee will intrinsically remain a convoluted two-dimensional curve. The fractal dimensions of movement paths are then bounded between a one-dimensional space (a line, $D = 1$) and a two-dimensional space (a surface, $D = 2$). The practical consequence of this specific property of movement paths is to call into question the validity of previous reports of fractal dimensions that fall beyond the $1 \leq D \leq 2$ limits discussed above for both two-dimensional ($D_c < 1$, Dowling et al. 2000; and $D_c > 2$, Bascompte and Vilà 1997) and three-dimensional ($D_c > 2$, Coughlin et al. 1992) analyses. As explored elsewhere (Chapter 7), these discrepancies might result from the lack of objective procedures to identify the scaling ranges, and the subsequent fractal dimensions of movement paths. All of the fractal dimensions estimated from *D. pulex* paths were always consistently significantly higher than 1 (linear movement, $p < 0.01$) and lower than 2 (Brownian motion, $p < 0.01$).

3.2.3 Cluster Dimension, D_c

3.2.3.1 Theory

Formally, the box dimension can be generalized to characterize the extent of self-similar spatial clustering in point patterns. This is of salient importance in ecology, where organisms can be regarded as discrete events distributed in two- and three-dimensional spaces. For instance, the distribution of trees in a forest, cows and sheep in a pasture, or phytoplankton and zooplankton in a water column can be regarded as points presenting different degrees of clustering. The cluster dimension, D_c, is conceptually equivalent to the box dimension D_b, and is defined rewriting Equation (3.13) as

$$N(\delta) = k\delta^{-D_c} \tag{3.16}$$

where δ is still the box size, $N(\delta)$ the number of boxes occupied by at least a single point (that is, an organism), k a constant, and D_c the cluster dimension, estimated as described in Section 3.2.2).

The cluster dimension D_c can also be computed using "counting disks" instead of boxes (Frontier 1987). Robertson et al. (1995) used a three-dimensional "cube-counting version" of the cluster dimension to study the distribution of earthquake hypocenters in space. As the one-to-one correspondence between the box dimension, D_b (Section 3.2.2) and the cluster dimension, D_c, might not be obvious from the literature, the reader should note that:

$$D_b = D_c \tag{3.17}$$

A variant of the above methods has been suggested by Hastings et al. (1992) in a study of the distribution of pancreatic islets. Assuming a Poisson distribution, the cumulative number of points $n(\delta)$ within a distance δ scales as:

$$n(\delta) = k\delta^{D_c} \tag{3.18}$$

where k is a constant.

Alternatively, King et al. (1989) suggested counting the number of points $n(\delta)$ within each grid unit of size δ and estimating the relative dispersion $RD(\delta)$ as the ratio between the standard deviation, $CV(\delta)$, and the mean, $\bar{x}(\delta)$, of grid counts; that is, $RD(\delta) = CV(\delta)/\bar{x}(\delta)$. This results in a power-law relationship between relative dispersion and the grid unit size δ defined as:

$$RD(\delta) = k\delta^{D_{c''}-1} \tag{3.19}$$

where $RD(\delta)$ is the relative dispersion at the spatial scale δ, and k is a constant. A set of points randomly distributed will have a fractal dimension $D_{c''} = 1.5$, while a value of $D_{c''} = 1.0$ reflects uniformity of the property over all length scales. Interestingly, the fractal dimension $D_{c''}$ can be used to quantify the spatial correlation r between clusters over the range of scales δ as (King et al. 1989):

$$r = 2^{3-2D_{c''}-1} \tag{3.20}$$

For $D_{c''} = 1.5$ (random pattern), Equation (3.20) leads to the minimal correlation $r = 0$, while the correlation is maximal ($r = 1$) for $D_{c''} = 1.0$.

3.2.3.2 Case Study: The Microscale Distribution of the Amphipod *Corophium Arenarium*

3.2.3.2.1 *Study Organism*

Corophium arenarium is a small (up to 7 mm) amphipod, an order of crustaceans in the subclass Malacostraca. *C. arenarium* lives in soft sand or mud on intertidal flats of the coasts of France, England, the Netherlands, and Germany. They usually form U-shaped tubes that extend down to 3 cm in summer and to 12 cm in winter to escape the freezing point. *C. arenarium* feeds on the organic matter deposited on the sediment surface, scratching along the surface with its elongated second pair of antennae, producing starlike patterns.

3.2.3.2.2 *Experimental Procedures and Data Analysis*

The study site is located in the Bay of Somme (France), at Le Crotoy (50°13′524 N, 1°36′506 E), which is the second-largest estuarine system, after the Seine estuary, and the largest sandy-muddy (72 km²) intertidal area on the French coasts of the eastern English Channel. The sampling site was chosen in a topographically homogeneous area, where the substrate grain size typically varied between 125 and 250 µm (modal size). The abundance of the amphipod *C. arenarium* was estimated nonintrusively from the number of opened burrows using digital images of every 20 cm × 20 cm subsection of a 1 m² quadrat (Figure 3.16A). Preliminary investigations showed a highly significant positive relationship ($r^2 = 0.95$, $p < 0.001$; Figure 3.16B) between the number of *C. arenarium* estimated from core samples sieved through a 500 µm screen and from the above-mentioned nonintrusive technique.

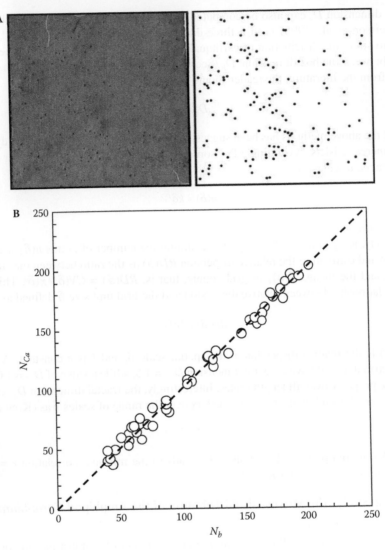

FIGURE 3.16 Nonintrusive photographic technique used to estimate the number of opened burrows (N_b) using digital images of every 20 cm × 20 cm subsection of a (A) 1 m² quadrat and (B) relationship between the number of opened burrows and the number of *C. arenarium* (N_{ca}) estimated from core-samples sieved through a 500 μm screen.

3.2.3.2.3   Results and Ecological Interpretations

The two-dimensional distribution of the 800 *C. arenarium* burrow openings found over one square meter (Figure 3.17A) was visually very distinct from the random point patterns simulated with the same number of data points (Figure 3.17B). In both cases, Equation (3.16) exhibited a highly significant linear behavior over two decades—that is, 1 cm to 1 m—resulting in cluster dimensions $D_c = 1.33 \pm 0.01$ (Figure 3.17C) and $D_c = 1.84 \pm 0.02$ (Figure 3.17D) for the empirical *C. arenarium* distribution and the simulated random point pattern, respectively. The cluster dimension of *C. arenarium* distribution is consistent with the fractal dimensions found for microphytobenthos biomass in the same environment, which range from 1.07 to 1.89, depending on the concentration threshold considered (Seuront 2005a); (see also Section 3.2.4.2; Figure 3.21). A fractal dimension $D_c = 1.33 \pm 0.01$ corresponds to the fractal dimension found for microphytobenthos

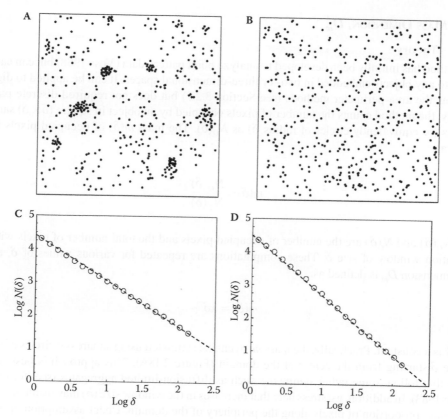

FIGURE 3.17 Two-dimensional microscale distribution of (A) the amphipod *Corophium arenarium* compared to (B) a random point pattern simulated with the same number of data points and the results of the corresponding box-counting method used to estimate the cluster dimension, D_c. The distribution of *C. arenarium* returned (C) a cluster dimension, $D_c = 1.33$, significantly different from (D) the dimension of the random point pattern, $D_c = 1.84$.

concentrations higher than 78 mg chlorophyll *a* m^{-2} (Seuront 2005a; for microphytobenthos biomass ranging from 42 to 114 mg chlorophyll *a* m^{-2}). This suggests that (1) *C. arenarium* might have a preferential concentration for the food they prey on, (2) the spatial distribution of food items (assessed through their fractal dimension) has the potential to drive the fractal distributions of their motile predators, and (3) the fractal dimension of predators might be used to infer the spatial distributions of their prey items, and *vice versa*.

3.2.3.3 Methodological Considerations: Constant Numbers or Constant Radius?

Using a two- or three-dimensional grid, at each node of the grid, the nearest neighbors are sampled in two or three dimensions, resulting in circular surfaces or spherical sampling volumes of either constant sample size or constant radius. Either approach is equally valid, and comparing the surface/volume of both is recommended to ensure that they are independent of the choice of the sampling method. By sampling a constant number of events at each node, the sample size, and hence uncertainty, is approximately constant, and the best spatial resolution possible at each node is achieved. In this case, the radii of sampling surfaces/volumes, or resolution, are inversely proportional to the local density of events and consequently are variable across a region. When using constant radii for sampling, the resolution does not vary spatially, but the sample size, and hence the uncertainty, does. It is necessary to exclude nodes where fewer than a minimum number of events are sampled, or one can use a cutoff, based on a maximum for the allowed uncertainty.

3.2.4 Mass Dimension, D_m

3.2.4.1 Theory

This method has initially been developed to analyze point pattern data (Voss 1988) but can easily be applied to any objects embedded in two- or three-dimensional spaces. It can be applied to digitized images as the area-perimeter methods (see Section 3.2.7) but does not required discrete patterns. Formally, the method counts the number of pixels occupied by an object in square ($\delta \times \delta$) sampling windows (or equivalently circles of radius δ) as $N_O(\delta)$. The mass $m(\delta)$ of occupied pixels is then defined as:

$$m(\delta) = \frac{N_O(\delta)}{N_T(\delta)} \tag{3.21}$$

where $N_O(\delta)$ and $N_t(\delta)$ are the number of occupied pixels and the total number of pixels within an observation window of size δ. These computations are repeated for various values of δ, and the mass dimension D_m is defined as:

$$m(\delta) = k\delta^{D_m} \tag{3.22}$$

where k is a constant. Practically, the mass $m(\delta)$ can be estimated using squares or circles of increasing size δ starting from the center of the domain (Figure 3.18A). This approach is best suited to objects that follow some radial symmetry, such as diffusion-limited aggregates (see, for example, Meakin 1983). In addition, we stress here that increases in the window size (δ) may result in exclusion of a greater proportion of pixels along the periphery of the domain. Under assumption of isotropy, a toroidal edge correction can nevertheless be applied to circumvent this problem. Alternatively, in the case of point-pattern data sets, calculating the mass $m(\delta)$ as the average mass in a number of squares or circles of radius δ (Figure 3.18B) is recommended. A surprising application of the mass dimension, D_m, to assess the existence of a "master plan" in the design of the Teotihuacan archaeological zone located 50 km northeast of Mexico City is provided in Box 3.5.

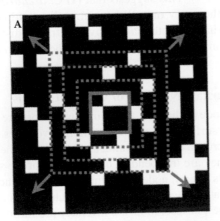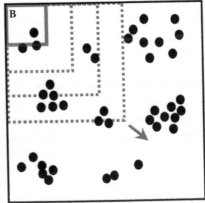

FIGURE 3.18 The mass dimension method. Using squares of increasing size starting from the center (A) or the side (B) of the domain under interest, one counts the number of occupied pixels (shown in black), and estimates the mass, $m(\delta)$ (see Equation 3.19). The slope of the linear behavior of $m(\delta)$ vs. δ in a log-log plot provides an estimate of the mass dimension, D_m (see Equation 3.22).

BOX 3.5   MASS FRACTAL DIMENSION IN ARCHEOLOGY AND ARCHITECTURE

The history of the Teotihuacan archaeological zone, located 50 km northeast of Mexico City, spans from 100 B.C. to 700 A.D. and is recognized to have greatly influenced civilizations in Mexico and Guatemala. The general design of this zone has been suggested as an integrated display of mathematical and geodetic information (Oleschko et al. 2000a). Teotihuacan was modified nearly continuously over more than 800 years. However, a range of anthropologists claim that the standardization in architectural orientation and construction, and the symmetry and proportionality in the spatial distribution of buildings, suggest that the major structures of Teotihuacan were laid out from its foundation according to a master plan intended to express a specific view of the world in material form (Sugiyama 1993). Using a range of gray-level radar images and aerial photographs of the Teotihuacan site, Oleschko et al. (2000a) refined those previous hypotheses, showing that the fractal dimension of the major monuments of Teotihuacan site were very similar, ranging from 1.8767 to 1.8993. In addition, those dimensions were very close to the dimension of the Sierpinski carpet, one of the best known theoretical fractals (see Figure 2.8A and the "negative" Sierpinski carpet, Figure 3.B5.1), 1.8928. The Sierpinski carpet was subsequently proposed as the model (that is, the master plan) for Teotihuacan (Oleschko et al. 2000a). Although it is difficult to explain the existence of a unique master plan for a site that continuously evolved over 8 centuries, this represents an illustration of fractal geometry not only being the geometry of nature but also representing the fractal structure of cities (see, for example, Batty and Longley 1994; Frankhauser 1994; Batty 1995), as well as a striking example of the similarity between a man-made structure and a theoretical fractal object created more than a millennium apart.

FIGURE 3.B5.1 Similarity between two-dimensional projections of the "negative" of the Sierpinski carpet (left) and the Great Compound, Ciudadela, of the Teotihuacan archeological site (right). (Modified from Oleschko et al., 2000a.)

3.2.4.2   Case Study: Microscale Distribution of Microphytobenthos Biomass

3.2.4.2.1   The Study Organism

Microphytobenthos are photosynthetic cells living within the surface layers of coastal sediments. They provide as much as 50% of the carbon fixed in some coastal systems (MacIntyre et al. 1996; Serôdio and Catarino 2000), especially in intertidal mudflats (MacIntyre et al. 1996; Barranguet 1997) and shallow subtidal locations (Miles and Sundbäck 2000; Glud et al. 2002). They are ecologically critical as a food resource (Blumenshine et al. 1997) and as "ecosystem architects," altering the erosion potential of coastal sediments (Rietmüller et al. 2000). The majority of the cells that can be found in the near-shore sediments, either attached to sand grains or rocks (epilithic cells) or living in the interstitial water (epipelic cells) are diatoms (Medlin 2006), characterized by robust and heavily silicified frustules (such as *Caloneis* sp., *Diploneis* sp.; Figure 3.19). Some of them are motile, and they secrete mucus that allows them to glide freely on the sediment. Vertical migration can be observed on both sandy and muddy flats, and exhibit diel rhythms. Microphytobenthos cells migrate to the surface of the sediment during daytime emersion, thus when photosynthesis and primary production occur (Janssen et al. 1999; Serôdio and Catarino 2000). The migration results in the formation of a biofilm, which consists of a dense layer of cells at the sediment surface (Paterson and Crawford 1986). While most cells migrate back into the sediment before the rising tide or at nightfall (Serôdio et al. 1997; Guarini et al. 2000), a portion of the biofilm may also remain on the sediment surface during the rising tide, leading to cell resuspension into the water column. Recent results have also revealed that the high diversity and rapid turnover of microphytobenthos populations make them ideal as a model system for the study of ecological theory (such as diversity vs. productivity issues) and aspects of ecosystem change (such as global warming; Riaux-Gobin 1997).

3.2.4.2.2 *Experimental Procedures and Data Analysis*

The study site, an intertidal sand flat in Wimereux (France), is typical of the hydrodynamically exposed sandy beach habitats that dominate the littoral zone along the French coast of the eastern English Channel. The sampling area (50°45′896 N, 1°36′364 E), located in the upper intertidal zone, did not exhibit any elevational gradient or sharp topographical features as ripple marks, high pinnacles, or deep surge channels, and was characterized by homogeneous medium-size sand (200 to 250 µm, modal size), typical of the surrounding sandy habitat (Seuront and Spilmont 2002). Air temperatures at the sampling site range from about 1 to 10°C in the winter to highs of about 10 to 25°C in the summer (Seuront 2005a). Water temperatures vary from 5°C to approximately 20°C depending on the season. Salinity is usually about 31% but can also vary with the season, being lower in late winter and early spring and higher in late summer and fall.

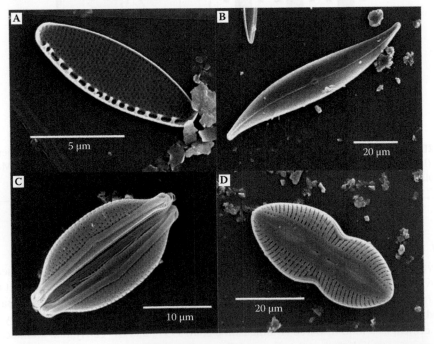

FIGURE 3.19 Electron microscopy photographs of microphytobenthic diatoms: (A) *Nitzchia* sp.; (B) *Gyrosigma* sp.; (C) *Amphora* sp.; and (D) *Diploneis* sp.

Measurements were performed at low tide, in the middle of the emersion period, on 27 September 2001. Samples were collected in a rigid 1-m² aluminum quadrat constructed from 225 1.9-cm² plastic cores resulting in an intersample distance of 6.67 cm. The cores were pushed into the sediment down to a depth of 1 cm, where the majority of the active cells are concentrated. This ensures that the observed spatial structure is not biased by any change in the spatial (that is, vertical) organization of the microphytobenthos during the sampling process. Sediment samples were then placed in 8 ml acetone, and pigments were extracted for 4 hours in the dark at 4°C (Seuront and Spilmont 2002; Seuront and Leterme 2006). After extraction, samples were centrifuged at 4000 rpm for 15 minutes. Chlorophyll a concentrations (mg) in the supernatant were determined by spectrophotometry following Lorenzen (1967). Chlorophyll a concentrations estimated in the supernatant have subsequently been expressed in terms of chlorophyll a per surface unit (mg m⁻²), taking into account the 1.9-cm² surface of the sampling unit.

3.2.4.2.3 Results

Microphytobenthos chlorophyll a concentration exhibits a very intermittent behavior, where sharp fluctuations occurring locally are clearly visible (Figure 3.20A). Chlorophyll a concentration ranged between 1.90 and 27.15 mg Chl.a m⁻², that is, 10.79 ± 4.15 mg m⁻² $(\bar{x} \pm SD)$. Results of descriptive analysis, including skewness and kurtosis estimates, show that the 225 microphytobenthos chlorophyll a concentration estimates are not normally distributed (Kolmogorov-Smirnov test, $p < 0.01$). Their frequency distribution rather exhibits a positively skewed behavior ($g_1 = 0.60$), reflecting a distribution characterized by a few dense patches and a wide range of low density patches. Finally,

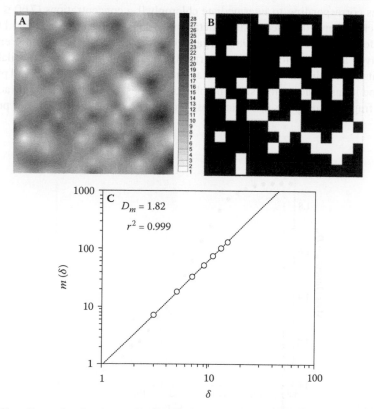

FIGURE 3.20 Two-dimensional microscale distribution of microphytobenthos chlorophyll a concentration (mg Chl.a m⁻²) illustrated (A) as a continuous pattern and (B) as a discrete pattern with concentrations $C > 10.79$ mg Chl.a m⁻² shown in black, and the scaling behavior of the log-log plot of $m(\delta)$ vs. δ for the distribution shown in (C).

the positive kurtosis shows a distribution that is more peaked than expected in the case of normality (g_2 = 1.83). As organisms foraging for food (for example, the amphipod *C. arenarium*; see Section 3.2.3) are likely to actively select areas of high chlorophyll *a* concentration, they may take advantage of positively skewed distribution. Equation (3.22) was then applied to the distribution pattern of chlorophyll *a* concentration higher than a given threshold (such as $C > 8.85$ mg Chl.*a* m⁻²; Figure 3.20B) to investigate the relative advantages that foraging organisms may have to actively select a specific range of food concentrations. The whole plot has size 15 × 15 in plot units, and each pixel represents a surface of 6.67 × 6.67 cm in the field. The log-log plot of $m(\delta)$ vs. δ exhibits a very clear scaling behavior, resulting in the mass dimension $D_m = 1.82$ for $C > 8.85$ mg Chl.*a* m⁻² (Figure 3.20C). More generally, the fractal dimensions D_m related to chlorophyll *a* concentrations ranging from 1.9 to 26.6 mg Chl.*a* m⁻² clearly decrease with increasing chlorophyll *a* concentration (Figure 3.21).

3.2.4.2.4 Ecological Interpretation

Low chlorophyll densities ($C \leq 6.65$ mg Chl.*a* m⁻²) are characterized by high fractal dimensions, $D_m = 1.89 \pm 0.01$. Such high dimensions (the maximum value that D_m can reach is $D_m = 2.00$) characterize very complex processes where short-range, local variability is highly developed and tends to obfuscate long-range trends; the variable is more evenly or regularly distributed (that is, less structured) in space. In other words, this indicates that the variation within a sampling unit is equal to the variation among the sampling units. Alternatively, the dimensions D_m related to high chlorophyll concentrations ($C \leq 6.65$ mg Chl.*a* m⁻²) are very low ($D_m = 1.03 \pm 0.02$) and cannot be distinguished from the lower D_m value ($D_m = 1$). This indicates that the variability of high microphytobenthos concentrations is characterized by the so-called random point pattern (Li 2000) (see Figure 2.9D). Finally, patches corresponding to intermediate chlorophyll *a* concentrations ($7.60 \leq C \leq 21.85$ mg Chl.*a* m⁻²) are characterized by decreasing fractal dimensions D_m from $D_m = 1.85$ to 1.07. According to the optimal foraging theory (Pyke 1984), organisms are expected to optimize the energy required to capture a given amount of food. This is particularly relevant here as food availability changes depending on its fractal dimension. A low fractal dimension relates to a smooth and predictable distribution of food items gathered in a small number of patches. In contrast, a high fractal dimension means rough, fragmented, space-filling and less predictable distribution. When a predator has no

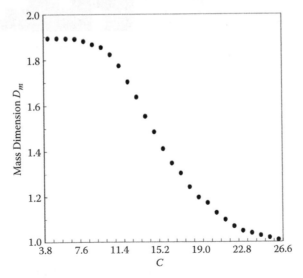

FIGURE 3.21 Mass fractal dimension D_m estimated from discrete patterns of increasing microphytobenthos chlorophyll *a* concentrations. The dimension D_m is estimated as the slope of the log-log plot of $m(\delta)$ vs. δ (Equation 3.20). (See Figure 3.20C.)

detection ability, prey distributions with high dimensions should be more efficient, as food availability (which is here equivalent to the predator–prey encounter rates) becomes proportional to the searched volume as fractal dimension increases. When a predator can remotely detect food items, prey distributions with low fractal dimension should be relatively better. Given the sensory abilities of intertidal organisms preying on microphytobenthos (see, for example, Erlandson and Kostylev 1995; Hutchinson et al. 2007), in the specific case studied here, microphytobenthos grazers should then derive maximum benefit from high concentration patches.

3.2.4.3 Comparing the Mass Dimension D_m to Other Fractal Dimensions

From Equations (3.9) and (3.10), and from Equations (3.19) and (3.20), it is readily seen that for strictly self-similar mathematical fractals, such as the Cantor dust (Figure 3.3) and the Sierpinski carpet and gasket (Figure 2.8), the mass dimension D_m is the same as the Hausdorff dimension D_H, and strictly speaking to any other fractal dimension related to Equations (3.9) and (3.10) in this section. However, for real-world fractals, there are significant differences. The mass dimension can nevertheless still be directly compared to the box dimension, D_b.

Consider the mass of occupied pixels in a window (that is, box) of size (δ) as:

$$m(\delta)_i = \frac{N_O(\delta)_i}{N_T(\delta)_i} \tag{3.23}$$

where $m(\delta)_i$ is the mass of the occupied pixels $N_O(\delta)$, and $N_T(\delta)_i$ the total number of pixels in the ith box of size δ. The average mass, $m(\delta)$, in $N(\delta)$ boxes of size δ is then:

$$m(\delta) = \frac{1}{N(\delta)} \sum_{i=1}^{N(\delta)} m(\delta)_i \tag{3.24}$$

Now, rethinking Equation (3.23) in terms of probabilistic arguments leads to expressing count frequencies as probability density function $p(\delta)$ following:

$$p(\delta) = \sum_{i=1}^{N} m(\delta)_i = 1 \tag{3.25}$$

One may note here that for a given value of δ, the mass $m(\delta)$ is expressed as the first moment (that is, the mean) of the probability distribution as:

$$m(\delta) = \sum_{i=1}^{N} N_O(\delta)_i\, p(\delta) \tag{3.26}$$

Using Equations (3.25) and (3.24) can be equivalently thought of as $m(\delta) = 1/N(\delta)$, leading to rewrite Equation (3.22) as:

$$N(\delta) = k\delta^{-D_m} \tag{3.27}$$

which is fully similar to the power-law relationship used to estimate the box dimension ($N(\delta) = k\delta^{-D_b}$), leading to:

$$D_m = D_b \tag{3.28}$$

and using Equation (3.17) to

$$D_m = D_b = D_c \tag{3.29}$$

The mass and box dimensions estimated for the distribution pattern of *Corophium arenarium* shown in Figure 3.17A are respectively $D_m = 1.29 \pm 0.02$ and $D_b = 1.28 \pm 0.01$, and cannot statistically be distinguished ($p > 0.05$) from the cluster dimension estimated in Section 3.2.3.2, $D_c = 1.28 \pm 0.02$. Similarly, the mass, box, and cluster dimensions estimated from the microscale distribution patterns of microphytobenthos chlorophyll biomass (Figure 3.20A) cannot be statistically distinguished whatever the chlorophyll a concentration C considered.

3.2.5 INFORMATION DIMENSION, D_i

3.2.5.1 Theory

The information dimension, D_i, can be conceptually related to the box dimension D_b and the cluster dimension D_c, because it is based on a count of occupied boxes of varying size δ (Figure 3.22). However, in the box and the cluster dimension estimates, a box is counted as occupied (Figure 3.22A,B,C) and enters the calculations of $N(\delta)$ (see Equations 3.12 and 3.14), regardless

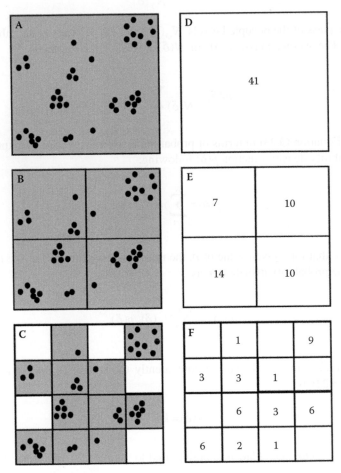

FIGURE 3.22 Schematic illustration of three successive steps of the box-dimension procedure, or equivalently the cluster dimension procedure (A, B, C) and information dimension procedure (D, E, F) using three different characteristic box sizes: δ_1(A, D), δ_2(B, E), and δ_3(C, F) defined as $\delta_1 = 2\delta_2 = 4\delta_3$.

of whether it contains one point or a relatively large number of points. The information dimension provides more details, as the number of points n_i within each occupied box are counted (Figure 3.22D,E,F), and is expressed as the relative frequency f_i:

$$f_i = \frac{n_i}{N} \tag{3.30}$$

where N is the total number of points in the set $\sum_{i=1}^{N(\delta)} f_i = 1$, and $N(\delta)$ is the number of occupied boxes of size δ. A weight is then assigned to each box; the boxes containing a greater number of points count more than boxes with fewer points. The information entropy, or Shannon entropy $H(\delta)$ at a scale δ is subsequently defined as:

$$H(\delta) = -\sum_{i=1}^{N(\delta)} f_i \log f_i \tag{3.31}$$

Consider now a uniformly distributed point pattern in one-, two-, and three-dimensional spaces. The number of points n_i and the frequencies f_i are thus proportional to δ^{-1}, δ^{-2}, and δ^{-3} and to δ^1, δ^2, and δ^3, respectively. Equations (3.12) and (3.14) are subsequently rewritten as;

$$H(\delta) = \log k - D_E \log \delta \tag{3.32}$$

where k is a constant. For nonuniformly distributed point patterns, the information dimension D_i is subsequently defined as:

$$H(\delta) = \log k - D_i \log \delta \tag{3.33}$$

where k is still a constant.

3.2.5.2 Comparing the Information Dimension D_i to Other Fractal Dimensions

Although the above-mentioned arguments lead us to consider the information dimension as a generalization of the box dimension, these dimensions are nevertheless not equal. Consider a uniform point pattern embedded in a D_E-dimensional space. All the frequencies f_i are equal, each $f_i = 1/N(\delta)$ and Equation (3.31) leads to $H(\delta) = \log N(\delta)$, that is, the maximum value of $H(\delta)$. All the other (that is, smaller) values of $H(\delta)$ thus quantify the nonuniformity of the point pattern or alternatively correct the dimension estimate by giving less weight to the boxes that contain relatively fewer points. In other words, if one defines the information dimension as:

$$D_i = \lim_{\delta \to 0} \frac{H(\delta)}{\log(1/\delta)} \tag{3.34}$$

and goes back to the formulation of the box dimension (Equation 3.15), it comes that:

$$D_i = D_b \tag{3.35}$$

for uniform point patterns, and

$$D_i < D_b \tag{3.36}$$

for nonuniform point patterns. The validity of Equations (3.33) and (3.34) is illustrated by the information dimension analysis of the microscale distribution of the amphipod *Corophium arenarium*

(Section 3.2.3; Figure 3.17A) and the related simulated uniform point pattern (Section 3.2.3) (Figure 3.17B). As predicted by Equation (3.35), the information and box dimensions of the uniform point pattern cannot be statistically distinguished ($p > 0.05$). In turn, the information dimension D_i is significantly different from the box dimension D_b for the point pattern of *C. arenarium* ($p < 0.01$). Farmer et al. (1983) refined the definitions of both box and information dimensions, stating that D_b is a "metric dimension" (that is, it depends only on metric scaling properties, δ), while D_i is a "probabilistic dimension" (that is, it depends on both metric, δ, and probabilistic, f_i, properties).

3.2.6 Correlation Dimension, D_{COR}

3.2.6.1 Theory

The correlation dimension is well adapted to the characterization of spatial clustering in point patterns and was initially introduced to characterize the dimension of strange attractors (see Section 6.1.3). This method, widely used in empirical analyses of dynamical systems (Grassberger and Procaccia 1983) and in cosmology (McCauley 2001), has, to our knowledge, never been used in ecology. A generalization of this method to the stochastic process has been applied by Ibanez (1986) and Seuront (1999, 2004) to plankton transects and time series, respectively (see Chapter 6). The correlation function of a point pattern, usually referred to as the correlation integral $C(\delta)$, is calculated as (Hentschel and Procaccia 1983):

$$C(\delta) = \frac{1}{N} \sum_{i=1}^{N} C_i(\delta) \tag{3.37}$$

where $C_i(\delta) = \frac{1}{N} \Sigma_{i \neq j=1}^{N} \theta(\delta - d_{i,j})$ is the number of distinct pairs of points in a circle (or equivalently a sphere if the point pattern is embedded in three dimensions) of radius δ, centered on the ith of N points, $d_{i,j}$ is the Euclidean distance between the ith and the jth points ($d_{i,j} = |x_i - x_j|$), and $\theta(\xi)$ is the Heaviside function, defined as $\theta(x) = 0$ and $\theta(x) = 1$ for $x < 0$ and $x \geq 0$, respectively. It is possible to take $0 < \delta < 1$. This means that the original dimensional variable δ for each point has been rescaled by dividing it by $\delta_{max,i}$, where $\delta_{max,i}$ is the value of the unscaled variable δ for which the radius δ, centered on point i, just touches the boundary of the data set (Figure 3.23).

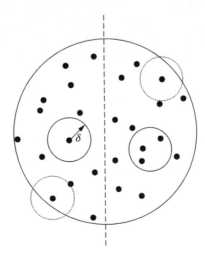

FIGURE 3.23 A point pattern, illustrating which clusters are used (solid circles) and not used (dashed circles) in the search for scale invariance.

Equation (3.37) can be rewritten more simply as:

$$C(\delta) = \frac{1}{N^2} N_{i,j}$$ (3.38)

where N is the total number of pairs of points in the set, and $N_{i,j}$ is the number of pairs with $d_{i,j} < \delta$. In other words, the correlation integral $C(\delta)$ represents the probability that the distance between a pair of randomly chosen points will be less than a distance δ apart. In the case of a uniform point pattern, the correlation integral scales with the distance δ as:

$$C(\delta) = k\delta^{D_E}$$ (3.39)

while for nonuniform distributions, $C(\delta)$ is given by:

$$C(\delta) = k\delta^{D_{cor}}$$ (3.40)

where k is a constant and the exponent D_{cor} is the correlation dimension, estimated as the slope of the log-log plot of $C(\delta)$ vs. r.

3.2.6.2 Comparing the Correlation Dimension D_{cor} to Other Fractal Dimensions

Considering the correlation dimension as:

$$D_{cor} = \lim_{\delta \to 0} \frac{\log C(\delta)}{\log(\delta)}$$ (3.41)

leads to:

$$D_{cor} = D_i = D_b$$ (3.42)

for uniform point patterns, and it is seen from the comparison of Equations (3.13), (3.32), and (3.39) that:

$$D_{cor} < D_i < D_b$$ (3.43)

for nonuniform distributions. The correlation dimension D_{cor}, information dimension D_i, and box dimension D_b returned by respectively applying Equations (3.36), (3.29), and (3.12) to the microscale point pattern of the amphipod *Corophium arenarium* (see Section 3.2.3, Figure 3.17A) and the related uniform point pattern simulated with the same number of data points (see Section 3.2.3, Figure 3.17B). As predicted by Equation (3.42), the correlation, information, and box dimensions of the uniform point pattern are not statistically different ($p > 0.05$) (Figure 3.24). On the other hand, the three dimensions D_{cor}, D_i, and D_b are significantly different ($p < 0.05$) for the point pattern of *C. arenarium*, with $D_{cor} = 1.15 \pm 0.01$, $D_{cor} = 1.22 \pm 0.02$, and $D_{cor} = 1.28 \pm 0.02$ (Figure 3.24), hence verifying Equation (3.43). Chapter 5 shows in detail how the correlation dimension D_{cor}, the information dimension D_i, and the box dimension D_b can also be related using a generalization of the entropy formulation introduced in Equation (3.31).

3.2.7 AREA-PERIMETER DIMENSIONS

Area-perimeter methods have generally been used to estimate the fractal dimension of objects coded as digitized images. Consider as an example a landscape consisting of a set of vegetation patches (Figure 3.25). These patches can be either monospecific (Figure 3.25A,B) or plurispecific (Figure 3.25C). Area-perimeter dimension methods can be used to determine the fractal dimension of a set of patches as a function of the complexity of their boundaries or their space-filling

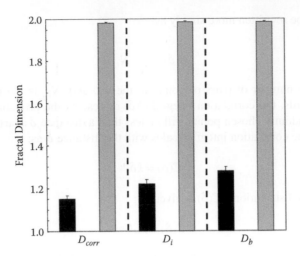

FIGURE 3.24 Correlation dimension D_{cor}, information dimension D_i, and box dimension D_b estimated for the microscale point pattern of the amphipod *Corophium arenarium* (see Figure 3.16) and a simulated uniform point pattern with the same number of data points.

character and also as a function of the quality of the surrounding patches. Three approaches are thus possible:

- *Perimeter-based*, to determine the extent that a patch perimeter fills the plane
- *Area-based*, to determine the extent that a patch fills the plane
- *Landscape-based*, to compare patch complexity in an environment characterized by patches of different types

Each of the three methods described in this section measures a fractal dimension, but it will be shown that their application and interpretation are quite different.

3.2.7.1 Perimeter Dimension, D_p

The perimeter dimension (D_p) method basically provides a measure of the perimeter-area ratio for a patch or a set of patches (Figure 3.25A). Consider an "ideal" circular patch (for example, Okubo 1980) in a two-dimensional space. The area A and the perimeter P are related as:

$$P = 2\pi r = r\sqrt{\pi A} \approx \sqrt{A}$$

$$A = \pi r^2 = \frac{P^2}{4\pi} \approx P^2 \tag{3.44}$$

where r is the radius of the patch, and "$\approx$" means proportionality. In case of a non-Euclidean patch structure, Equation (3.44) leads to the following perimeter-area relationship:

$$P = kA^{D_p/2} \tag{3.45}$$

where k is a constant, the area A is the number of pixels needed to cover a given object, the perimeter P is the number of pixel edges, and D_p is the perimeter dimension. For a single patch, the perimeter dimension is thus simply written as:

$$D_p = 2\log P / \log A \tag{3.46}$$

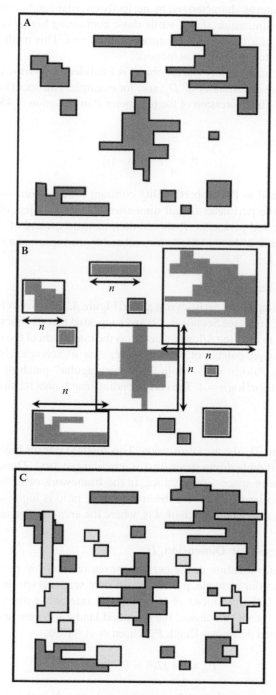

FIGURE 3.25 Area-perimeter dimensions. (A) Perimeter dimension D_p is based on estimates of patch areas (light gray) and perimeter (indicated by a black solid line). (B) Area dimension D_a is based on estimates of patch areas and largest horizontal or vertical dimension (n). (C) Landscape/seascape dimension D_s is a generalization of the perimeter dimension to patches of different species, here shown by different gray intensities.

Alternatively, the slope of the log-log area-perimeter plot for a set of objects gives a fractal dimension (Burrough 1986). Patterns characterized by perfectly circular patches (that is, a low perimeter-area ratio) have a fractal dimension $D_p = 1$, while those containing highly convoluted patches (that is, a high perimeter-area ratio) have a fractal dimension $D_p \to 2$. This method is thus well adapted to studies focusing on ecotonal boundaries (edges).

However, a digitized image of any fractal object is Euclidean by virtue of its projection onto a grid, and thus leads to biased estimates of D_p; see, for example, Falconer (1993), and Section 7.2.1. As a remedial procedure, the expression of the perimeter P in Equation (3.45) needs to be modified as (Olsen et al. 1993):

$$P_1 = \frac{1}{4}[P + 2(A-1)] \tag{3.47}$$

The ratio 1:4 is determined as the proportionality constant for pixelized systems, and $2(A-1)$ is the maximum value of the perimeter fractal dimension D_p of an object of area A considered on a square grid. The corrected perimeter P_1 should then be used instead of P when estimating D_p from Equation (3.45). Some additional limitations related to the representation of digital images are discussed in Section 7.2.1.

3.2.7.2   Area Dimension, D_a

To quantify the space-filling characteristics of a patch (Figure 3.25B), Voss (1988) modified the standard box-counting framework (see Section 3.2.2) and suggested that the fractal dimension of a patch can be instead measured as $D_a = \log A / \log L$, where L is the maximum of the row and column lengths of the pixelized patch. Square patches of size n ($L = n$, $A = n^2$) completely fill a two-dimensional space (that is, $D_a = \log n^2 / \log n = 2$), while for a "rectangular" patch of length n and width 1 ($L = n$, $A = n$), $D_a = \log n / \log n = 1$. The corresponding length-area relationship is written as:

$$A = kL^{D_a} \tag{3.48}$$

where k is a constant, and D_a the area dimension. This method returns high fractal dimensions for patches that best fill their embedding space (that is, circular patches). Thus, in studies focusing on acquisition and retention of space—for instance, in the framework of invading species (Kooi and Kooijman 2000)—a circular patch (where the area-to-edge ratio is high) would be more likely to retain space than a thin, convoluted patch (that is, where the area-to-edge ratio is low).

3.2.7.3   Landscape/Seascape Dimension, D_s

This framework is a generalization of the perimeter-area relationship provided above to a set of patches of different types (for example, patches of different seagrass species; Figure 3.25C) to estimate the extent to which the perimeter of a given patch interacts with neighboring patch types. The related measure of fractal dimension, the so-called landscape/seascape dimension D_s, can be estimated using a modified perimeter length P_m (Olsen et al. 1993):

$$P_m = P + [2(A - 1)(N/(N - 1))] \tag{3.49}$$

where P and A are respectively the patch perimeter and area, N is the number of adjacent patch types, and N_t is the number of all patch types in the landscape/seascape. The related fractal dimension for the seascape is subsequently estimated by substituting Equation (3.48) into the perimeter-area relationship given by Equation (3.45). This method can be extremely useful to estimate the impact of intra- and interspecific competition for space and resources—for instance, in the framework of zoobenthic community establishment and maintenance (Boström and Bonsdorff 2000; Hovel et al. 2002).

3.2.7.4 Fractal Dimensions, Areas, and Perimeters

Although the perimeter and the area fractal dimensions introduced in Section 3.2.7 specifically describe the relationships between the perimeter P and the area A of a patch, they can more generally be used to quantify the structure of any rough object such as proteins, biological aggregates obtained from natural environments (for example, marine snow) or bioreactors (for example, bioflocculated microbial aggregates generated by the activated sludge process), inorganic colloidal aggregates (for example, clays, alum, ferric hydroxides), and growth patterns of inorganic and organic systems (for example, cluster formation, dendritic growth, diffusion-limited aggregation). However, it appears that many fractal relationships involving the perimeter or the area of an object significantly differ from the concepts introduced above. These relationships are reviewed hereafter and discussed in relation to the related fractal dimensions.

3.2.7.4.1 Fractal Structure of Surfaces

3.2.7.4.1.1 On the Fractal Surface Dimension of Proteins

The characteristic roughness and corrugation of protein surfaces are of extreme biological relevance in their function, including (1) the association of different subunits; (2) the recognition, diffusion, and binding of a ligan; and (3) the release of products. Typically, the surface areas of proteins is defined by the area accessible to a probe sphere (Lee and Richards 1971; Richards 1977) and the related fractal surface dimension D_s, estimated as

$$D_s = 2 - \frac{d \log A}{d \log R_p} \tag{3.50}$$

or equivalently

$$A = k R_p^{2-D_s} \tag{3.51}$$

where k is a constant, A the molecular surface area, and R_P the probe radius (Lewis and Rees 1985). Note that Equation (3.51) is the strict equivalent of Equation (3.6), and consequently of the box-counting dimension D_b described in Section 3.2.2. This approach has been used to investigate the fractal surface dimension of three enzymes (lysozyme, ribonuclease A, and superoxide dismutase), which was $D_b = 2.44$ on average for scales ranging from 0.1 and 0.35 nm (Lewis and Rees 1985). The variation in D_s over the protein surfaces revealed high fractal dimensions ($D_s > 2.5$) for surface regions of lysozyme characterized by elevated reactivity, while lower D_s values ($D_s \in [2.3 - 2.5]$) were obtained from regions located near the activity surface of the enzyme (Lewis and Rees 1985). The region of greatest surface roughness corresponded to the dimmer interface and to the subunit interface in superoxide dismutase and ribonuclease A, respectively. Regions involved in the formation of tight complexes and permanent binding (for example, interfaces between subunits, antibody-combining regions) appeared to be more irregular than average ($D_s > 2.4$). In contrast, regions of proteins that interact transiently with ligans and cannot tolerate formation of stable complexes (for example, active sites) appear to be smoother than average. Despite the very narrow range of scales used to estimate the fractal dimensions and in the absence of any discussion related to the obvious changes in the values of D_s with the size of the probe (Lewis and Rees 1985; see their Figure 2b), this work indicates that increased roughness favors strong bounds, thus relating the fractal structure to specific functions and suggesting that fractal dimension could actually be used to predict protein functional sites, as functional surfaces are much rougher than protein surfaces in general (Pettit and Bowie 1999). Note that Equation (3.50) has further been modified to account for statistical errors resulting from local variations in roughness at specific sites on a protein surface as

$$f_i = 2 - \frac{d \log \sum_j A_j}{d \log R_P} \tag{3.52}$$

where f_i is the smoothed atomic fractal dimension for atom i, A_j the contact area (Richards 1977) of atom j, and the sum accounts for all neighbor atoms j within 5 Å of atom i (Pettit and Bowie 1999) (Figure 3.26).

More generally, investigations of the global fractal surface dimensions of 14 proteins indicated that most of them were bounded between 2.1 and 2.2, which suggests that the surface dimension of proteins may have a physical basis and control some biologically relevant processes of proteins (Elber 1989). This might be the case for substrate diffusion to and along the protein surface. Molecules close to the surface are therefore captured at a rate that increases exponentially with D_s. In contrast, the slower the substrate migrates along the surface to the active site, the higher the D_s is (Pfeifer et al. 1985). However, while high D_s accelerate the diffusion rates to the surface, they slow it down on the surface (Pfeifer et al. 1985). As a consequence, the overall activity of proteins is likely to result from the balance between efficient diffusion to the surface and efficient trapping on the surface by active sites. The value of the global fractal surface dimensions obtained for a range of proteins, $D_s \approx 2.2$ (Lewis and Rees 1985; Aqvist and Tapia 1987; Elber 1989), might then represent the optimum surface complexity needed to achieve enzymatic activity.

3.2.7.4.1.2 On the Fractal Surface Dimension of Vegetation and Soil

The roughness of surface areas of vegetation and of soil pores is of particular biological and ecological relevance, as it defines the habitat space available to species and individuals at their characteristic scales. This has been used to relate body size to population density in fractal habitat (Morse et al. 1985; Kampichler and Hauser 1993).

In a study of the impact of human activities on the shape of remnant riparian forest patches in Iowa (Rex and Malanson 1990), Equation (3.45) has been rewritten as:

$$C = \frac{P}{A^{D_P/2}} \tag{3.53}$$

where C is defined as the patch shape, P the patch perimeter, A the patch area, and D_P the perimeter dimension. Equation (3.53) implies that if a patch has a smooth outline (that is, low D_P) the shape C will be large. In contrast, a very rough patch (that is, high D_P) results in a smaller patch shape for constant perimeter and area. Note that Equation (3.53) is strictly equivalent to Equation (3.45); that is, $C = k$. The prefactor (that is, constant) k is barely taken into account in fractal analysis, mainly due to early criticisms stating that it did not appear important and that

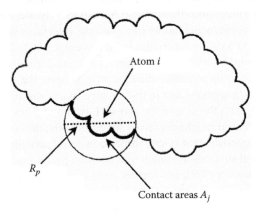

FIGURE 3.26 Smoothed atomic fractal dimension f_i. f_i is used to map local variations in roughness of protein surfaces based on a sphere of radius R_p centered on each atom and used to estimate the contact areas A_j of neighboring atoms.

a geometrical framework was lacking (Longley and Batty 1989). However, patch shape significantly affects ecological processes such as distribution and abundance of organisms (Hamazaki 1996; Hawrot and Niemi 1996; Muriel and Grez 2002), predator–prey interactions (Orrock et al. 2003), response to habitat fragmentation (Collinge and Palmer 2002), and species richness (Heegaard et al. 2007). Patch shape, but more generally the prefactors k of any scaling relationships and their relationships with fractal dimensions, should consequently be more thoroughly investigated.

Besides the widely reported fractal organization of vegetation patches (Palmer 1988; DeCola 1989; Rex and Malanson 1990; Williamson and Lawton 1991; van Hees 1994; deJong and Burrough 1996; Cantero et al. 1998; Despland 2003; Alados et al. 2005) and their influence on foraging organisms (Russel et al. 1992; Ritchie 1998; Etzenhouser et al. 1998; Cuddington and Yodzis 2002; Hoddle 2003; Phillips et al. 2004; Nams and Bourgeois 2004), plant surfaces were also found to be fractal, thus influencing the type, size, and abundance of arthropods living on them (see, for example, Morse et al. 1985; Gunnarsson 1992; Shorrocks et al. 1991; Jeffries 1993). Applying the box-counting approach described in Section 3.2.2 to a selection of woody plants, Morse et al. (1985) found box dimensions D_b ranging between 1.28 and 1.79. Based on the assumption that animals perceive and use vegetation surfaces in a way proportional to their body length, replacing the box size δ by the body length L in Equation (3.13) leads to:

$$N(L) \propto L^{-D_b} \tag{3.54}$$

where $N(L)$ can be equivalently thought of as the number of boxes of size L or the number of body lengths necessary to cover the vegetation surface. From Equation (3.54), it comes that for $D_b = 1.5$ the area perceived by an organism 3 mm in length is 3.16-fold greater than the same area perceived by a 30-mm organism, hence the available surface area increases with decreasing body length. This increase in available space for smaller organisms was then combined with the way in which metabolic rate r scales with body length (that is, $r \propto W^{0.75} = (L^3)^{0.75}$, where W is the body weight; Peters 1983; Schmidt-Nielsen 1984) to make predictions about the distribution of body lengths of animals living on vegetation. As population densities scale reciprocally of metabolic rates (Peters 1983), the number of arthropods $N(L)$ of size L living on a vegetation of fractal surface dimension, D_b, scales as:

$$N(L) \propto L^{-2.25} \tag{3.55}$$

A 10-fold decrease in body length results in a 178-fold increase in the density of organisms. The 3.16-fold increase in the surface area available predicted for a 10-fold decrease in body length results in an expected increase of 560-fold in the number of individuals. The predictions of Morse et al. (1985) have been confirmed for both terrestrial and aquatic invertebrates (Gunnarsson 1992; Shorrocks et al. 1991; Jeffries 1993). They are nevertheless likely to significantly underestimate the number of animals found on the vegetation surface of known fractal dimension D_b estimated from the two-dimensional projection of a three-dimensional structure; see Section 3.2.2.4. Note that under the assumption that all the individuals present on the vegetation surface can be collected, Equation (3.54) allows us to predict the box dimension D_b of the three-dimensional surface of the vegetation from the slope of the log-log plot of the number of arthropods $N(L)$ versus their body length L. This is a consequence of the applicability of Equation (3.13) to two- and three-dimensional structures (see Section 3.2.2.4; Seuront et al., 2004a). This is also consistent with the divergence between the fractal dimension estimated by Morse et al. (1985) from the two-dimensional projections of three-dimensional vegetation surfaces—that is, $D_b \in [1.28 - 1.79]$—and the dimension D_b ($D_b = 2.78$) predicted from Equation (3.13) using data collected by pyrethrum knockdown of a tree canopy.

Fractal dimensions have also been used as a measure of the roughness of soil pore walls to estimate the available pore area for microarthropods of different sizes and to predict the relative abundance of various size classes and body sizes of soil microarthropod communities (Kampichler and Hauser 1993). Photographic images of thin sections of soil and of pores actually habitable by microarthropods (that is, pores having an area larger than 0.003 mm²) were analyzed as two-dimensional patches. The fractal dimension of the patch perimeter was assessed using the perimeter dimension D_P (see Section 3.2.7.1, Equation 3.45), and from Equations (3.1) and (3.6), it shows that the related pore-surface dimension D_s can be calculated as:

$$D_s = D_P + 1 \tag{3.56}$$

D_s varied slightly around 2.32. A decrease of an order of magnitude of body length would increase habitat space approximately four times and results in a fourfold increase in the density of microarthropods in a given pore area (Kampichler and Hauser 1993). The relation between the size of organisms and habitat availability through considerations of habitat spatial complexity is critical for population dynamics of soil microorganisms and for the dynamics of cycling and transport of nutrients via the soil microbial population. The persistence of soil bacterial communities then depends on the existence of refuge sites in soil, that is, niches of microscale structure accessible to bacteria but excluding predatory protozoa. Bacteria tend to reside on the surface of pore walls where they may be subjected to predation by protozoa (see, for example, Coleman and Crossley 2004). As a consequence, information related to the fractal properties of soil structure (see also Section 3.2.7.4.1) allows the prediction of (1) the magnitude of the area of refuge sites for prey species of various sizes, and (2) the fraction of the potential habitat of the prey species that is accessible to predators (Crawford et al. 1993a). For instance, only half of the potential habitable area of the bacteria has been estimated to be inaccessible to predation by protozoa in a soil of fractal dimension 2.36 (Crawford et al. 1993b). The fractal nature of soil is also likely to influence the motility of organisms as a function of their size. From fractal data of soil structure and measurements of the diffusion rate to body size, Fujikawa (1994) estimated that, despite their low motility, large species could move more rapidly than expected through soil because the effect of constricting pore necks is to limit the effective tortuosity, thus permitting a greater mobility to an otherwise more slowly moving large predator.

Fractal geometry provides a valuable theoretical and quantitative framework with which to analyze, infer, and forecast the role of vegetation and soil structure in the space-time dynamics of invertebrates and microbial organisms and the related processes involved in matter cycling and transport.

3.2.7.4.2 Fractal Structure of Aggregates

3.2.7.4.2.1 On the Fractal Structure of Protein Aggregates

The aggregation of monomers and the formation of clusters are of central interest in biology and polymer and colloid chemistry, including the physical chemistry of macromolecules (for example, Antonietti 2003; Antonietti and Förster 2003; Antonietti and Tauer 2003; Sun 2004). As stressed earlier, for protein surface, fractal properties of protein aggregates can provide valuable insight into the structural basis and a better understanding of protein–protein interactions and aggregation in biomembranes, thus allowing a better assessment of their functional role. The fractal dimension of protein aggregates is typically estimated as:

$$N \propto \left(\frac{r}{R_0} \right)^D \tag{3.57}$$

where N is the number of monomers inside a radius r from the center of the aggregate, and R_0 is the monomer radius (Mandelbrot 1983). Note that the fractal dimension D estimated from Equation

(3.57) is conceptually similar to the cluster dimension D_c, introduced in Equation (3.18). More generally, the characteristic radius, R_i, of an aggregate is related to the number of monomers, i, in a cluster as:

$$R_i = R_0 i^\beta \tag{3.58}$$

where the cluster exponent $\beta = 1/D$. The cluster resulting from the heat-induced aggregation kinetics of immunoglobulin followed Equation (3.58) with $D = 2.56 \pm 0.3$ (Feder et al. 1984), a value significantly lower than the space-filling dimension $D = 3$.

The fractal dimension of aggregating protein systems have also been estimated using small angle X-ray, neutron scattering, and light scattering techniques (for example, Horne 1987, 1989a, 1989b; Khlebtsov and Melnikov 1994; Schuler et al. 1999). These procedures measure the mean scattered intensity, I, as a function of scattering angle and subsequently as a function of the magnitude of the scattering vector Q, or momentum transfer vector. For mass fractals (Gouyet 1992; Pfeifer and Ober 1989), there is a power-law relationship between the scattered intensity, I, and the magnitude of the momentum vector, Q (Schmidt 1989):

$$I(Q) \propto Q^{-D_m} \tag{3.59}$$

where the mass fractal dimension $D_m \leq 3$ and Q is a function of the scattering angle θ, that is, $Q = (4\pi n/\lambda)\sin(\theta/2)$, where n and λ are the refractive index of the medium and the wavelength of the laser light, respectively. Equation (3.59) is valid if the cluster size is large compared with the primary particle size, that is, $R_0 < Q^{-1} < r$ (Raper and Amal 1993). In contrast, for surface fractals, Equation (3.59) rewrites as:

$$I(Q) \propto Q^{-(6-D_s)} \tag{3.60}$$

where the surface fractal dimension D_s is in the range $3 \leq 6 - D_s \leq 4$ (Schmidt 1989). Equation (3.57) has been successfully applied to ramified clusters of α-elastin obtained by reversible aggregation of a nondispersed elastin solution upon increasing temperature ($D_m = 2.24$; Tamburro and Guantieri 1991) and casein aggregates under different conditions of temperature, $D_m \in [2.11 - 2.44]$ (Vétier et al. 1997, 2003). These values are comparable with the values reported for renneted casein and casein aggregation induced by ethanol, $D_m = 2.40$ and 2.33, respectively (Horne 1987, 1989a, 1989b) and for acidified sodium caseinate aggregates, $D_m = 2.27$ (Bremer and Walstra 1989).

Fluorescence-resonance energy transfer was used to estimate the fractal properties of the contour of membrane protein aggregates (Dewey and Datta 1989). This method lies on the general expressions relating energy transfer from a donor to acceptors randomly distributed on a fractal structure (Klafter and Blumen 1984). For multiple donors and multiple acceptors, the ratio of quantum yields of donor in the presence, Q_P, and absence, Q_A, of acceptor is given by:

$$\frac{Q_P}{Q_a} = 1 - N_{A,B} \left\langle (R_0/R)^6 \right\rangle \tag{3.61}$$

where $N_{A,B}$ is the total number of donors in the presence of acceptors equivalent to the efficiency of the energy transfer between donors and acceptors, R_0 the protein diameter, and R the distance between donor and acceptor. Because $N_{A,B}$ cannot be controlled experimentally, it must be related to N_A, the total number of acceptor molecules (Dewey and Datta 1989). First, consider the surface area, S_A, of the acceptor molecules. Equation (3.45) thus leads to:

$$P \propto S_A^{D_p/2} \tag{3.62}$$

where P is the perimeter of the acceptor molecules. The surface area is assumed to be directly proportional to N_A, the total number of acceptor molecules, and to the surface density of acceptors, σ. Because:

$$N_{A,B} = P/\delta \tag{3.63}$$

where δ is the width of the protein, it becomes:

$$N_{A,B} \propto N_A^{D_p/2} \tag{3.64}$$

Combining Equation (3.61) and Equation (3.64), it finally becomes:

$$E \propto \sigma^{D_p/2} \tag{3.65}$$

The value of D_p obtained for bacteriorhodopsin ($D_p = 1.6$; Dewey and Datta 1989) is comparable to the value $D_p = 1.56$ found for lattice animals (Jullien and Kolb 1984; Kolb and Jullien 1984; Brown and Ball 1985) and $D_p = 1.55$ for chemical cluster-cluster aggregates (Stauffer 1979). These D_p values were consistent with favorable and relatively nonspecific protein–protein interactions, which probably produced extended aggregate structures. In contrast, the higher fractal dimension ($D_p = 1.80$) found for calcium-ATPase was close to values obtained for percolation clusters (Stauffer 1979). This may suggest that purified ATPase vesicles have such a high protein-to-lipid ratio, leading to an almost totally contracted network (that is, filling uniformly a two-dimensional space).

3.2.7.4.2.2   On the Fractal Structure of Marine Snow Aggregates

The interior of deep lakes and oceans is far from the pristine vision of an idyllic deep blue sea that may intuitively come into mind (Figure 3.27A). Instead, it is often characterized by the occurrence of marine snow (Figure 3.27B). Marine snow refers to the continuous flow of inorganic and organic detritus falling from the productive euphotic zone down toward the ocean's interior, and somehow similar to snow on land. It is formed by the collision and subsequent coagulation of large macromolecules such as the transparent exopolymer particles secreted as a waste product by bacteria and phytoplankton, and a variety of inorganic and organic materials such as dying or dead organisms (either plants or animals), secretion and excretion of organisms, terrigenic dust particles, and resuspended sediment. Although marine snow can sometimes be identified as having been formed from specific

FIGURE 3.27 Classical view of the pristine waters of the ocean interior (A) compared to a more realistic view illustrating the occurrence of marine snow particles (B). **(See color insert.)**

particles in the water column (for example, fecal pellets, larvacean houses, or diatoms), it may also be highly amorphous and not recognizable as having any distinct origin (Alldredge and Gotschalk 1988, 1990). Marine snow aggregates grow over time as they sink through coagulation with other particles and aggregates at a rate controlled by turbulent diffusion and differential sedimentation (Kiørboe and Hansen 1993; Kiørboe et al. 1994, 1996, 1998; Kiørboe 1997; Li and Logan 1997a, 1997b) (Figure 3.28) and can reach up to several centimeters in diameter, sinking for weeks before reaching the ocean floor (Alldredge and Gotschalk 1988). Marine snow is a critical component of the ocean biogeochemical cycles through the transport of organic and inorganic material from the water column to the ocean floor (Azam and Long 2001). Little is still known about the geometrical properties of different types of aggregates. These properties can, however, affect the coagulation and sedimentation of these particles (Jiang and Logan 1991). In addition, as the geometry of aggregates in general—and marine snow aggregates in particular—conditions the surface and space available to microorganisms (for example, heterotrophic bacteria, micro- and mesozooplankton) that colonize their surface and degrade their organic content as inorganic compounds (Figure 3.29), a thorough assessment of their properties is critical to infer their overall contribution to biogeochemical cycles (Jackson 1990; Hill 1992; Riebesell and Wolf-Gladrow 1992).

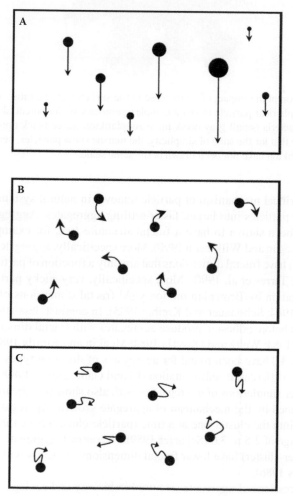

FIGURE 3.28 Physical processes contributing to particle collisions. (A) Differential sinking, (B) turbulence, and (C) Brownian motion.

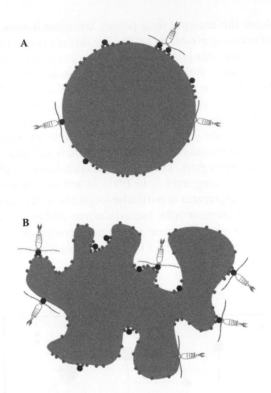

FIGURE 3.29 Illustration of the impact of an increase in the surface/volume ratio of a marine snow particle from (A) an idealized spherical particle to (B) a complex geometry on its potential to host microorganisms such as heterotrophic bacteria (small gray dots), microzooplankton (large black dots) and mesozooplankton (that is, copepods). Note that for the sake of simplicity, the marine snow particles, heterotrophic bacteria, and micro- and mesozooplankton have not been drawn at the same scale.

Coagulation is a critical mechanism of particle removal in natural systems, transforming many small, slowly settling particles into larger, faster-settling aggregates. Aggregate properties formed by coagulation have been shown to have a fractal structure (see, for example, Witten and Cates 1986; Meakin 1988; Logan and Wilkinson 1990). More specifically, aggregates formed by Brownian motion (Figure 3.28C) have fractal dimensions that are only a function of particle stickiness (Meakin 1988; Lin et al. 1989; Torres et al. 1990). More specifically, very sticky particles that undergo fast cluster-cluster aggregation by Brownian motion yield fractal dimensions of ca. 1.8 (Jullien et al. 1984; Schaefer et al. 1984; Schonauer and Kreibig 1985). In contrast, less-sticky particles that need to collide many times before adhesion produce aggregates with fractal dimensions between 1.9 and 2.2 (Jullien and Kolb 1984; Weitz and Oliveria 1984; Meakin and Family 1987). Fractal dimensions ranging from 1.0 and 3.0 have been found for aggregates of different types of particles formed by either shear motion or differential sedimentation (Li and Ganczarczyk 1989; Logan and Wilkinson 1990, 1991). Computer simulations of aggregate growth also show that the magnitude of the fractal dimension is determined by the mechanism of aggregate growth. Aggregates formed through the addition of particles into the cluster one at a time (particle-cluster) have three-dimensional fractal dimensions in the range of 2.5 to 3.0 (Schaefer 1989). In contrast, aggregates formed through collision of clusters (cluster-cluster) have lower fractal dimensions, typically with D ranging from 1.6 to 2.2 (Witten and Cates 1986).

Two standard expressions of aggregate size used in biology and ecology are (1) the average distance between two points in the outline of a particle (that is, the statistical diameter; Herdan 1953), and (2) the diameter of a circle having the same area as the projected image of the particle, when viewed in the direction perpendicular to the plane of greatest stability (that is, the circular, or spherical, equivalent diameter, [SED]) (Herdan 1953). Shape factors (Box 3.6), including fractal dimension, have further been introduced as size-independent features calculated from geometric dimensions.

BOX 3.6   SHAPE FACTORS IN PARTICLE IMAGE ANALYSIS

The different morphological features of particles and aggregates can be measured using shape factors as size-independent descriptors calculated from basic geometric dimensions. Four shape factors are defined hereafter following Yonekawa et al. (1996):

Elongation, E:

$$E = \frac{1}{w} \qquad (3.B6.1)$$

Circularity, C:

$$C = \frac{4a\pi}{p^2} \qquad (3.B6.2)$$

Roundness, R:

$$R = \frac{4a}{l^2\pi} \qquad (3.B6.3)$$

Porosity, P:

$$P = \frac{a-a_d}{a} \times 100 \qquad (3.B6.4)$$

where a is the total area of the particle (the integral of the particle surface excluding enclosed holes), a_d the detected area (the area of the particle taking into account the holes), l the maximum distance between two points located on the boundary of the particle/aggregate, w the width (diameter orthogonal to the length), l_f the Feret length (diameter parallel to the y axis), w_f the Feret width (width parallel to the x axis), and p the perimeter (the sum of distances between midpoints of the vectors forming the boundary (Figure 3.B6.1).

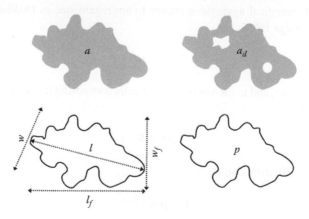

FIGURE 3.B6.1 The morphological features of a particle used to define shape factors. Length l, width w, Feret length l_f, Feret width w_f, perimeter p, total area a, and detected area a_d.

These factors are used to characterize the shapes of particles, regardless of their sizes. Shape factors have been used to discriminate, for example, rock particles (Schäfer and Teyssen 1987) and plant leaves (Yonekawa et al. 1996). Examples of shape factors are given in Box 3.6. Applying Equations (3.9) and (3.10) to photographs of suspended particulate matter, Billiones et al. (1999) found fractal dimensions D_d bounded between 1.04 and 1.37 and between 1.25 and 1.86 for detritus originated from monocotyledon and dicotyledon plants, respectively. The observed difference in D_d values could be due to the different venation patterns on the leaves of the two groups of plants, the monocotyls having parallel venation while the dicotyls having net venation (Muller 1979). Assuming that during the fragmentation of the leaves into detrital particles, the breaking up follows along the lines of the venation pattern: Monocotyl detritus will have the tendency to break up into somewhat rectangular shapes, following parallel lines of venation, while dicotyls will have more irregular borders, like the fringes of a torn net. The ability to discriminate detritus of different sources is highly relevant in aquatic environments where a large portion of the detritus is of terrestrial plant origin (Pomeroy 1980).

More specifically, the n-dimensional fractal dimension D_n of aggregates relates how aggregate properties vary with their characteristic length, in which n is the Euclidean dimension the object is embedded in (see Section 2.2). Fractal dimensions of aggregates embedded in one-, two-, and three-dimensions are:

$$P \propto l^{D_1} \tag{3.66}$$

$$A \propto l^{D_2} \tag{3.67}$$

$$N_p \propto l^{D_3} \tag{3.68}$$

where l is the maximum aggregate size, P the aggregate perimeter, A the projected aggregate area, and N_p the number of particles in the aggregate. Since these three fractal dimensions characterize how the aggregate properties change with size, their magnitude is related to aggregate morphology. Aggregates with highly irregular perimeters have fractal dimension D_1 greater than unity. Similarly, aggregates tend to become more porous with increasing size, resulting in two- and three-dimensional fractal dimensions that are usually less than their corresponding integer (that is, space-filling) values; that is, $D_2 < 2$ and $D_3 < 3$. This approach led to identifying different fractal dimensions among different types of aggregates and with the type of fluid motion forming the aggregates (Logan and Wilkinson 1990).

The porosity, p, of biological aggregates relates to aggregate size as (Alldredge and Gotschalk 1988; Logan and Alldredge 1989):

$$1 - p = al^b \tag{3.69}$$

where a and b are constants, and to the volume, V_c, of cells composing the aggregate and the volume, V, of the aggregate as:

$$1 - p = NV_c / V \tag{3.70}$$

Equations (3.68) and (3.70) then lead to:

$$1 - p \propto l^{D_3 - 3} \tag{3.71}$$

where Equations (3.69) and (3.71) lead to $D_3 = 3 + b$ (see also Table 3.4). Using porosity data determined directly from gravimetry assuming an average particle density (Alldredge and

TABLE 3.4

Relationships between the Physical and Fractal Properties of Aggregates

Aggregate Property	Equation	Scaling Relationship
Solid volume[a] (v)	$v = \psi^{D_3/3} \xi_0 l_0^{3-D_3} l^{D_3}$	$v \propto l^{D_3}$
Encased volume[b] (v_c)	$v_c = \xi l^3$	$v_c \propto l^3$
Mass (m)	$m = \rho_0 \psi^{D_3/3} \xi_0 l_0^{3-D_3} l^{D_3}$	$m \propto l^{D_3}$
Density (ρ)	$\rho = \rho_0 \psi^{D_3/3} \left(\xi_0/\xi\right)\left(l/l_0\right)^{D_3-3}$	$\rho \propto l^{D_3-3}$
Porosity (ε)	$\varepsilon = 1 - \psi^{D_3/3} \left(\xi_0/\xi\right)\left(l/l_0\right)^{D_3-3}$	$\varepsilon \propto l^{D_3-3}$
Settling velocity (u)	$u = \left[\dfrac{2g\xi_0(\rho_0 - \rho_w)\psi^{D_3/3} l_0^{1+D_2-D_3} v^{-b}}{a\rho_w \xi_2} l^{D_3+b-D_2}\right]^{1/(2-b)}$	$u \propto l^{(D_3+b-D_2)/(2-b)}$

Source: Modified from Logan and Kilps (1995).

a ξ_0 and ξ_0 are the packing and shape factors of primary particles in the aggregate of size l_0 and density ρ_0; ζ and ξ are the packing and shape factors of the aggregate; g the gravitational constant; ρ_w the density of water; $\psi = \zeta\xi/\xi_0$; and a and b are constants relating the drag coefficient of spheres C_D to the Reynolds number Re. For Re ≤ 0.1, $a = 24$ and $b = 1$, and for $0.1 < \text{Re} < 10$, $a = 29$ and $b = 0.871$. The Reynolds number of an aggregate of size l is defined as Re $= ul/v$, where v is the fluid kinematic viscosity, and relates to the drag coefficient C_D as $C_D = a\,\text{Re}^{-b}$. (See Jiang and Logan [1991] for more details.)

b   The difference between the solid volume and the encased volume is that the latter includes both the volume of particles and the volume of pores.

Gotschalk 1988; Logan and Alldredge 1989), Equation (3.71) returned $D_3 = 1.39 \pm 0.15$ for general marine snow aggregates, $D_3 = 1.52 \pm 0.19$ for diatom aggregates, and $D_3 = 1.8 \pm 0.3$ for microbial aggregates of the rod-shaped, gram-negative bacterium *Zooglea ramigera* (Logan and Wilkinson 1990). Assuming that $A \propto l^{D_2}$ (Equation 3.67), fractal dimensions have also been estimated from the settling velocity, u, of aggregates using the scaling relationship:

$$u \propto l^{D_3+1-D_2} \tag{3.72}$$

as $D_3 = 1.26 \pm 0.06$ for marine snow aggregates (Logan and Wilkinson 1990). From digitized *in situ* photographs of a variety of marine snow aggregates, Kilps et al. (1994) found that the lowest fractal dimensions D_2 ($D_2 = 1.28 \pm 0.11$) were found for aggregates composed predominantly of a single type of particle (for example, diatoms or fecal pellets) containing a large amount of miscellaneous debris. Marine snow formed of fecal pellets ($D_2 = 1.34 \pm 0.16$), nonidentifiable amorphous particles ($D_2 = 1.63 \pm 0.72$), and diatoms ($D_2 = 1.86 \pm 0.13$) had increasingly larger fractal dimensions.

Meakin (1988) demonstrated that when the three-dimensional fractal dimension $D_3 \leq 2$, then $D_2 = D_3$. As a consequence, Equations (3.67) and (3.68) lead to:

$$A \propto N_p^{\phi} \tag{3.73}$$

where $\phi = D_2/D_3$. Equation (3.73) can be used to test whether $\phi = 1$ when $D_3 \leq 2$, or alternatively to test whether $D_3 \leq 2$ when $\phi = 1$. Note that the two- and three-dimensional fractal dimensions $D_2 = D_3$ and $D_2 = D_3$ can also be calculated using a series of scaling relationships involving aggregate volume, mass, density, porosity, and settling velocity (Table 3.4). Based on the consistency in the slopes of particle size distributions observed in the ocean and in laminar shear devices during coagulation of clay particles (Hunt 1982), Jiang and Logan (1991) showed that for shear coagulation the three-dimensional fractal dimension D_3 could also be described as:

$$s(l) = -\frac{1}{2}\left(1 + \frac{3}{D_3}\right) \tag{3.74}$$

$$s(v) = -\frac{3}{2}\left(1 + \frac{1}{D_3}\right) \tag{3.75}$$

where $s(l)$ is the slope of a discrete particle size distribution, $n(l)$, as a function of maximum aggregate length l, and $s(v)$ is the slope of a discrete particle size distribution, $n(v)$, as a function of solid volume, v. The equations obtained for coagulation by differentiated sedimentation are shown in Table 3.5. The particle size and volume distribution are given by (Jiang and Logan 1991):

$$n(l) = A_l l^{s(l)} \tag{3.76}$$

$$n(v) = A_v v^{s(v)} \tag{3.77}$$

where A_l and A_v are proportionality constants. Using the expression for the solid volume of an aggregate (Table 3.4), Equations (3.74) and (3.75) lead to:

$$A_l l^{s(l)} = \Psi^{D_3/3} \xi_0 l_0^{3-D_3} l^{D_3-1} D_3 A_v v^{s(v)} \tag{3.78}$$

TABLE 3.5

Three-Dimensional Fractal Dimension D_3 of Particles Expressed as a Function of Length and Volume Distributions and the Coagulation Mechanisms

Basis	Distribution Type	Mechanism		
		Shear	Differential Sedimentation	Mechanism Independent
Steady state	Discrete length	$-2s(l) - 5$	$\dfrac{s(l)(2b-4) + 3b + D_2 - 8}{3-b}$	
	Cumulative length	$-2S(l) - 3$	$\dfrac{S(l)(2b-4) + b + D_2 - 4}{3-b}$	
	Discrete volume	$\dfrac{-3}{2s(v)+3}$	$\dfrac{D_2 + b - 4}{s(v)(4-2b) - 3b + 7}$	
	Cumulative volume	$\dfrac{-3}{2S(v)+1}$	$\dfrac{D_2 + b - 4}{S(v)(4-2b) - b + 3}$	
Non-steady state	Discrete, both volume and length			$\dfrac{s(l)+1}{s(v)+1}$
	Cumulative, both volume and length			$\dfrac{S(l)}{S(v)}$

Source: Modified from Logan and Kilps (1995).

or equivalently:

$$l^{s(l)} \propto l^{D_3-1} l^{D_3 s(v)} \tag{3.79}$$

The three-dimensional fractal dimension D_3 can then be expressed from Equation (3.79) independently of the coagulation mechanism as:

$$D_3 = \frac{s(l)+1}{s(v)+1} \tag{3.80}$$

Note that a similar relationship can be derived using the slopes obtained from log-log plots of cumulative size distributions (Table 3.5):

$$D_3 = \frac{S(l)}{S(v)} \tag{3.81}$$

where $S(l)$ and $S(v)$ are the slopes of the cumulative size distributions $N(l)$ and $N(v)$ based on particle maximum length l and solid volume v, respectively. Equation (3.81) has been successfully used to estimate the fractal dimension $D_3(D_3 = 1.99 \pm 0.08; \bar{x} \pm \text{SD})$ of *Escherichia coli* flocs (Tang et al. 2001). Calculation of fractal dimension does not require an assumption of steady-state conditions if (1) both the size and the volume distributions are known for the same population of particles, and (2) the fractal dimension is unchanged over the distribution considered. As a consequence, discrepancies between the values of D_3 returned by Equations (3.74) and (3.75) would indicate that the system under consideration was not at steady state, and that the fractal dimensions should be more reliably estimated using the non-steady-state equations, that is, Equations (3.80) and (3.81). Table 3.5 provides estimates of steady-state and non-steady-state conditions for both shear and differential sedimentation as coagulation mechanisms.

3.2.7.4.2.3 *On the Fractal Structure of Soil Aggregates*

Although used relatively early to investigate various soil properties (Jenkins and Watts 1968; Serra 1968; Webster and Butler 1976; Campbell 1978; Burgess and Webster 1980), fractal concepts and the fractal dimension were formally introduced to soil science by Burrough (1981, 1983a, 1983b), who showed that the nature of the spatial variation of some soil properties, including pH, bulk density, percentage of clay, sand, and silt, and sodium content could be described by examining their fractal dimension; see also Pachepsky et al. (2000a, 2000b) for a review of the applications of fractals in soil science. Fractals have subsequently been applied to a wide range of soil properties and processes (Tyler and Wheatcraft 1990, 1992; Perfect and Kay 1991; Turcotte 1991; Rieu and Sposito 1991; Brakensiek et al. 1992; Rawls et al. 1993). One of the most critical purposes of soil morphological analysis is to describe the soil aggregate and associated pore morphologies, continuities, and sizes (Bullock et al. 1985). This has reached a new dimension with the recognition that, because soil is both a fragmented material and a porous medium, a fractal representation may be particularly suitable (Rieu and Sposito 1991). Two main streams can be identified in "fractal soil science": approaches dealing with the complexity of soil aggregate structure and those dealing with soil aggregate-size distributions.

Soil structure has typically been investigated using thin section photographs (Figure 3.30) that were digitized and analyzed in terms of fractal surfaces (that is, pore-solid interface) (Pfeifer and Avnir 1983; Friesen and Mikula 1987; Davis 1989; Toledo et al. 1990; Bartoli et al. 1991; Crawford et al. 1993a, 1993b) and fractal mass dimension D_m (Bartoli et al. 1991; Young and Crawford 1991; Anderson and McBratney 1995; Anderson et al. 1996, 1998; Oleschko et al. 1997; Giménez et al. 1998, 2002; Millán and Orellana 2001; Bird et al. 2006) (see Section 3.2.4 and Equation 3.22). However, because soils

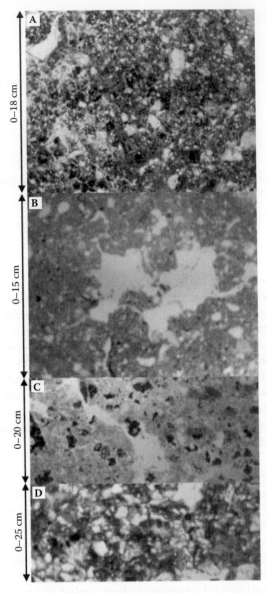

FIGURE 3.30 Examples of the thin section images typically used to estimate the solid mass dimension, D_{ms}, and the porous mass dimension, D_{mp}. Images are from the first few centimeters of (A) Melanic Andosol developed from volcanic ashes; (B) Eutric Vertisol originated from deposition of different types of alluvium derived from basalt and other igneous extrusive rocks; (C) Texcoco Lake deposits, lacustrine clay accumulation; and (D) Tepetates, hardened soil of volcanic origin, from Mexico. (Modified from Oleschko et al., 2000.)

are intrinsically porous and do not have a uniform internal mass distribution, soil structure can also be characterized from digitized images on the basis of the space-filling ability of both solid and pore networks, respectively. Two mass dimensions can then be defined, the solid mass dimension, D_{ms}, and the porous mass dimension, D_{mp}. Note that the porous mass fractal dimension, D_{mp}, has rarely been reported in the literature (Katz and Thompson 1985; Ghilardi et al. 1993). Using a similar approach on vertical sections of Melanic Andosol (developed from volcanic ashes), Eutric Vertisol (originated from deposition of different types of alluvium derived from basalt and other igneous extrusive rocks),

Texcoco Lake deposits (lacustrine clay accumulation), and Tepetates (hardened soil of volcanic origin), Oleschko et al. (2000b) found that fractal dimensions discriminated between different soil structures (that is, materials with contrasting genesis) through significant correlations with soil properties (that is, bulk density, dielectric constant) and depth. However, for many soil properties affecting water movement and root growth, the pore size distribution pattern has much greater importance than total porosity, bulk density, and mass fractal dimensions. This led a range of studies to investigate both theoretically and practically fractal pore size distributions (Friesen and Mikula 1987; Ahl and Niemeyer 1989; Tyler and Wheatcraft 1990; Rieu and Sposito 1991; Perrier et al. 1996) and particle size distributions (Tyler and Wheatcraft 1989, 1992; Wu et al. 1993).

The fractal characterization of aggregate size distribution of fragmented soil material was originally introduced to assess the influence of cropping and wetting treatments on aggregate fragmentation (Perfect and Kay 1991; Rasiah et al. 1992) using the number-size relation (Turcotte 1986) and mass-size relation (Tyler and Wheatcraft 1992). The related number-size and mass-size fractal dimensions belong to a specific class of fractal processes that are investigated in Chapter 5. Specifically, the reader is referred to Section 5.4 for more details on the fragmentation and mass-size fractal dimensions.

3.2.8 RAMIFICATION DIMENSION, D_r

3.2.8.1 Theory

Consider a tree trunk of diameter $d(1)$ bifurcating into two main branches with diameters $d_1(2)$ and $d_2(2)$. Assuming a constant bifurcation ratio, a general relationship can be proposed to connect the diameter of two successive bifurcations (Rouse and Ince 1963):

$$d(1)^\Delta = d_1(2)^\Delta + d_2(2)^\Delta \tag{3.82}$$

where $\Delta = 2$ for the confluence of two rivers, and $d(1)$, $d_1(2)$, and $d_2(2)$ are the river widths (Schroeder 1991), $\Delta = 2.7$ over 20 bifurcations in mammalian vascular systems (Suwa and Takahashi 1971), and $\Delta = 3$ over 15 bifurcations in human bronchial tree (Thompson 1961). Equation (3.82) can be generalized following Equations (3.1) and (3.2) as:

$$d(n) = kn^{-D_r} \tag{3.83}$$

and

$$L(n) = kn^{1-D_r} \tag{3.84}$$

where $d(n)$ and $L(n)$ are the mean tube diameter and tube length after n ramifications, k is a constant, and D_r is the ramification dimension; see, for example, West and Goldberger (1987) and Crawford and Young (1990). Extensive studies of the relationship between fractal geometry and allometric scaling of organisms can be found in West et al. (1997, 1999) and Enquist et al. (1998, 1999).

3.2.8.2 Fractal Nature of Growth Patterns

A nearly infinite multitude of forms is found in living organisms. Many natural objects, in contrast with man-made objects, show at first sight a high degree of irregularity, nonsmoothness, and fragmentation. As a consequence, the description of indeterminate growth forms of many organisms—including canopies, root systems, sessile marine invertebrates, and microbial growth patterns and morphologies—is often only achievable in qualitative terms. The lack of ability to describe growth forms accurately has hampered the use of external morphology as a diagnostic character and has made interpretation of the interaction between the growth form and the environment difficult to achieve. Fractal analysis can then be thought of as a convenient alternative to describing the complexity of growth patterns from a dynamic perspective. This is briefly illustrated hereafter from previous studies conducted on microbial and fungal growth and morphology and plant root systems.

3.2.8.2.1 On the Fractal Nature of Microbial and Fungal Growth and Morphology

Fractal analysis has been introduced to microbiology as a tool to describe growth patterns and morphology of a variety of microorganisms under a wide spectrum of growth conditions. This is particularly relevant because of the critical role that microbial growth patterns and morphologies play in, for example, pathogenicity, metabolic activity, and enzyme production. Fractal geometry is thus expected to provide a powerful tool for the geometric, pattern-oriented description and measure of irregularity of the complex structure of bacteria and mycelia, which are still widely described in empirical terms such as diffuse, compact, smooth, and rough. However, simple, nonequilibrium, probabilistic growth models resulting from computer simulations can also lead to complex structures that mimic certain types of biological morphologies and exhibit fractal properties (Meakin 1986). The first evidence of fractal growth in microbial systems was reported for the enteric pathogenic bacterium *Serratia marcescens* growing on a minimal-nutrient agar medium (that is, Davis or Vogel-Bonner agar medium) after a week at 30°C (Matsuyama et al. 1989; Fujikawa and Matsushita 1989). In contrast, it formed a typical round colony (that is, nonfractal) on normal nutrient agar after 1 day of culture (Matsuyama et al. 1989; Matsuyama and Matsushita 1992). The role of nutritional conditions on the induction of microbial fractal colony growth has been confirmed on the soil bacterium *Bacillus subtilis* (Fujikawa and Matsushita 1989, 1991; Matsushita and Fujikawa 1990; Fujikawa 1994). More specifically, it appears that the box dimension $D_b (D_b = 1.73 \pm 0.02)$ of *Bacillus subtilis* colonies is very close to the value ($D_b = 1.70$) expected from two-dimensional simulations of diffusion-limited aggregation (DLA) patterns (Figure 3.31) (Meakin 1986). It is also stressed that the macroscopic growth patterns of *B. subtilis* exhibit clear macroscopic similarities with the modeled DLA (Figure 3.32A), including (1) the repulsion between two neighboring colonies (Figure 3.32B), (2) the tendency of the colony to grow toward the nutrient (Figure 3.32C), and (3) the appearance of a screening effect of protruding main branches against inner ones with elapsing time from

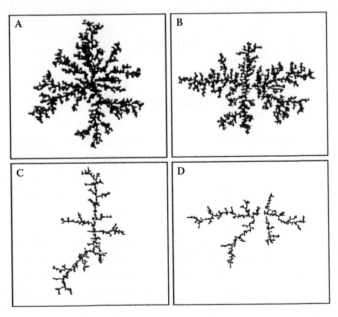

FIGURE 3.31 Two-dimensional simulations of clusters generated by a diffusion-limited process from one central point with growth exponents (A) 1.0 and (C) 2.0, and from two points with growth exponents (B) 1.0 and (D) 2.0. (Modified from Meakin, 1986.)

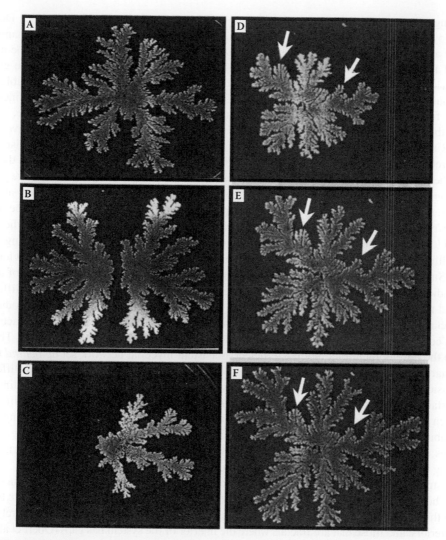

FIGURE 3.32 Illustration of diffusion-limited aggregation in (A) a *Bacillus subtilis* colony pattern, showing the repulsion between (B) two neighboring colonies, (C) the tendency of the colony to grow toward a nutrient gradient, and the appearance of a screening effect of protruding main branches against inner ones (identified by the arrows) with elapsing time from inoculation. (Modified from Matsushita and Fujikawa, 1990.)

inoculation (Figure 3.32D,E,F) (Matsushita and Fujikawa 1990; Fujikawa and Matsushita 1991). Those results have later been generalized to various strains of Enterobacteriaceae (that is, *Proteus mirabilis*, *Citrobacter freundii*, *Escherichia coli*, *Salmonella anatum*, *Salmonella typhimurium*, and *Klebsiella pneumoniae*), which all form fractal colonies after relatively long incubations in minimal-nutrient agar enriched with 0.4% glucose (Matsuyama and Matsushita 1992). The related fractal dimensions were in the range 1.7 to 1.8, similar to the one of the DLA model.

Mycelia have also widely been described in terms of fractal dimensions on soil (Bolton and Boddy 1993; Donnelly et al. 1995; Abdalla and Boddy 1996; Donnelly and Boddy 1997a, 1997b, 1998; Wells et al. 1997) and agar (Ritz and Crawford 1990, 1991; Crawford et al. 1993a; Mihail et al. 1994, 1995; Baar et al. 1997). Fractal dimensions, in turn, are used to quantify the extent to which mycelia permeate space in relation to extent of the system. Fractal geometry has then been used to quantify interspecific differences in mycelial morphology and relate these to habitat (Donnelly et al. 1995; Mihail et al. 1995), and intraspecific changes induced by introducing new carbon resources or competing fungi to established or establishing mycelia systems (Bolton and Boddy 1993; Donnelly and Boddy 1997b, 1998). For instance, fractal analysis has been used to assess the mechanisms of resource acquisition and the adaptation of *Trichoderma viride* colony morphology to the nutrient status of the substrate (Ritz and Crawford 1991). Under low-nutritional conditions, the colonies formed a low fractal-dimensional morphology by distributing as little hyphal mass as possible across a maximal area. In contrast, under elevated nutrient concentration, the fractal dimension increased, with the fungus filling the space as effectively as possible to exploit fully the substrate, suggesting that the fractal dimension reflected a compromise between exploitative and explorative growth forms (Ritz and Crawford 1991). The response of *T. viride* colonies to a heterogeneous distribution of resources indicates that the fractal dimension of the colony structure that developed in the direction of the nutrient source did not differ significantly from 2, whereas that of the structure growing away from the nutrient source had a dimension significantly lower than 2 (Crawford et al. 1993). However, a greater amount of hyphal mass was measured in the direction away from the nutrient source (Crawford et al. 1993a). Those results suggest that (1) the space-filling capacity of the pattern adjusted to the heterogeneous levels of nutrition, and (2) the processes controlling branching (that is, space-filling efficiency) and the phenomenon of mass distribution were independent (Crawford et al. 1993a). Subsequent studies focusing on the ecological significance of the fractal nature of mycelia have studied development in nonsterile soils. These studies have revealed interspecific differences in fractal morphology, especially at initial stages of outgrowth from resources. Some produce surface fractal systems while others produce mass fractal systems, though with time as surface fractal systems cover a large area they become increasingly mass fractal (Donnelly et al. 1995). This may indicate the development of a biomass-efficient, persistent mycelial network set up *behind* the foraging margin. Significantly, differences in morphology appear to be associated with differences in extension rate, with more aggregated systems (that is, mass fractal systems) extending faster than surface fractal systems (Donnelly et al. 1995). Morphological and physiological differences have been related to resource specificity, broad-fronted, slowly extending systems utilizing diverse locally abundant resources, while narrow-fronted rapidly extending systems utilize bulky, disparate resources. These contrasting strategies have also been described for clonal plants, the former strategy being termed "phalangeal" and the latter "guerrilla" (Schmid and Harper 1985).

This stresses the need to be able to quantify both the space filling occurring at mycelia margins (that is, the search fronts) and within the system. This effectively allows us to discriminate between systems that are only fractal at their boundaries (that is, surface/border fractal) having entirely plane-filled interiors, and those that are fractals where the interior of the system has gaps (Obert et al. 1990). This is when it becomes critical to estimate two complementary fractal dimensions, the interior and the border fractal dimensions, D_{b_i} and D_{b_b}, as described in Section 3.2.2.1. Using this approach, Boddy et al. (1999) found distinct temporal patterns for D_{b_i} and D_{b_b} during the development of the mycelial systems of *Hypholoma fasciculare* but relatively similar patterns for *Phallus impudicus*. This suggests that (1) interior and the border fractal dimensions are critical to understanding the dynamics of growing microbial and fungal structures, and (2) the intrinsic dynamics of search fronts and space-filling properties may differ at the intra- and interspecific levels.

3.2.8.2.2   On the Fractal Nature of Plant-Root Systems

Fractal geometry is a relatively new approach to the analysis of root system architecture and was first introduced by Tatsumi et al. (1989). Several studies have since demonstrated that the fractal dimension increases as root systems grow and become larger (Fitter and Stickland 1992; Lynch and van Beem 1993) and also between plants of equal age but different size (Eghball et al. 1993; Berntson 1994; Lynch and van Beem 1993). However, fractal dimension ontogenically increases during early growth and then levels off (Fitter and Stickland 1992), suggesting that consideration of fractal dimension as an estimate of root system size is appropriate only during initial growth. The fractal dimension of root systems has also been shown to be positively related to the density of roots (Berntson 1994), to vary significantly between different species and genotypes (Berntson et al.

1998; Nielsen et al. 1999), and to increase significantly with nutrient supply (Eghball et al. 1993; Berntson 1994). More specifically, fractal dimension of the common bean (*Phaseolus vulgaris*) root system was found to correlate with root phosphate content, suggesting fractal dimension to be a possible indicator for root phosphate uptake (Nielsen et al. 1999). Fractal geometry may then offer improved ways to quantify and summarize root system complexity as well as yield ecological and physiological insights into the functional relevance of specific architectural patterns (Tatsumi et al. 1989; Berntson 1994; Lynch and van Beem 1993; Nielsen et al. 1997).

The relative simplicity of the roots branching system, especially when compared to the DLA-type of growth observed for microbial and fungal colonies, allowed the introduction of a variety of models meant to link root architecture to root length and biomass. As described in Section 3.2.8.1, the self-similarity principle applied to a tree-root system predicts that roots follow the same bifurcation pattern from proximal roots to the smallest transport roots. The basic parameters of fractal root models describe the ratio of the sum of root cross-sectional areas after a bifurcation to the cross-sectional area before bifurcation, α, and the distribution of the cross-sectional areas after bifurcation, q (Spek and van Noordwijk 1994; van Noordwijk et al. 1994). Independency of ratios α and q on root diameters has been observed in tropical legume trees over a large range of diameters (Van Noordwijk and Purnomosidhi 1995; Ozier-Lafontaine et al. 1999). However, large variability within the whole root system was observed, which affected the precision of the root length and biomass estimates and the architecture generated by the model. West et al. (1999) presented a general fractal allometric model for vascular plants that takes into account the tapering of the water-conducting vessels. Model parameters are derived from fractal geometry and hydrodynamics rather than from empirical observations. A general fractal root model derived from previous modeling attempts (Van Noordwijk and Purnomosidhi 1995; Ozier-Lafontaine et al. 1999; West et al. 1999) with the ability of describing root systems with different branching properties (that is, number of new segments at each ramification, root ramification angle) is described in Salas et al. (2004); see also Box 3.7.

BOX 3.7   FRACTAL ROOT MODEL

Root systems are described as networks of connected links whose length and diameter are root-order dependent (Van Noordwijk and Purnomosidhi 1995; Ozier-Lafontaine et al. 1999). At a given bifurcation level, a root segment (order n, segment i) is divided into several new segments that form the next higher order (order $n + 1$, segment j). A recursive algorithm is then applied until the final ramification of the network (that is, roots of minimum diameter d_m) is reached. The scaling factor α is defined as the ratio of the square of root diameter before bifurcation (d_b^2) to the sum of squares of the diameters of bifurcating roots (Σd_a^2):

$$\alpha = d_b^2 / \sum d_a^2 \qquad (3.B7.1)$$

and the allocation factor of root cross-sectional area, q, is given as:

$$q = \max(d_a^2) / \sum d_a^2 \qquad (3.B7.2)$$

First, the relative frequency distribution of both vertical and horizontal bifurcation angles in 0.175 rad (10°) classes was generated from field data. From this distribution, the cumulative frequency range that corresponds to each class was computed. The cumulative frequency was then compared to a random number bounded between 0 and 100, and the angle within the probability range of which the random number corresponded was used as the angle of the following bifurcation. This algorithm provides reliable estimates of the root bifurcation angles independently of the form and density of the angle distribution, and no fitting of a theoretical distribution to data is necessary.

3.2.9 SURFACE DIMENSIONS

Although the methods related to the estimations of the fractal dimensions of surfaces were initially
designed to studies of topographic surface (see, for example, Goodchild 1980; Mandelbrot 1983), in
ecology a surface can also be thought of as the two-dimensional distribution of a given descriptor,
the "elevation" being thus referred to as the values of the descriptor. This can be illustrated by the
"mountain shape" aspect of two-dimensional microscale distributions of bacterioplankton abundance (Figure 3.33A) and microphytobenthos biomass (Figure 3.33B).

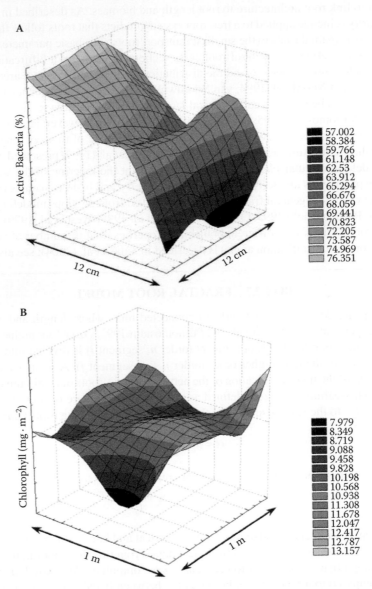

FIGURE 3.33 Two-dimensional "mountain shape" patterns of (A) microscale distributions of active bacterioplankton (data courtesy of Dr. J. Seymour) and (B) microphytoplankton biomass. The active bacterioplankton
was estimated through flow cytometric analysis of 1 ml samples taken every 1 cm using a spring-loaded micro-
sampler ($n = 100$); see Seymour et al. (2004) for more details. The microphytobenthos biomass was estimated
through estimates of chlorophyll a concentrations in 1 cm deep sediment core samples every 5 cm on a 1 m²
surface ($n = 225$). (For further details see Seuront and Spilmont, 2000; Seuront and Leterme, 2006.)

3.2.9.1 Transect Dimension, D_t

The method is very simple to implement and very intuitive. First, consider a surface (Figure 3.34A) characterizing the topographic elevation or the values of a descriptor $X(i, j)$ on a plane. Second, consider a series of one-dimensional transects taken from that surface (Figure 3.34B). One can then apply a one-dimensional variant of the cluster algorithm (see Section 3.2.3) to each transect; for

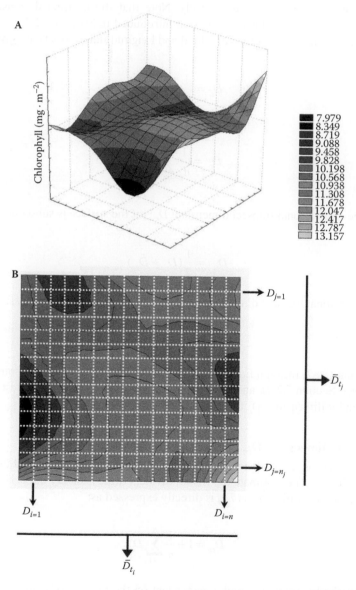

FIGURE 3.34 The transect dimension. Schematic illustration of the way to estimate (A) the two-dimensional fractal dimension of a pattern from (B) the one-dimensional fractal dimensions of horizontal and vertical subsections of the initial pattern. The two-dimensional fractal dimension is estimated as the mean of both the horizontal and vertical dimensions.

each box of length δ, the mean values $x_i(\delta)$ and $x_j(\delta)$ of the descriptor $X(i, j)$ along longitudinal and latitudinal transects are recorded and expressed, modifying Equation (3.34) as:

$$x_i(\delta) = k\delta^{D_{t_i}} \qquad i = 1,\ldots,n_i \qquad\qquad (3.85)$$

$$x_j(\delta) = k\delta^{D_{t_j}} \qquad j = 1,\ldots,n_j \qquad\qquad (3.86)$$

where k is a constant, n_i and n_j are the numbers of lines and columns of the Euclidean domain supporting the descriptor $X(i, j)$, and D_{t_i} and D_{t_j} the transect dimensions of the ith and jth longitudinal and latitudinal transects, respectively. Note that the transect dimensions D_{t_i} and D_{t_j} are conceptually similar to the cluster dimension introduced in Section 3.2.3.

The mean fractal dimensions of the latitudinal and longitudinal transects are given as:

$$\bar{D}_{t_i} = \frac{1}{n_i}\sum_{i=1}^{n_i} D_{t_i} \qquad\qquad (3.87)$$

$$\bar{D}_{t_j} = \frac{1}{n_j}\sum_{j=1}^{n_j} D_{t_j} \qquad\qquad (3.88)$$

The resulting one-dimensional transect dimension $\bar{D}_{t_{i,j}}$ of the surface is subsequently given as:

$$\bar{D}_{t_{i,j}} = \frac{1}{2}(\bar{D}_{t_i} + \bar{D}_{t_j}) \qquad\qquad (3.89)$$

and the two-dimensional transect dimension of the surface is finally estimated as:

$$\bar{D}_{t_{i,j}} = 1 + \bar{D}_{t_{i,j}} \qquad\qquad (3.90)$$

Note that Equation (3.87) through Equation (3.89) are relevant only when the isotropy condition is fully satisfied (see Section 7.2.2), and then requires an appropriate statistical test of homogeneity between the n_i and n_j dimensions D_{t_i} an D_{t_j}; see Zar (1996).

3.2.9.2 Contour Dimension, D_{co}

This method is based on a conversion of a surface plot to a contour map, and using the dividers method (see Section 3.2.1) to estimate the fractal dividers dimension D_{d_i} of each of the n contours (Figure 3.35). The contour dimension D_{co} is directly expressed as:

$$D_{co} = 1 + \frac{1}{n}\sum_{i=1}^{n} D_{d_i} \qquad\qquad (3.91)$$

However, this method is implicitly highly dependent on the number of isolines used in the computation of the fractal dimension D_{co}. In particular, when applied to a topographic surface, the consideration of more isolines should result in an increase of the resolved details and in subsequent modifications of the fractal dimension estimates.

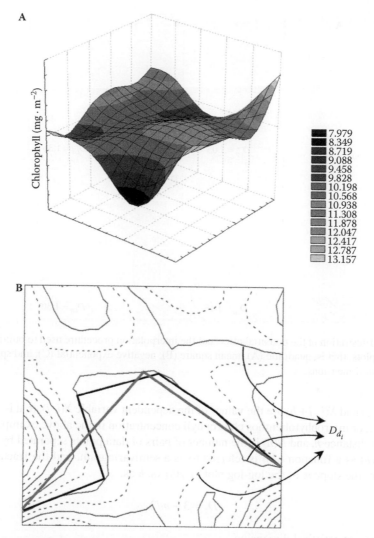

FIGURE 3.35 The contour dimension. Schematic illustration of the way to estimate (A) the two-dimensional fractal dimension of a pattern from (B) the one-dimensional fractal dimensions of its n contours. The two-dimensional fractal dimension is estimated as the mean of one-dimensional contour dimensions. The gray and black broken lines indicate two successive steps of the divider method.

On the other hand, in the specific case of the two-dimensional distribution of a descriptor $X(i, j)$, the fractal dimension estimates returned by this method will be strongly influenced by the interpolation procedure used to build the two-dimensional contour plots, which might themselves appear as being very different (Figure 3.36). As a consequence, any fractal dimension estimated using the contour method should make explicit reference to the interpolation procedure used to build the two-dimensional contours.

3.2.9.3 Geostatistical Dimension, D_g

The geostatistical dimension D_g is a two-dimensional generalization of the variogram dimension introduced in Section 4.2.8 to self-similar patterns. The fractal dimension of a landscape surface is based on the semivariance $\gamma(h)$ defined as (Huang and Turcotte 1989):

$$\gamma(h) = \frac{1}{4N(h)} \sum_{i=1}^{N(h)} \sum_{j=1}^{N(h)} [X(i, j) - X(i+h, j)^2 + X(i, j) - X(i, j+h)^2] \tag{3.92}$$

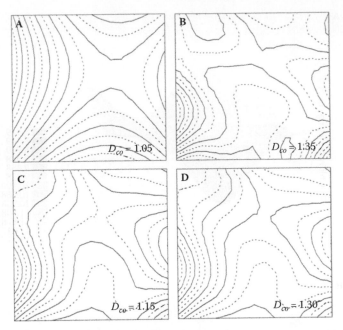

FIGURE 3.36 Illustration of the potential effects of the interpolation procedure used to build the two-dimensional contour plots, that is, quadratic (A), mean square (B), negative exponential (C), and splines (D), on the resulting contour dimensions.

where $X(i + h, j)$ and $X(i, j + h)$ are the values of the dependent variable $X(i, j)$, that is, bacterioplankton abundance, or microphytobenthos chlorophyll concentration in the above examples, at locations separated by a distance h, and $N(h)$ is the number of pairs of data points separated by the distance h. The plot of $\gamma(h)$ as a function of h is referred to as a semivariogram, and the fractal dimension is estimated from the slope m of the log-log plot of $\gamma(h)$ vs. h as:

$$D_g = 3 - m/2 \tag{3.93}$$

where D_g is the geostatistical dimension.

3.2.9.4   Elevation Dimension, D_e

This method, based on the mean absolute elevation difference $|\Delta h|$ between two points separated by a distance δ, has been specifically developed to estimate the fractal dimension of topographic surfaces (Polidori et al. 1991). The related fractal dimension D_e is given as:

$$|\Delta h| = k\delta^{3-D_e} \tag{3.94}$$

where k is a constant. Considering the two-dimensional distribution of a descriptor $X(i, j)$, Equation (3.94) is rewritten as:

$$|\Delta X(i, j)_\delta| = k\delta^{3-D_e} \tag{3.95}$$

where k is a constant and $|\Delta X(i, j)_\delta|$ is the mean absolute difference between the values of $X(i, j)$ separated by a distance δ in the x-y plane and expressed as:

$$\left|\Delta X(i, j)_\delta\right| = \frac{1}{N_\delta}\left|(X(i, j) - X(i, j + \delta)) + (X(i, j) - X(i + \delta, j))\right| \tag{3.96}$$

Although the geostatistical and the elevation dimensions can provide a valuable estimate of the two-dimensional fractal structure of a descriptor $X(i, j)$—for example, bacterioplankton abundance or microphytobenthos biomass—it is straightforward to see that Equations (3.92) and (3.96) imply different forms of spatial averaging. These two methods are then implicitly based on the stationarity hypothesis discussed in Section 7.2.3 and cannot provide any information regarding potential differences in the local fractal structures. Such information is available using the transect and the contour dimensions. The transect and the contour dimensions and the geostatistical and elevation dimensions can be referred to as *local fractal dimensions* and *global fractal dimensions*, respectively.

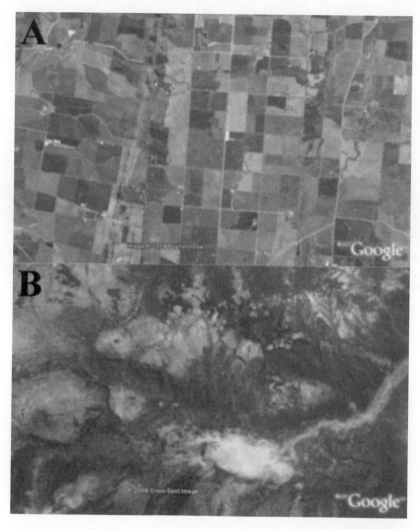

FIGURE 1.7 South Australian landscapes, with (A) and without (B) anthropogenic influences. Both pictures were taken from an altitude of 20 km.

FIGURE 1.9 Contrast existing between the geometry of a man-made surface, a brick wall (A) and (B) the bark patterns of the white fig (*Ficus virens*), (C) the English oak (*Quercus robur*), and (D) the cotton palm (*Washingtonia filifera*).

FIGURE 2.3 Nested structure perceptible in the geometry of clouds. At increasing resolution, the local and global structures remain very similar.

FIGURE 3.11 Salt-marsh environments where *H. haswellianus* burrow morphology was investigated in (A) Goolwa, (B) Torrens Island, (C) Middle Beach, and (D) Port Noarlunga. (Courtesy of G. Katrak, Flinders University, Australia.)

FIGURE 3.27 Classical view of the pristine waters of the ocean interior (A) compared to a more realistic view illustrating the occurrence of marine snow particles (B).

FIGURE 4.14 The Australian nectar-feeding passerines (A) New Holland honeyeater (*Phylidonyris novae-hollandiae*) and (B) red wattlebird (*Anthochaera carunculata*).

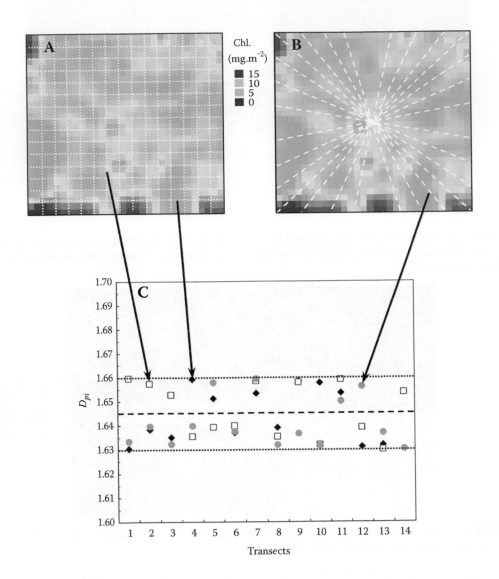

FIGURE 7.9 Spatial isotropy of fractal dimension estimates. From a two-dimensional pattern of microphy-tobenthos biomass (recorded in the Bay of Somme, April 25, 2002; Seuront and Spilmont, unpublished data) one-dimensional fractal dimensions have been estimated for horizontal and vertical sections (A) and different diagonal sections (B) of the initial pattern. The resulting patch-intensity dimensions (see Section 5.2) cannot be regarded as significantly different from the two-dimensional estimate (C; $p > 0.05$), showing the isotropic character of the initial distribution. The dashed and dotted lines represent the two-dimensional patch-intensity dimension and its 95% confidence interval, respectively.

4 Self-Affine Fractals

4.1 SEVERAL STEPS TOWARD SELF-AFFINITY

4.1.1 DEFINITIONS

Consider, for example, a turbulent velocity component measured as a function of time at a single point. As shown in Figure 4.1, it looks rough, like the boundary of a random fractal, but with the difference that the two axes correspond to different physical quantities (velocity and time) that are intrinsically different. Different units can be chosen for the two axes to make the trace look either very steep or nearly smooth. Similarly, from a scalar field in two dimensions, one can construct a mountain-like structure in which the height at each point equals the magnitude of the local concentration (see Figure 3.33). Again, the height and the two spatial coordinates are intrinsically different and independent, so that the mountain can be made to look jagged or relatively smooth depending on the choice of units for the two quantities. In general, whenever different quantities involved in such constructions scale differently, the notion of self-similarity contained in Equations (3.1) through (3.3) will not be adequate; to describe these phenomena, one needs the more versatile machinery of *self-affinity*.

An affine transformation is one that transforms a set S of points at positions $\vec{x} = (x_1, x_2, \ldots, x_{D_E})$ into a new set $\delta(S)$ with points at $\delta(\vec{x}) = (\delta_1 x_1, \delta_2 x_2, \ldots, \delta_{D_E} x_{D_E})$, where the scale ratios $(\delta_1, \delta_2, \ldots, \delta_{D_E})$ are all different. A bounded set S is *self-affine* when S is the union of N nonoverlapping subsets, each of which is identical (under translations and rotations) to $\delta(S)$. In other words, if a subset of a pattern is similar to the whole under an affine transformation, the pattern is said to be self-affine. In addition, S is *statistically self-affine* when S is the union of N distinct subsets, each of which is identical in distribution to $\delta(S)$. One cannot, however, define fractal dimension using Equation (3.1) through Equation (3.3) for even the simplest self-affine fractal curve. If one evaluates this dimension mechanically, pretending the curve in Figure 4.1 to be like a coastline, the value depends on the expansion used for one quantity relative to the other. If, for example, the time scale is stretched enough to render the signal to appear as a collection of smooth increments, it is intuitively clear that the dimension (called the global dimension) will be unity. If, on the other hand, the ordinate is stretched over a wide range of values, one can define the usual fractal dimension according to Equation (3.1). This is the so-called local dimension of the self-affine fractal. Although it has been pointed out that more than one dimension is necessary to characterize self-affine fractals (Mandelbrot 1986), and thus referring to the concept of multifractals studied more thoroughly hereafter (see Chapter 8), one must note that the fractal dimension for self-affine fractals is not as easily defined as with self-similar ones.

4.1.2 FRACTIONAL BROWNIAN MOTION

First, one needs to introduce the concept of fractional Brownian motion (fBm) (Mandelbrot and Wallis 1969; Mandelbrot 1977, 1983), which can be thought of as a generalization of the so-called concept of Brownian motion that played such an important role in both physics and mathematics. A fractional Brownian motion, $B_H(x)$, is a single valued function of one variable, x (that

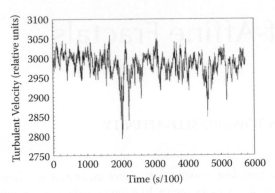

FIGURE 4.1 A time series of microscale turbulent velocity fluctuations, as an illustration of a self-affine fractal.

is, time or space) defined by its increments $B_H(x_i) - B_H(x_{i-1})$ that have a Gaussian distribution with variance

$$\langle |B_H(x_i) - B_H(x_{i-1})|^2 \rangle = k |x_i - x_{i-1}|^{2H} \tag{4.1}$$

where k is a constant, the angle brackets "$\langle\rangle$" denote ensemble averaging, and the parameter H, $0 < H < 1$. Practically, Equation (4.1) means that its mean square increments depend only on the difference $(t_i - t_{i-1})$. In case of $H = 0.5$, Equation (4.1) recovers the Brownian motion where $\Delta B^2 = k\Delta t$. More formally, for any three time $(x_{i-1}, x_i,$ and $x_{i+1})$ such that $x_{i-1} < x_i < x_{i+1}$, $\Delta B_1 = B_H(x_i) - B_H(x_{i-1})$ is statistically independent of $\Delta B_2 = B_H(x_{i+1}) - B_H(x_i)$ for $H = 0.5$. This means that at every stage and at every scale of Δt, all directions of displacement are equally likely. For $H > 0.5$ and $H < 0.5$, the increments are positively and negatively correlated, respectively. More specifically, if $H > 0.5$, the increments of the displacement may be roughly thought of as overlapping each other, above time increments that do not overlap. Such a process may be said to be positively correlated, or persistent, in the sense that a particle moving in some direction at time t will tend to move in the same direction regardless of Δt. Alternatively, if $H < 0.5$, the process is said to be negatively correlated, or antipersistent.

It can be seen that Equation (4.1) is qualitatively similar to a power law. Any change by a factor of δ in the scale t will change ΔB_H by a factor of δ^H as:

$$\langle \Delta B_H(\delta x)^2 \rangle = k\delta^{2H} \langle \Delta B_H(x)^2 \rangle \tag{4.2}$$

where k is a constant. Practically, Equation (4.1) introduces a major difference to the self-similar power law; see, for example, Equations (3.1) and (3.2). Indeed, a fractional Brownian trace requires different scaling factors in the two coordinates (δ for x, and δ^H for B_H). Each value of x corresponds to only one value of B_H, while any value of B_H may occur at multiple values of x. This specific non-uniform scaling provides an additional definition to *self-affinity*.

4.1.3 DIMENSION OF SELF-AFFINE FRACTALS

Consider a trace of $B(x)$ covering a time span $\Delta t = 1$ and a vertical range $B(x) = 1$. $B(x)$ is statistically self-affine when t is scaled by δ and $B(x)$ is scaled by δ^H. Divide now the time span into N equal intervals, each with $\Delta t = 1/N$. Each of these intervals contains one portion of $B(x)$ with vertical range

$\Delta B = \Delta t^H$. Since $0 < H < 1$, each of these new sections will have a large vertical and horizontal size ratio and the occupied portion of each interval will be covered by $\Delta B/\Delta t = (1/N^H)/(1/N) = N/N^H$ elements of size $\delta_n = 1/N$. The number of length elements required to cover the trace goes from 1 to $N(\delta_n)$ as:

$$N(\delta_n) = N \times N / N^H = 1/\delta_n^{2-H} \tag{4.3}$$

Finally, comparing Equations (3.2) and (4.3) leads to the fractal dimension D_F:

$$D_F = 2 - H \tag{4.4}$$

A Brownian motion will thus have a fractal dimension $D_F = 1.5$.

Fractional Brownian motion, illustrated here in a one-dimensional framework, can be generalized to higher dimensions, namely to self-affine surface and volume. For instance, replacing the single variable x by the coordinates (x, y) in a plane leads to considering the resulting fBm $B_H(x, y)$ as the surface altitude at position (x, y). In analogy with Equation (4.1), the increments $B_H(x_i, y_i) - B_H(x_{i-1}, y_{i-1})$ of $B_H(x, y)$ have a Gaussian distribution with variance:

$$\left\langle | B_H(x_i, y_i) - B_H(x_{i-1}, y_{i-1}) |^2 \right\rangle = k[(x_i - x_{i-1})^2 + (y_i - y_{i-1})^2]^H \tag{4.5}$$

From Equation (4.4), the fractal dimension of a fractal landscape can be derived as:

$$D_F = 3 - H \tag{4.6}$$

Note that the intersection of a vertical plane with the surface $B_H(x, y)$—that is, the altitude fluctuations of a mountain sheep following any straight path in the (x, y) plane—is a self-affine fractional Brownian motion fully similar to those observed in Figure 4.2 with $D_F = 2 - H$. Alternatively, the intersection of a horizontal plane with the surface $B_H(x, y)$—that is, the coastline of a mountain lake—has a fractal dimension $D_F = 3 - H$, but since the two coordinates x and y are equivalent, the coastlines of $B_H(x, y)$ are self-similar, not self-affine. A fractal volume—that is, a fractal cloud—has the fractal dimension:

$$D_F = 4 - H \tag{4.7}$$

Generally speaking, a self-affine fractional Brownian function, $B_H(\vec{x})$, in a D_E Euclidean space satisfies:

$$\left\langle | B_H(\vec{x_i}) - B_H(\vec{x_{i-1}}) |^2 \right\rangle = k | \vec{x_i} - \vec{x_{i-1}} |^{2H} \tag{4.8}$$

with $\vec{x} = (x_1, x_2, \ldots, x_{D_E})$, and its fractal dimension is written as:

$$D_F = D_E + 1 - H \tag{4.9}$$

Finally, the intersection of $B_H(\vec{x})$ with a D_E-dimensional object form a self-similar fractal with dimension $D_F = D_E - H$.

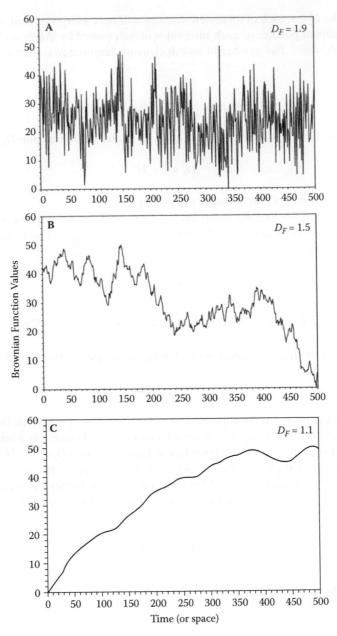

FIGURE 4.2 Self-affine fractional Brownian motion (fBm) characterized by different fractal dimensions. Note that the case $D_F = 1.5$ corresponds to the basic Brownian motion.

4.1.4   1/f Noise, Self-Affinity, and Fractal Dimensions

Self-affine functions have been defined (via fractional Brownian motion) as being a function of a single variable x that can be regarded as either space or time, mainly because changes in space have many of the same similarities at different scales as changes in time. However, unpredictable changes of any quantity $x(t)$ varying in time t are known as noise. More specifically, Montroll and Badger (1974), Montroll and Shlesinger (1982), and West and Shlesinger (1990) have reported a number of examples showing that when the spectral density $E_Q(f)$ (that is, an estimate of the mean square fluctuations at frequency f, and consequently of the variations over a time scale of order 1/f)

of certain data are presented on log-log plots, the data appear as a straight line over a certain range. Beyond that range, the straight line assumes the shape of a curve according to an inverse power law of the form $E_Q(f) \approx 1/f^\beta$, where f is the frequency and β a positive exponent referred to as the spectral exponent. In particular, the $1/f^\beta$ law, referred to as scaling $1/f$ *noise* (Mandelbrot 1983), can serve as a powerful tool to describe music, speech, and a wide variety of noise. For instance, studying different compositions such as the First Brandenburg Concerto and Scott Joplin rags, Voss and Clark (1975, 1978) found that composition having a frequency generated by $1/f$ sources sounded pleasing, while those generated by $1/f^2$ sounded too correlated, and those sounds generated from white noise, namely by $1/f^0$ sources, sounded too random. The spectral density of $1/f$ noise thus varies with a predictability between white noise ($1/f^0$, no correlation in time) and Brownian motion ($1/f^2$, no variability between increments; see Section 4.1.2). More generally, the so-called $1/f$ noise has been observed in a wide variety of phenomena in nature, ranging from earthquakes (Bak and Tang 1989; Carlson and Langer 1989), turbulence (Gollub and Benson 1980), cosmology (Chen and Bak 1989), relaxation in nonperiodic solids (Evangelou and Economou 1990), ionization of excited hydrogen atoms (Jensen 1990), microcirculatory control of blood flow (Intaglieta and Breit 1991), and human interbeat dynamics (Nunes Amaral et al. 1998; Ivanov et al. 1999) to complex systems involving a large number of interacting subunits that display "free will," such as city growth (Makse et al. 1995) and economics (Mantegna and Stanley 1995). An illustration of different $1/f$ noises, together with the related power spectra, is given in Figure 4.3.

Both white noise ($1/f^0$, no correlation in time) and Brownian motion ($1/f^2$, no correlation between increments) are well understood in terms of mathematical physics. On the other hand, the origin of $1/f^\beta$ noise, which represents the most common type of noise found in nature, nevertheless remains a mystery after almost a century of investigations. The universality of $1/f$ noise suggests that it does not represent a consequence of particular physical interactions but instead is a general manifestation of complex dynamical systems that have remarkably similar critical components, perhaps because the "interaction parts" between the constituent subunits in such extremely complex systems dominate the observed cooperative behavior more than the detailed properties of the subunits themselves (Stanley 1995). From a mathematical point of view, this universality may be attributed to a very rich random statistical ensemble that has typical configurations dominating over the usual mean values (West and Shlesinger 1989).

4.1.5 Fractional Brownian Motion, Fractional Gaussian Noise, and Fractal Analysis

Fractional Gaussian noise (fGn) represents another family of self-affine processes, defined as the series of successive increments in an fBm. An fBm signal is nonstationary with stationary increments. The increments, $y(t) = x(t) - x(t-1)$, of a nonstationary fBm signal $x(t)$ yield a stationary fGn signal and *vice versa*. Fractional Gaussian noise and fractional Brownian motion signals are then interconvertible: When an fGn is cumulatively summed, the resultant series constitutes an fBm, and when an fBm is differenced, the resultant constitutes an fGn. Each fBm is then related to a specific fGn, and both are characterized by the same H exponent. These two processes, however, possess fundamentally different properties: fBm is nonstationary with time-dependent variance, while fGn is a stationary process with a constant mean and variance expected over time. Examples of fBm and fGn corresponding to three values of H are presented in Figure 4.4. The H exponent can be assessed from an fBm series as well as from the corresponding fGn, but because of the different properties of these processes, the methods of estimation are necessarily different. The dichotomy between fGn and fBm motivated a systematic evaluation of fractal analysis methods (Caccia et al. 1997; Cannon et al. 1997; Eke et al. 2000, 2002) that showed that most methods gave acceptable estimates of the Hurst exponent H when applied to a given class (fGn or fBm) but led to inconsistent results for the other. The first step in a fractal analysis is to identify the class to which the analyzed data set belongs, fGn or fBm (Figure 4.5). The Hurst exponent H can subsequently be properly estimated, using a method relevant for the identified class. The nature of $1/f^\beta$ noises described can here be very

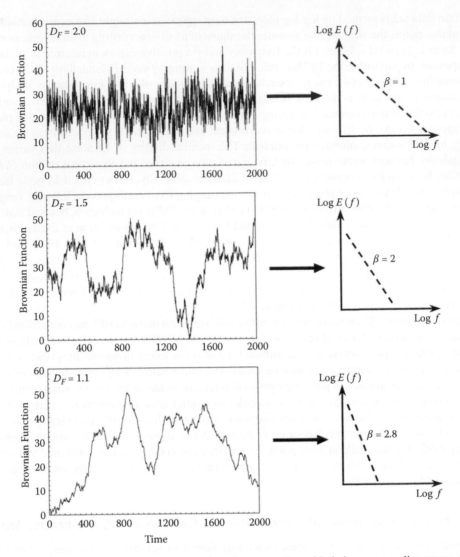

FIGURE 4.3 Self-affine fractional Brownian motion, shown together with their corresponding power spectra.

useful as a diagnostic tool to identify the class of a given signal. More specifically, signals with $-1 < \beta < 1$ have constant variance at all values of x, which classifies them as stationary signals, that is, fractional Gaussian noise. In contrast, signals with $1 < \beta < 3$ are nonstationary (that is, fractional Brownian motion) because their variances increase with x, such that (Mandelbrot and van Ness 1968; Beran 1994):

$$\text{var}\,[B(x)] \propto x^{2H} \tag{4.10}$$

Those fractional Brownian motions have a spectral slope β_{fbm} defined as:

$$\beta_{fbm} = \beta_{fGn} - 2 \tag{4.11}$$

Fractional Brownian motions and fractional Gaussian noises characterized by the same Hurst exponent H will then have different spectral exponents β. The dichotomy between fractional Brownian

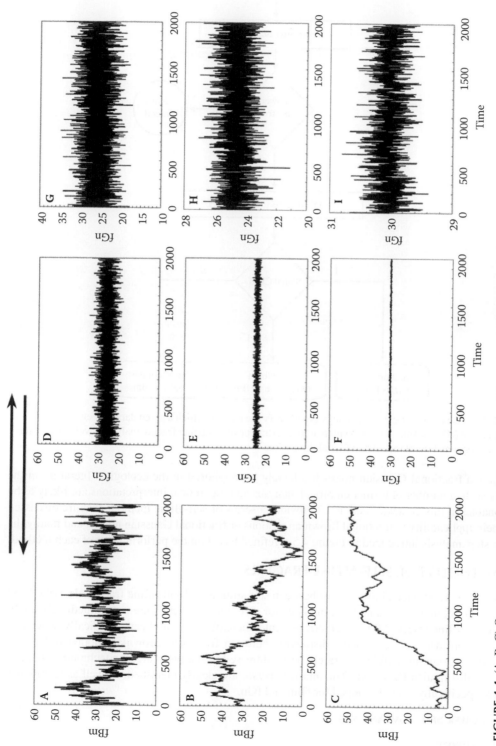

FIGURE 4.4 (A, B, C) One-to-one correspondence between fractional Brownian motion (fBm) and (D, E, F) fractional Gaussian noise (fGn). Three signals are shown in each signal class with $H = 0.25$ (A, D, G) a Hurst exponent $H = 0.50$; (B, E, H) and $H = 0.75$ (C, F, I). The case $H = 0.50$ is the special case of uncorrelated white noise, that is, Brownian motion.

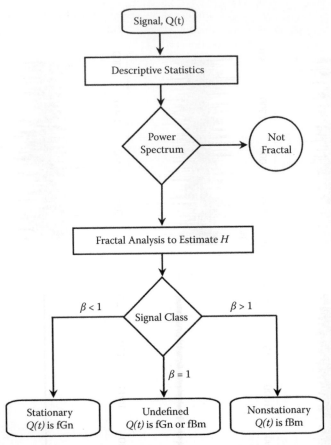

FIGURE 4.5 Use of power spectrum analysis to identify the class to which a given data set belongs, fractional Gaussian noise (fGn) or fractional Brownian motion (fBm), prior to further fractal analysis devoted to estimate the Hurst exponent H, and the related fractal dimension.

motions and fractional Gaussian noises has largely been ignored in the ecological literature in the past. As such, a number of former empirical analyses and theoretical interpretations are likely to be questionable. In this context, Section 4.2 provides a series of self-affine methods that are discussed upon their applicability to fractional Brownian motions or fractional Gaussian noises, and finally the step-by-step analysis introduced in Figure 4.5 is refined based on the performances of each method.

4.2 METHODS FOR SELF-AFFINE FRACTALS

Because of the fundamental differences between self-similar and self-affine fractals described and discussed in Sections 4.1.3 and 4.1.4, methods for self-similar fractals are not immediately applicable to self-affine traces. All the methods presented hereafter have all been specifically developed to analyze self-affine traces. As the distinction between fractional Brownian motion (fBm) and fractional Gaussian noise (fGn) has rarely been addressed in the literature, the techniques that can be used to distinguish fBm from fGn, and *vice versa*, are clearly identified, as well as techniques that have specifically been developed for fBm and fGn.

4.2.1 POWER SPECTRUM ANALYSIS

4.2.1.1 Theory

Power spectrum analysis (PSA) is probably the most extensively used technique to detect both spatial and temporal patterns in aquatic ecology since the seminal work of Platt and Denman in the early seventies (Platt 1972; Platt and Denman 1975; Denman and Platt 1976; Denman et al. 1977) but has still seldom been related to the concept of fractal dimension (Seuront et al. 2002). Formally speaking, a power spectrum is defined as the square of the amplitude of the Fourier transform of a time series and can thus be regarded as an expression of the variance of the underlying process at different spatial or temporal scales. In practice, the power spectral density $E(x)$ is given by:

$$E(x) \propto x^{-\beta} \qquad (4.12)$$

where x is the frequency f (s^{-1}; $f = 1/t$, where t is time) or the wave number k (m^{-1}; $k = 1/l$, where l is space) for temporal and spatial self-affine processes, respectively. The spectral exponent β is related to the Hurst exponent H as:

$$H = \frac{\beta - 1}{2} \qquad (4.13)$$

for fractional Brownian motions, and

$$H = \frac{\beta + 1}{2} \qquad (4.14)$$

for fractional Gaussian noises. Fractional Brownian motions characterized by $\beta < 2$ (that is, $H < 0.5$) and $\beta > 2$ ($H > 0.5$) are respectively antipersistent and persistent. In contrast, antipersistent and persistent fractional Gaussian noises are characterized by $\beta < 0$ and $\beta > 0$, respectively. This subsequently leads to expressing the Fourier dimension D_{FFT} as (Schroeder 1991):

$$D_{FFT} = D_E + 1 - H \qquad (4.15)$$

where D_E is the Euclidean dimension of the embedding space. Following Equation (4.9), Equation (4.13) and Equation (4.14) can also be rewritten as:

$$D_{FFT} = D_E + 1 - \frac{(\beta - 1)}{2} \qquad (4.16)$$

for fractional Brownian motions, and

$$D_{FFT} = D_E + 1 - \frac{(\beta + 1)}{2} \qquad (4.17)$$

for fractional Gaussian noises.

Practically, the power spectral density $E(x)$ of a signal $x(t)$ is estimated by the fast Fourier transform (FFT) (Aho et al. 1974; Horowitz and Sahni 1978; Burrus and Parks 1985; Kreyszig 1988), which uses complex exponentials in place of the equivalent sine and cosine terms of a traditional Fourier series (for example, Bloomfield 2000). This provides the spectral density $E(x)$ at frequencies $f (x = f)$ or wave numbers $k (x = k)$ increasing by a factor of 2^n, where n is a positive integer. Note that series whose length is smaller than 2^n may be extended to a length 2^n by adding zeros to the end of the series. Although this "zero padding" procedure shifts the apparent fundamental frequency, it does not distort the spectrum. To improve the consistency of spectral estimates for fBm, it is recommended to proceed successively to parabolic windowing and bridge detrending of the fBm signals before running FFT analysis. The parabolic window for a series of length n is a function that multiplies each value in the series and is given as (Fougere 1985):

$$w(i) = 1 - \left(\frac{2i}{n+1} - 1 \right)^2 \qquad (4.18)$$

where $i = 1,\ldots, n$. Bridge detrending (Cannon et al. 1997) is done by substracting from the data the line connecting the first and last points of the series.

4.2.1.2 Spectral Analysis in Aquatic Sciences

In aquatic ecology, the origin of $1/f$ noise can be traced to the seminal work of Platt (1972), who showed for the very first time that the spectral density of fluorescence (that is, a proxy of phytoplankton biomass) exhibits an inverse power law of the form $E_F(f) \approx 1/f^{5/3}$ (Figure 4.6A). The spectral exponent fairly similar to the theoretical value ($\beta = 5/3$) expected for purely passive scalar advected by three-dimensional and isotropic turbulent processes indicates that the phytoplankton is fully controlled by physical processes over a wide range of scales (that is, from a meter to thousands of meters). This has subsequently been verified in many environments, ranging from lakes (Powell et al. 1975; Abbott et al. 1982), coastal and open ocean waters (Weber et al. 1986; Seuront et al. 1996a, 1996b, 1999; Lovejoy et al. 2001), and estuaries (Lekan and Wilson 1978). Conceptually similar results have been found from remote sensing observations of sea-surface temperature and chlorophyll (Gower et al. 1980; Barales and Trees 1987; Denman and Abbott 1988, 1994; Smith et al. 1988), with chlorophyll exhibiting a $1/f^3$ behavior and thus fully controlled by two-dimensional

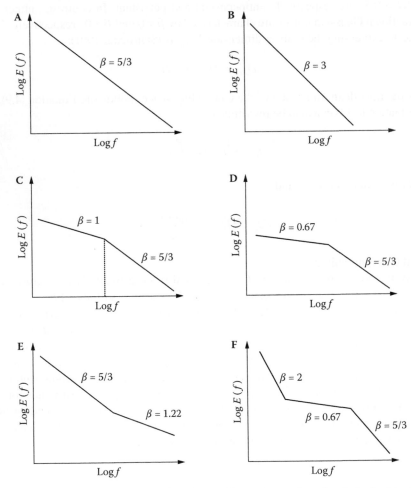

FIGURE 4.6 Schematic illustration of $1/f$ noise observed in marine ecology for phytoplankton biomass fluctuations (see text for explanation).

turbulence (Figure 4.6B). It is stressed here that if a $1/f^3$ is expected for the energy spectrum of quasi-geostrophic flows (Kraichnan 1967; Charney 1971), a $1/f$ has nevertheless been suggested for inert particles such as phytoplankton cells when their distribution is driven by the enstrophy cascade instead of the energy cascade (Lesieur and Sadourny 1981; Bennett and Denman 1985). Note that the $\beta = 5/3$ ($H = 1/3$ and $D_{FFT} = 5/3$; Equation 4.13 and Equation 4.16) spectral exponent expected in the case of turbulent velocity fluctuations as well as a purely passive scalar advected by turbulent fluid motion is then indicative of an antipersistent memory in turbulence-driven processes.

Alternatively, both empirical and theoretical investigations have demonstrated that the spectral exponent β could be whitened or reddened, that is, a decrease or increase in β values (Denman and Platt 1976; Denman et al. 1977; Powell and Okubo 1994). On the basis of both dimensional analysis and modeling approaches, potential changes in the phytoplankton $1/f^\beta$ noise have been attributed to the combination of turbulence, phytoplankton growth, and predator–prey relationships. Thus, in the absence of predation, a phytoplankton species with both negative and nil growth rates is characterized by a spectral exponent $\beta = 5/3$ whatever the scales (Figure 4.6A). When the growth rate is positive, two spectral exponents should be expected, $\beta = 5/3$ and $\beta = 1$, below and above the critical scale where growth dynamics overcome the diffusive dynamics of turbulence (Figure 4.6C). These theoretical scale breakings have been observed only a few times for temporal and spatial scales compatible with phytoplankton growth dynamics (Powell et al. 1975; Lekan and Wilson 1978; Abbott et al. 1982; Weber et al. 1986). More recently, three kinds of transitions have been reported (Figure 4.6):

1. A transition from $\beta = 5/3$ at small scales ($t < 25$ s) and $\beta = 0.68$ at larger scales (Seuront et al. 1996a; Seuront 1999) (Figure 4.6D).
2. A transition from $\beta = 5/3$ at large scale ($t > 160$ s) to $\beta = 1.22$ at smaller scales (Seuront 1998, 1999) (Figure 4.6E).
3. A transition from $\beta = 5/3$ at small scales ($t > 20$ s), to $\beta = 0.67$ over intermediate scales ($20 < t < 1000$ s) and to $\beta = 1.96$ for larger scales (Seuront et al. 1999) (Figure 4.6F).

However, the temporal and spatial scales involved are far too small to be related to growth dynamics and have rather been related to coagulation processes ($\beta = 5/3 \rightarrow \beta = 0.67$; cases (1) and (2); Seuront et al. 1996a, 1999), zooplankton grazing pressure ($\beta = 5/3 \rightarrow \beta = 1.22$; Seuront 1999; Lovejoy et al. 2001), and to the presence of a frontal area ($\beta = 0.67 \rightarrow \beta = 1.96$; case (3); Seuront et al. 1999). Finally, the introduction of predation reddened the spectral exponent from $\beta = 5/3$ to $\beta = 3$ for a three-dimensional turbulence, and whitened the spectral exponent from $\beta = 3$ to $\beta = 1$ for a two-dimensional turbulence (Powell and Okubo 1994). To our knowledge, only the latter case has been verified from remote sensing observations of sea-surface chlorophyll concentrations (Barale and Trees 1987; Smith et al. 1988). More generally, $1/f^\beta$ noise have also been identified in the distribution of nutrients (Seuront et al. 2002); $\beta \in [1.20 - 1.69]$) and zooplankton abundance (Seuront and Lagadeuc 2001) ($\beta = 1.42$).

4.2.1.3 Case Study: Eulerian and Lagrangian Scalar Fluctuations in Turbulent Flows

Sessile and motile organisms intrinsically perceive their environments in a Eulerian and Lagrangian framework, respectively (Figure 4.7). More specifically, sessile organisms will only perceive environmental fluctuations from a Eulerian perspective. In contrast, motile organisms will perceive environmental fluctuations occurring at scales smaller and larger than they in Eulerian and Lagrangian ways, respectively (Figure 4.7). As a consequence, these two frameworks need to be thoroughly investigated and understood to critically assess the impact of environmental fluctuations on their biology and ecology. In this context, the theoretical scaling relations expected for Eulerian and Lagrangian fluctuations of turbulent velocity and passive scalars are briefly reviewed hereafter before being tested using oceanic biophysical time series recorded from a fixed and a drifting platform used to mimic respectively the perception of sessile and free-living organisms.

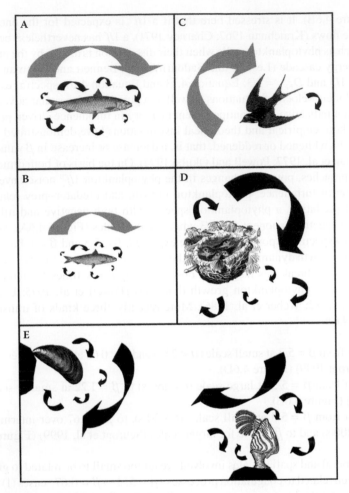

FIGURE 4.7 Lagrangian and Eulerian perceptions of the environment by living organisms. For (A, B) swimming and (C) flying organisms, the perception of their environment is intrinsically linked to their size. Velocity fluctuations larger than the organisms are then perceived in a Lagrangian way (gray arrows), while velocity fluctuations smaller than the organisms will be perceived in a Eulerian way (black arrows). In contrast, nonmotile organisms always perceive their environment in a Eulerian way (D, E), black arrows. (Modified from Seuront, 2008.)

4.2.1.3.1 Eulerian and Lagrangian Scaling Relations for Velocity and Passive Scalars

Scaling relations for turbulent velocity and passive scalars (originally temperature) fields have been expressed in Eulerian turbulence using the energy flux ε as (Kolmogorov 1941; Obukhov 1941):

$$\varepsilon \approx \frac{(\Delta V_l)^3}{l} \tag{4.19}$$

and the scalar variance flux χ as (Obukhov 1941, 1949; Corrsin 1951):

$$\chi \approx \frac{(\Delta S_l)^2 (\Delta V_l)}{l} \tag{4.20}$$

where $\Delta V_l = |V(x + l) - V(x)|$ and $\Delta S_l = |S(x + l) - S(x)|$ are the velocity shear and passive scalar gradients at scale l and $\Delta V_l/l$ is the inverse of the local eddy turnover time. These scaling relations were originally considered in the framework of homogeneous turbulence; that is, the fluxes ε and χ were considered as homogeneous, exhibiting no scale dependence. The statistics of turbulent velocity and passive scalar fluctuations were then regarded as universal and determined by the mean dissipation rate ε and the mean scalar variance flux χ. Consequently, a unique exponent was required for the velocity and passive scalar, the so-called 1/3 law in physical space:

$$\Delta V_l \approx l^{1/3} \tag{4.21}$$

$$\Delta S_l \approx l^{1/3} \tag{4.22}$$

In Fourier space, assuming local isotropy and three-dimensional homogeneity of turbulence in the inertial subrange, Equations (4.21) and (4.22) can be rewritten to describe the velocity fluctuations and the fluctuations of a passive scalar using the spectral densities $E_V(k)$ and $E_S(k)$ as:

$$E_V(k) \approx k^{-\beta_V} \tag{4.23}$$

$$E_S(k) \approx k^{-\beta_S} \tag{4.24}$$

where k is either the frequency (Hz) or the wave number (m^{-1}) whether velocity and passive scalar fluctuations are considered in time or in space, and β_V and β_S are characteristic spectral exponents defined as $\beta_V = \beta_S = 5/3$. This has been verified for velocity and temperature fluctuations in the atmosphere (Gurvich 1960; Grant et al. 1962), in the ocean (Seuront et al. 1996a, 1996b, 1999; Lovejoy et al. 2001), and in laboratory experiments (see, for example, Baumert et al. 2005), and for phytoplankton biomass in a variety of marine environments (Seuront et al. 1996a, 1996b, 1999; Currie and Roff 2006; Yamazaki et al. 2006).

In a Lagrangian framework, the scaling relations given by Equations (4.19) and (4.20) are now a function of the time between observations instead of the spatial separation considered in the Eulerian framework. Replacing $\Delta V_l/l$ by $1/t$ in Equations (4.19) and (4.20) leads to (Inoue 1952a, 1952b; Monin and Yaglom 1975):

$$\varepsilon \approx \frac{\Delta V_l^2}{t} \tag{4.25}$$

and the scalar variance flux χ as:

$$\chi \approx \frac{\Delta S_l^2}{t} \tag{4.26}$$

where $\Delta V_t = |V(\tau + t) - V(\tau)|$ and $\Delta S_t = |S(\tau + t) - S(\tau)|$ are the velocity shear and passive scalar gradients for an element of fluid at the scale t. In Fourier space, Equations (4.25) and (4.26) directly lead to:

$$E_V(k) \approx k^{-\beta_V} \tag{4.27}$$

$$E_S(k) \approx k^{-\beta_S} \tag{4.28}$$

where β_V and β_S are characteristic spectral exponents defined as $\beta_V = \beta_S = 2$. Note that for a given data set, the difference in spectral slope can then be used to identify Lagrangian and Eulerian regimes in atmospheric and oceanic data.

4.2.1.3.2   Eulerian Sampling Procedure

Sampling was carried out from two anchor stations located in the inshore (9/23/1997) and offshore (9/23/1997) waters of the eastern English Channel (Figure 4.8). Time series of temperature, salinity, and *in vivo* fluorescence (that is, a proxy for phytoplankton biomass) were recorded using a Sea-Bird 25 Sealogger CTD probe and a Sea Tech fluorometer at a frequency of 2 Hz. This led to 28,590 and 28,777 data points available for analysis from inshore and offshore waters, respectively.

In coastal waters, the power spectra of temperature and salinity both fully agree with the theoretical $\beta = 5/3$ expectations over four decades (Figure 4.9A,B). In contrast, *in vivo* fluorescence power spectrum clearly flattens for frequencies smaller than 0.01 Hz (that is, 100 seconds) with $\beta = 0.34$ (Figure 4.9C). For frequencies larger than 0.01 Hz, the fluorescence spectrum follows a power law with $\beta = 1.71$, which cannot be statistically distinguished from $\beta = 5/3$ ($p > 0.05$). This shows that fluorescence fluctuations belong to the classes of fractional Gaussian noise and fractional Brownian motion for scales smaller and larger than 100 seconds, respectively. Temperature and salinity fluctuations are, however, fully compatible with fractional Brownian motions. More specifically, this indicates that temperature and salinity can be considered as passive scalars advected by turbulent flows, while phytoplankton biomass can only be thought of as a passive scalar for scales larger than 100 seconds. For scales smaller than 100 seconds, the flattening of the phytoplankton power spectrum suggests that biological activity overcomes turbulent diffusion. The fractal dimensions and Hurst exponents for temperature and salinity, and fluorescence for scales larger than 100 seconds, are $D_{FFT} = 5/3$ and $H = 1/3$ using Equations (4.13) and (4.15). In contrast, Equations (4.14) and (4.15) lead to $D_{FFT} = 1.33$ and $H = 0.67$ for fluorescence for scales smaller than 100 seconds. This suggests that biological activity modifies the intrinsic properties of fluorescence signals from an antipersistent fractional Brownian motion to a persistent fractional Gaussian noise.

In offshore waters, the temperature and salinity power spectra still exhibit power-law behaviors over four decades but with different spectral exponents (Figure 4.9D,E); the temperature spectrum is in perfect agreement with $\beta = 5/3$ (Figure 4.9D) while the salinity one exhibits a $\beta = 7/5$ behavior ($\beta = 1.4$) (Figure 4.9E), characteristic of a buoyancy area where salinity fluctuations are controlled by gravity rather than by temperature fluctuations (Nozdrin 1974; Monin and Ozmidov 1985). The power spectrum observed for *in vivo* fluorescence unambiguously followed a scaling behavior over

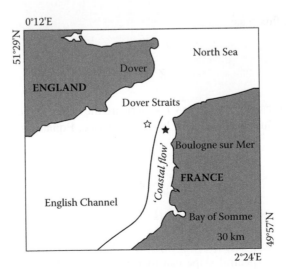

FIGURE 4.8 Study area and location of the sampling stations in the inshore (black star) and offshore (open star) waters of the eastern English Channel.

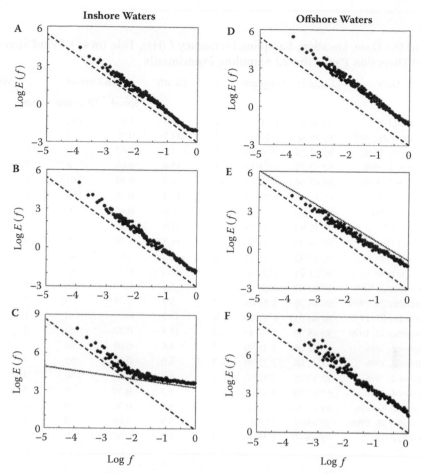

FIGURE 4.9 Power spectra of (A, D) temperature, (B, E) salinity, and (C, F) *in vivo* fluorescence recorded in the inshore and offshore waters of the eastern English Channel in a Eulerian framework (that is, a fixed location). Dashed lines are the $\beta = 5/3$ theoretical slope expected in case of purely passive scalars advected by turbulent flows. The dotted lines have a slope of $\beta = 7/5$ for (E) salinity and $\beta = 0.34$ for (C) *in vivo* fluorescence.

the whole range of available frequencies with $\beta = 5/3$ (Figure 4.9F). The resulting fractal dimensions and Hurst exponents for temperature and fluorescence ($D_{FFT} = 5/3$ and $H = 1/3$), and for salinity ($D_{FFT} = 1.80$ and $H = 0.2$) using Equation (4.13) and Equation (4.15) then suggest that in the buoyancy regime salinity fluctuations are more negatively correlated than temperature and phytoplankton biomass fluctuations.

4.2.1.3.3 Lagrangian Sampling Procedure

Sampling was carried out between March 1995 and December 1996 adrift in the coastal waters of the eastern English Channel at different depths, and in different tidal and meteorological conditions (Table 4.1) aboard the N/O *Sepia II* (CNRS-INSU), and on April 2, 1998, aboard the N/O *Côte de la Manche* (CNRS-INSU). During each sampling experiment (Table 4.1), physical parameters (temperature, salinity, light transmission) and *in vivo* fluorescence (that is, a proxy of phytoplankton biomass) were simultaneously recorded at 1 to 2 Hz from a single depth with a SBE 25 Sealogger CTD and a Sea Tech fluorometer, respectively. The present analysis is based on 22 time series labeled from S1 to S22 (Table 4.1) that contain temperature, salinity, light transmission, and *in vivo* fluorescence data, that is, 1,431,084 data points.

TABLE 4.1
Summary of the Date, Location, Sampling Frequency f (Hz), Tide (m s^{-1}), Wind Speed (m s^{-1}), and Direction (°) for the 22 Sampling Experiments

Code	Date	Latitude	Longitude	f	Depth	Tidal Current		Wind	
						Speed	Direction	Speed	Direction
S1	March 30,1995	50°40′00	1°31′00	2	15.7	1.00	180	7	120
S2	November 29,1995	50°48′56	1°29′45	2	14.7	0.22	90	5	22
S3	January 19, 1996	50°52′24	1°34′93	1	7.6	1.12	0	6	230
S4	February 1, 1996	50°42′73	1°27′49	1	15.8	0.60	0	3	220
S5	February 22,1996	50°43′09	1°32′30	1	6.2	0.88	180	4	90
S6	March 28,1996	50°45′56	1°33′82	2	10.4	0.28	180	8	90
S7	April 26, 1996	50°55′26	1°32′64	2	16.8	0.45	90	3	170
S8	May 28, 1996	50°49′93	1°32′93	2	15.6	0.75	0	3	90
S9	June 3, 1996	50°49′35	1°31′62	2	16.1	1.50	0	6	130
S10	June 19, 1996	50°42′42	1°28′5I	2	10.5	1.01	180	1	260
S11	June 25, 1996	50°51′21	1°29′65	2	21.2	0.91	180	1	330
S12	September 4, 1996	50°40′53	1°30′63	2	5.9	0.99	180	5	310
S13	September 25, 1996	50°44′73	1°33′05	2	6.2	0.39	0	5	210
S14	September 25, 1996	50°44′91	1°33′19	2	11.2	0.39	0	5	210
S15	September 25, 1996	50°45′39	1°33′45	2	15.8	0.39	0	5	210
S16	October 2, 1996	50°42′08	1°32′90	2	6.6	0.69	180	1	100
S17	October 2, 1996	50°42′08	1°32′90	2	6.0	0.69	180	1	100
S18	October 2, 1996	50°42′08	1°32′90	2	6.2	0.69	180	4	40
S19	October 8, 1996	50°47′79	1°33′68	2	6.3	0.50	0	1	220
S20	December 5, 1996	50°45′50	1°32′85	2	6.6	0.30	0	2	190
S21	December 18, 1996	50°45′46	1°32′87	2	5.9	0.60	0	6	210
S22	April 2, 1998	50°45′00	1°33′50	2	6.0	0.60	0	2	100

Samples of the double logarithmic power spectra for the studied time series together with their best fitting lines are given in Figure 4.10. The power spectra present a mixed behavior with two scaling tendencies, the change of behavior of the power spectra occurring for frequencies f ranging from 0.02 to 0.11 Hz (0.06 ± 0.03 Hz, $\bar{x}\pm$SD), which are associated with characteristic time scales t ranging from about 9 to 54 seconds (23.3 ± 12.9 sec, $\bar{x}\pm$SD). Within each data set, the transition scales are very similar whatever the variables in question, as shown by the weak dispersion of the estimates of the transition frequencies f (SD$_f = 0.005\pm0.001$).

Those temporal transition scales can be associated with spatial scales using Taylor's hypothesis of frozen turbulence (Taylor 1938), which basically states that temporal and spatial averages t and l, respectively, can be related by a constant velocity V, $l = V \cdot t$. Then, using the mean tidal drift observed during each field experiment (Table 4.1), we estimate that the associated length scales were ~12 meters (12.1 ± 1.6 m, $\bar{x}\pm$SD) for sampling experiments S1 to S21 and 24.6 meters for sampling experiment S22. These length scales are close to the size of the ships used during the sampling experiment, that is, 12.5 m and 24.9 m for N/O *Sepia II* and N/O *Côte de la Manche*, respectively. These results thus confirm and generalize the results obtained by Seuront et al. (1996b) from a single sampling experiment conducted at the end of March 1995 during a period of spring tide (Table 4.2; Figure 4.11).

In order to interpret this change of behavior of the power spectra, remember that the measurements were taken from a boat adrift in the channel. This means that for the high-frequency range

TABLE 4.2
Temporal and Spatial Transition Scales between Eulerian and Lagrangian Regimes

Time Series	Time (s)[1]	Time (s)[2]	Space (m)[1]	Space (m)[2]
*	12.30	12.70	12.30	12.70
S1	12.59	12.70	12.59	12.70
S2	53.70	54.00	11.81	11.88
S3	10.23	10.50	11.46	11.76
S4	21.38	21.00	12.83	12.60
S5	12.88	12.67	11.34	11.16
S6	44.67	44.00	12.51	12.32
S7	25.70	26.00	11.57	11.70
S8	16.22	16.50	12.16	12.38
S9	8.91	9.00	13.37	13.50
S10	12.30	12.00	12.43	12.12
S11	14.13	14.00	12.85	12.74
S12	12.30	12.50	12.18	12.38
S13	33.33	33.50	13.00	13.07
S14	30.90	31.00	12.05	12.09
S15	33.88	34.00	13.21	13.26
S16	18.62	19.00	12.85	13.11
S17	17.78	18.00	12.27	12.42
S18	19.50	20.00	13.45	13.80
S19	26.92	27.00	13.46	13.50
S20	40.74	41.00	12.22	12.30
S21	19.50	19.50	11.70	11.70
S22	44.68	44.50	24.58	24.48

Note: The association between temporal and spatial transition scales has been done via the Taylor's hypothesis of frozen turbulence.

[1] Power spectra.

[2] Structure functions.

Source: From Seuront et al. (1996b).

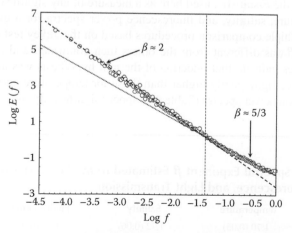

FIGURE 4.10 Power spectrum of temperature fluctuations recorded in the inshore waters of the eastern English Channel from a drifting boat 12.5 m long. Two scaling regimes with $E(f) \approx f^{-2}$ and $E(f) \approx f^{-5/3}$ respectively occur above and below a critical time scale $t = 22$ seconds. The high and low frequency regimes correspond to Eulerian and Lagrangian regimes, respectively.

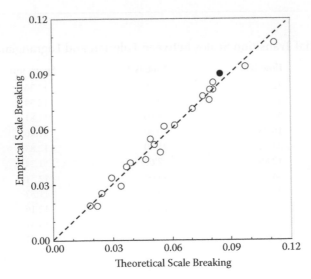

FIGURE 4.11 Theoretical vs. empirical scale breakings (Hz) between Eulerian and Lagrangian scales. Theoretical scale breaking has been estimated by multiplying the size of the boat used during the sampling experiment by the mean tidal drift observed during each experiment. Empirical scale breaking has been estimated as the mean transition scale from the power spectra of temperature, salinity, and *in vivo* fluorescence. The result obtained by Seuront et al. (1996b) is shown for comparison (black dot).

of the measurements we can consider the boat as not moving, so the measurements correspond to a fixed-point procedure, that is, Eulerian sampling. This is confirmed by the slopes of the small-scale temperature, salinity, and *in vivo* fluorescence power spectra (Table 4.3), which were not significantly different (Kruskal-Wallis test, $P > 0.05$) and cannot be statistically distinguished from the theoretical spectral value $\beta = 5/3$ (Binomial test, $p > 0.05$) (Siegel and Castellan 1988) expected in the case of an isotropic three-dimensional homogeneous turbulence (Obukhov 1949; Corrsin 1951). Nevertheless, for each sampling experiment, an analysis of covariance (Zar 1996) has been conducted for the slopes of the power spectra for temperature, salinity, and *in vivo* fluorescence. It is found that the chlorophyll *a* concentration exhibits highly significant positive correlation ($p = 0.01$) with the F-statistic, used here as a measure of any significant difference between the slope of temperature, salinity, and fluorescence power spectra for a given sampling experiment. Subsequent multiple comparison procedures based on the Tukey test (Zar 1996) conducted to determine which β was different from the others then confirmed and specified the previous results. These analyses indicate that rejection of the null hypothesis was always due to β values for *in vivo* fluorescence significantly higher than those for temperature and salinity. On the contrary, light transmission power spectra (Table 4.3) appear significantly smaller than the theoretical

TABLE 4.3

Mean Values of the Spectral Exponent β Estimated from Time Series of Temperature, Salinity, *In Vivo* Fluorescence, and Light Transmission

	Temperature	Salinity	Fluorescence	Transmission
E	1.70 (0.05)	1.72 (0.06)	1.69 (0.03)	1.31 (0.36)
L	2.03 (0.05)	2.03 (0.04)	1.03 (0.24)	1.02 (0.24)

Note: The numbers in parentheses are the standard deviations; Eulerian (E; $n = 22$) and Lagrangian (L; $n = 13$) scales.

$\beta = 5/3$ value ($p < 0.05$). Finally, one may also note here that analyses of covariance showed that the 22 spectral exponents β were not all equal for each parameter, indicating potential differential spectral structures of the variables in question at the space-time scales of the whole sampling experiment.

For frequencies smaller than the observed scale breakings, the inertia of the boat becomes negligible and the measurements are effectively taken following the flows, that is, in a Lagrangian framework. One may note here that we had to average the original time series up to the Eulerian/Lagrangian transition scale (Table 4.2) in order to be in the Lagrangian scales. In that way, our characterization of the Lagrangian behavior of time series of temperature, salinity, light transmission, and *in vivo* fluorescence is based on time series exhibiting at least 256 data points (that is, the lower recommended bound for a data set to lead to reliable spectral analysis). In the following, we then focused on 13 time series S1–S3, S5, S7, S9, S10–S12, S14, S19, and S21–S22. The previously described transition is also confirmed by the similar scaling behaviors exhibited by temperature and salinity time series (Table 4.3), which cannot be statistically distinguished from the theoretical slope $\beta = 2$ (Binomial test, $p > 0.05$) expected in the case of purely passive scalars advected by Lagrangian fluid motions (Monin and Yaglom 1975). On the contrary, *in vivo* fluorescence and light transmission spectral exponents show very specific behaviors (Table 4.3) that cannot be statistically distinguished (Wilcoxon-Mann-Whitney U-test, $p > 0.05$). Moreover, analyses of covariance concluded that the 13 spectral exponents β were not all equal ($p < 0.05$) for both light transmission and fluorescence power spectra.

4.2.2 Detrended Fluctuation Analysis

4.2.2.1 Theory

Detrended fluctuation analysis (DFA) is an elegant tool to quantify simply and reliably the correlations found in both stationary and nonstationary data (Peng et al. 1992, 1993, 1994). Compared to more traditional analyses used to estimate the degree of correlation in temporal signals, such as power spectrum analysis (see Section 4.2.1), Hurst analysis (Section 4.2.5), and autocorrelation analysis (Section 4.2.7), the advantage of the DFA method is that it can accurately quantify the correlation property of signals masked by polynomial trends (Hu et al. 2001; Chen et al. 2002). Here, a temporal signal (note that the method described hereafter is also applicable to a spatial signal $x(l)$), is integrated by computing for each t the accumulated departure from the mean of the whole series:

$$X(t) = \sum_{i=1}^{N} [x(i) - \bar{x}]$$ (4.29)

where N is the length of the data set. This integrated series is divided into nonoverlapping intervals of length n. In each interval, a least-squares regression representing the trend in the interval is fitted to the data (Figure 4.12). The series $X(t)$ is then locally detrended by substracting the theoretical values $X_n(t)$ given by the regression. For a given interval length n, the characteristic size of fluctuations for this integrated and detrended series is calculated by:

$$F(n) = \sqrt{\frac{1}{N} \sum_{t=1}^{N} [X(t) - X_n(t)]^2}$$ (4.30)

This computation is repeated over all possible interval lengths. Note, however, that the width of the regression windows ranged from six data points and to $N/4$ (Peng et al. 1994). The degree of correlation in the signal is finally quantified as:

$$F(l) \propto l^{\alpha_{DFA}} \qquad (4.31)$$

where the exponent α is estimated as the slope of the linear trend of the log-log plot of $F(l)$ versus l. The use of a linear regression on the log-transformed data instead of nonlinear regression on the raw data is recommended, as the residual error will be distributed as a quadratic and the minimum error is not guaranteed. This is not the case with nonlinear regression (Seuront and Spilmont 2002).

As with power spectrum analysis, DFA allows the distinction between fGn and fBm signals; fGn correspond to α_{DFA} exponents ranging from 0 to 1, and fBm to exponents bounded between 1 and 2

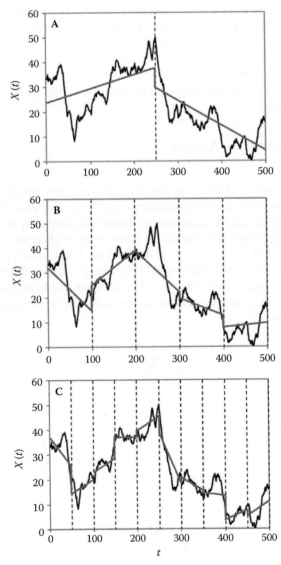

FIGURE 4.12 Three successive steps of the implementation of the detrended fluctuation analysis (DFA). DFA is applied to intervals of size (A), 100 (B), and 50 (C). The gray lines are the best least-squares fit to the data in each interval.

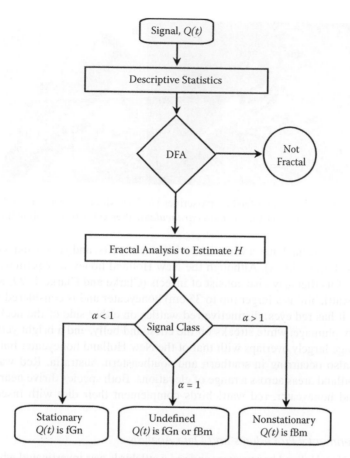

FIGURE 4.13 Use of detrended fluctuation analysis (DFA) to identify the class to which a data set belongs, fractional Gaussian noise (fGn) or fractional Brownian motion (fBm), prior to further fractal analysis devoted to estimate the Hurst exponent H, and the related fractal dimension.

(Figure 4.13). Note that the correspondence between α and H is:

$$H = \alpha_{DFA} - 1 \qquad (4.32)$$

for fractional Brownian motions, and

$$H = \alpha_{DFA} \qquad (4.33)$$

for fractional Gaussian noises. The corresponding fractal dimensions, D_{DFA}, are given by (Equation 4.9):

$$D_{DFA} = 2 - H \qquad (4.34)$$

Note here that combining Equations (4.32) and (4.33) with Equations (4.13) and (4.14) leads to relate the exponents α and β as $\alpha = (\beta + 1)/2$, and reciprocally $\beta = 2\alpha - 1$.

4.2.2.2 Case Study: Assessing Stress in Interacting Bird Species

4.2.2.2.1 The Study Organisms

The New Holland honeyeater (*Phylidonyris novaehollandiae*; Figure 4.14A) and the red wattle-bird (*Anthochaera carunculata*; Figure 4.14B) are Australian nectar-feeding passerines. The New Holland honeyeater is a medium-sized (up to 18 cm) very active bird found throughout southern

FIGURE 4.14 The Australian nectar-feeding passerines (A) New Holland honeyeater (*Phylidonyris novae-hollandiae*) and (B) red wattlebird (*Anthochaera carunculata*). **(See color insert following page 80.)**

and southeastern Australia. It has a white iris, white facial tufts, and yellow markings on its wings and tail feathers (Figure 4.14A). Although the New Holland honeyeater primarily feeds on nectar, a large part of its diet may also consist of insects (Clarke and Clarke 1999; Kleindorfer et al. 2006). The red wattlebird is a larger (up to 35 cm) honeyeater and is considered a dominant honeyeater species. It has red eyes, distinctive red wattles on either side of the neck (Figure 4.14B), gray-brown body plumage, white streaks on the chest and belly, and a bright yellow patch on the abdomen. Its range largely overlaps with that of the New Holland honeyeater but extends slightly further inland, also occurring in southern and southeastern Australia. Red wattlebirds inhabit coastal and woodland areas across a range of elevations. Both species thrive near habitation. Like the New Holland honeyeater, red wattlebirds complement their diet with insects, berries, and other fruits.

4.2.2.2.2 *Experimental Procedures and Data Analysis*

The behavior of New Holland honeyeaters and red wattlebirds was investigated when the birds were eating the nectar of aloe flowers separately (Figure 4.15A,B) and simultaneously (Figure 4.15C); the objectives of this work were to investigate if their feeding behavior exhibited any long-term correlation and if the presence of another species would affect the feeding behavior of a given species. The behavior of those two species was recorded with a digital camera (DV Sony DCR-PC120E) at a rate of 25 frame s^{-1} on July 22, 2007, in West Beach, South Australia. The behaviors of New Holland honeyeaters and red wattlebirds typically were very similar and consisted of either eating the nectar with their beak in the flower or scanning their surroundings. The nature of the behavioral sequences observed in the New Holland honeyeaters and the red wattlebirds was assessed through the construction of a binary sequence $z_t(i)$ for each behavioral activity i taken from continuous observations (Figure 4.16A,B). When a specific activity was observed, $z_t(i) = 1$, and $z_t(i) = -1$ otherwise. Here $z_t(i) = 1$ and $z_t(i) = -1$ when the beak of either the New Holland honeyeater or the red wattlebird was respectively inside and outside the flower. This generated binary sequences $z_t(i)$ taken at 0.04-second time intervals t for each behavioral activity. Behavior sequence random walks $w_i(t)$ were subsequently obtained as

$$w_i(t) = \sum_{i=1}^{N} z_t(i) \qquad (4.35)$$

where N is the number of behavioral observations. Equation (4.35) provides a graphical representation to calculate the degree of correlation in the behavioral time series (Figure 4.16C,D). Note the clear difference between the behavioral $z_i(t)$ and $w_i(t)$ shown in Figure 4.16 and Figure 4.17 and those obtained in case of a purely random process (Box 4.1).

BOX 4.1 PURE RANDOMNESS: COIN FLIPPING

The archetypical example of a purely random process is the flipping coin experiment in which the probability of obtaining the obverse (heads) or the reverse (tails) is the same (that is, 0.5). In addition, the probability of obtaining one event is totally independent of the previous event (tails or heads); the coin-flipping process does not have memory.

The subsequent detrended fluctuation analysis of $w_i(t)$ results in $\alpha_{DFA} = 1.50$, hence $H = 0.50$. The sequential random walk $w_i(t)$ resulting from successive coin flipping is then a Brownian motion; that is, no form of persistence ($H > 0.50$) or antipersistence ($H < 0.50$) exists in the signal.

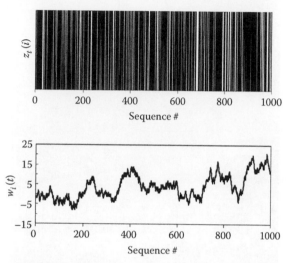

FIGURE 4.B1.1. Binary sequence obtained from 1000 iterations of a coin-flipping experiment, where $z_t(i) = 1$ for tails (black vertical lines, top panel) and $z_t(i) = -1$ for heads (white vertical lines, top panel), and the resulting sequential random walk $w_i(t) = \Sigma_{t=1}^N z_t(i)$ (bottom panel).

4.2.2.2.3 Results

Behavior sequence random walks $w_i(t)$ obtained for New Holland honeyeaters (Figure 4.17A) and red wattlebirds (Figure 4.17B) were visually clearly different, as was New Holland honeyeater behavior in the absence (Figure 4.17B) and presence of red wattlebirds (Figure 4.17C). Specifically, applying Equations (4.30) and (4.31) to the behavior sequence random walks shown in Figure 4.17 leads to values of the exponent $\alpha_{DFA} = 1.65$ for New Holland honeyeaters and $\alpha_{DFA} = 1.72$ for red wattlebirds. The values of α_{DFA} obtained for the red wattlebird were not affected by the presence of the New Holland honeyeater. In contrast, the exponent α_{DFA} significantly decreased down to $\alpha_{DFA} = 1.48$ for the New Holland honeyeater in the presence of red wattlebirds.

4.2.2.2.4 Ecological Interpretation

The estimates of α_{DFA} for the sequential behavior of the New Holland honeyeaters and the red wattlebirds were consistently larger than 1, indicating that the behavior sequence random walks $w_i(t)$ consistently belong to the class of fractional Brownian motions. Specifically, the Hurst exponents H estimated from Equation (4.32) lead to values higher than $H = 0.5$ for the New Holland honeyeater ($H = 0.65$) and the red wattlebird ($H = 0.72$). Both species then exhibit a persistent behavior; that is, a feeding bout is more likely to be followed by a feeding bout than by nonfeeding bout.

FIGURE 4.15 The New Holland honeyeater (*Phylidonyris novaehollandiae*) and red wattlebird (*Anthochaera carunculata*) foraging around aloe flowers separately (A, B), and simultaneously (C).

In contrast, in the presence of the red wattlebird, the exponent H of the New Holland honeyeater decreased to $H = 0.48$ and is nonsignificantly different from the special case $H = 0.50$ expected for Brownian motion, which indicates no correlations in the behavior sequence $w_i(t)$; in other words, the probability of feeding bouts is independent of the probability of nonfeeding bouts. Those results are consistent with previous behavioral studies using DFA to assess the health and stress from binary behavioral sequences of wild chimpanzees (Alados and Huffman 2000) and captive chickens (María et al. 2004). The exponents α_{DFA} were consistently lower in sick chimpanzees than in healthy ones, and decreased with chicken stress. The present results then suggest that the presence

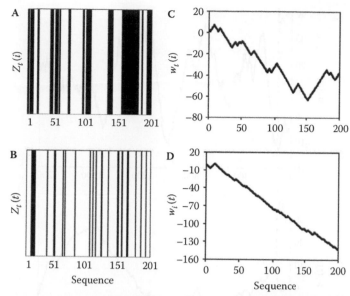

FIGURE 4.16 Examples of binary behavioral sequences $z_t(i)$ recorded for the New Holland honeyeater in the (A) absence and (B) presence of the red wattlebird and respectively visualized as black and white vertical lines for $z_t(i) = 1$ (bird's beak inside aloe flower) and $z_t(i) = -1$ (bird's beak outside aloe flower). The binary sequences $z_t(i)$ are subsequently used to build behavior sequence random walks $w_i(t)$ as $w_i(t) = \Sigma_{i=1}^{N} z_t(i)$, where N is the number of behavioral sequences.

of the large red wattlebirds increases the level of stress in the New Holland honeyeaters, while the significantly higher α_{DFA} observed for the red wattlebirds suggests an overall higher behavioral persistence than might be related to their larger size. This is also consistent with other results based on the use of cumulative frequency analyses and spectral analysis showing that the sequential behavior of parasited Spanish ibex, *Capra pyrenaica*, which returned lower exponents in cumulative frequency analyses (see Chapter 5) and spectral analysis than nonparasited ones (Alados et al. 1996). A cumulative frequency analysis of the move duration observed in the marine zooplankton species *Centropages hamatus* also showed a decrease in scaling exponent under conditions of naphthalene contamination (Seuront and Leterme 2007). As a conclusion, detrended fluctuation analysis of sequential behavior in particular, and scaling analysis of sequential behavior in general, provide an effective noninvasive method to record and evaluate the general state of health and related stress of captive and wild animals.

Note that in their analysis of wild chimpanzee behavioral sequences, Alados and Huffman (2000) seem to have mixed up the values of α_{DFA} expected for fractional Brownian motions and fractional Gaussian noises, as well as the concepts related to α_{DFA} and H. They indeed claim that "$\alpha = 1/2$ indicates no correlation in the sequence (white noise), and $\alpha \neq 1/2$ indicates long-range power law correlations. If α exceeds 1/2, the sequence is persistent; if $\alpha < 1/2$ the sequence is anti-persistent." However, as discussed in Section 4.2.2.1, this is only true for fGn in which case $\alpha_{DFA} = H$ and is bounded between 0 and 1. Over the 20 values of α_{DFA} provided in their Table 3, 19 are greater than 1 and range between 1.114 and 1.436, suggesting that their random walks $w_i(t)$ belong to the family of fractional Brownian motions. As such, Equation (4.32) leads to a reinterpretation of the sequential behavior of wild chimpanzees as being antipersistent (that is, $\alpha_{DFA} < 1.5$ and $H < 0.5$) and not persistent as originally claimed (Alados and Huffman 2000). This stresses the critical need to identify unambiguously the nature of a signal to be analyzed to ensure the results of the analysis are relevant and meaningful.

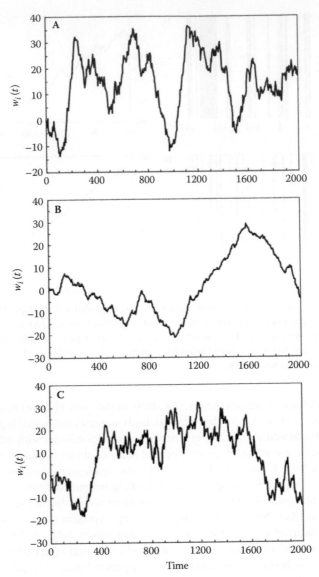

FIGURE 4.17 Behavior sequence random walks $w_i(t)$ obtained for (A) the New Holland honeyeater and (B) the red wattlebird, and (C) for the New Holland honeyeater in the presence of red wattlebird.

4.2.3 SCALED WINDOWED VARIANCE ANALYSIS

4.2.3.1 Theory

Scaled windowed variance (SWV) analyses, also referred to as roughness-length analyses, are applicable only to fBm signals. Those methods are based on dividing a temporal signal $x(t)$ into nonoverlapping windows of size τ and computing the standard deviation, $SD(i)$, in each of these windows as:

$$SD(i) = \left[\frac{1}{\tau - 1} \sum_{t=1}^{\tau} [x(t) - \overline{x}_i]^2 \right]^{1/2} \tag{4.36}$$

where $\bar{x}_i$ is the average within each interval. The average standard deviation, $SD(\tau)$, for each window size τ is further expressed as:

$$SD(\tau) = \frac{1}{n} \sum_{i=1}^{n} SD(i) \qquad (4.37)$$

where n is the number of nonoverlapping windows of size τ. This procedure is iterated for all possible window sizes. For a fractal signal $x(t)$, $SD(\tau)$ scale with τ as:

$$SD(\tau) \propto \tau^H \qquad (4.38)$$

where H is the Hurst exponent described above. The subsequent fractal dimension D_{SWV} is derived from Equation (4.9) as:

$$D_{SWV} = 2 - H \qquad (4.39)$$

Initially introduced by Mandelbrot (1985), this method applies no correction to the trends present within a given window and potentially leads to bias in H estimates (Schmittbuhl et al. 1995; Turcott and Teich 1995). Trends in the signal seen within a given window can, however, be corrected either by subtracting a linearly estimated trend (that is, line-detrended scaled windowed variance analysis, ldSWV) or the values of a line bridging the first and last values of the signal (bridge-detrended scaled windowed variance analysis [bdSWV]) (Cannon et al. 1997). Note that line detrending and bridge detrending are conceptually similar to the detrending procedure used in detrended fluctuation analysis (Section 4.2.2) and illustrated in Figure 4.12.

4.2.3.2 Case Study: Temporal Distribution of the Calanoid Copepod *Temora Longicornis*

4.2.3.2.1   The Study Organism

Copepods are the largest and most diversified group of crustaceans, they are the most numerous metazoans (that is, multicelled organisms) in the aquatic communities, and they are considered the most plentiful group on Earth, outnumbering even the insects, which include more species but fewer individuals. As an example, considering a mean density of just one copepod per liter of the overall volume of the open ocean (that is, 1347×10^6 km^3) would suggest a total world population in the order of 1.35×10^{21} (Boxshall 1998). They include over 14,000 species, 2,300 genera, and 210 families, a surely underestimated number. Their habitat ranges from freshwater to hypersaline conditions, from subterranean caves to water collected in bromeliad leaves or leaf litter on the ground, from streams, rivers, and lakes to the sediment layer in the open ocean, from the highest mountains to the deepest ocean trenches, and from the cold polar ice–water interface to the hot active hydrothermal vents. Copepods may be free-living, symbiotic, internal, or external parasites on almost every phylum of animals in water. The usual length of adults is 1 to 2 mm, but adults of some species may be as small as 0.2 mm and others may be as large as 10 mm. Copepods are the dominant forms of the marine plankton that constitute the secondary producers in the marine environments. As such, they constitute a fundamental step in the oceanic food chain, linking microscopic algal cells to juvenile fishes and whales, and play a pivotal role in the functioning of marine systems and biogeochemical cycles (Roemich and McGowan 1995).

More specifically, the calanoid copepod *Temora longicornis* (Figure 4.18A) is a very abundant and nearly ubiquitous species in coastal waters. It is also of great ecological significance in many areas as it represents 35 to 70% of the total copepod population in the Southern Bight of the North Sea (Daan 1989) and in the eastern English Channel (Seuront 2005c); in Long Island Sound (USA), it is able to remove up to 49% of the daily primary production (Dam and Peterson 1993).

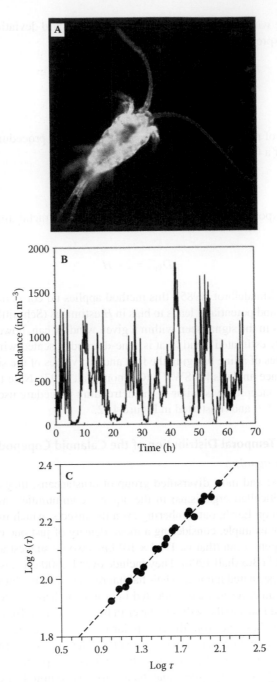

FIGURE 4.18 The calanoid copepod *Temora longicornis* (A), a high-frequency time series of the abundance of adult male and female *T. longicornis* sampled in the inshore waters of the eastern English Channel (B), and the related log-log plot of $S(\tau)$ vs. τ (Equation 4.38) resulting from the scaled windowed variance analysis.

4.2.3.2.2   *Experimental Procedures and Data Analysis*

Sampling was conducted in the coastal waters of the eastern English Channel (see Figure 4.8) for 66 hours. Water was continuously sampled from a depth of 10 m through a weighted seawater intake and directly brought through a 200-μm mesh plankton net using a Flight pump with a 300 l-min^{-1}

output, connected to a 10-cm-diameter plastic tubing. Every 3 minutes, filtered organisms were collected and immediately preserved in a 10% formaldehyde solution. This resulted in a total of 1321 samples. Adult males and females from each sample were subsequently enumerated under a dissecting microscope. The resulting time series (Figure 4.18B) exhibits significant semidiurnal tidal cycle (that is, 12.5 hours) but was also characterized by very violent and erratic fluctuations. A previous power spectrum analysis showed a very strong scaling behavior over more than two decades, with $\beta = 1.42$ (Seuront and Lagadeuc 2001), thus suggesting that the temporal distribution of *T. longicornis* belongs to the family of fractional Brownian motions. After detrending, the scaling properties of the time series shown in Figure 4.18B were investigated using Equation (4.38) and the fractal dimension derived from Equation (4.39).

4.2.3.2.3 Results and Discussion

The log-log plot of $s(\tau)$ versus τ shows a unique scaling regime over the whole range of available scales (Figure 4.18C), with $H = 0.37$, thus $D_{SWV} = 1.64$. The presence of a unique scaling regime suggests that the same process, or similar processes, is responsible for the scaling structure of the abundance of *T. longicornis* for time scales ranging from 6 minutes to 66 hours. Using Taylor's hypothesis of frozen turbulence, the related spatial scales range between 92 m and 120 km. Note that applying rescaled range (R/S) analysis (see Section 4.2.6) to the same data set returned a very similar value for the Hurst exponent; that is, $H = 0.34$, thus $D_H = 1.66$. In contrast, the Hurst exponent derived from power spectrum analysis led to a significantly lower value for H, $H = 0.21$ (that is, $D_{FFT} = 1.79$). This is in agreement with previous studies that showed the potential differences in the H values returned by R/S analysis (Section 4.2.6), power spectrum analysis (Section 4.2.1), roughness-length analysis (Section 4.2.3), variogram analysis (Section 4.2.8), and wavelet analysis (Section 4.2.8); see, for example, Mulligan (2004) for a review. This issue is addressed hereafter in Section 4.2.10.

4.2.3.2.4 Ecological Interpretation

The fractal dimensions obtained for the temporal distribution of *Temora longicornis*, bounded between 1.76 and 1.79, are higher than those expected for passive scalar advected by three-dimensional turbulence (see Section 4.2.1). Those fractal dimensions are also lower than those found for phytoplankton distribution from *in situ* time series of *in vivo* fluorescence ($D \in [1.61 - 1.67]$; Seuront et al. 1996a, 1996b, 1999) and from satellite images of sea-surface chlorophyll patterns ($D \in [0.98 - 1.69]$; Denman and Abbott 1988, 1994; Smith et al. 1988). This can be related to the differences existing between phytoplankton and zooplankton in terms of size and motility, suggesting that copepod behaviors such as diel migration, phototaxis, rheotaxis, social behaviors, and predation pressure—behaviors relevant at the space and time scales of the present study—induce larger fractal dimensions (that is, a flatter power spectrum and weaker scale dependence) in comparison with phytoplankton. This is consistent with numerical experiments based on simple predator–prey formulations considered in a turbulent frame that demonstrated that the interactions between diel vertical migration and turbulent shear could lead to a flatter zooplankton power spectrum (Steele and Henderson 1992). Similar conclusions were reached by Powell and Okubo (1994) from their study of interacting plankton populations in two-dimensional turbulence. In addition, the fractal dimension estimated here from the distribution of *T. longicornis* is very similar to that estimated for the oceanic copepod *Neocalanus cristatus* abundance transects from the subarctic Pacific, $D = 1.80$ (Tsuda 1995), over a similar range of scales (that is, between tens of meters and over 100 kilometers), suggesting that the distribution of zooplankton species could be very similar independent of their surrounding environments. This is also consistent with the white spectra ($\beta = 0$) found for total zooplankton density in the St. Laurence estuary (Currie and Roff 2006), leading to fractal dimension $D \rightarrow 2$.

4.2.4 Signal Summation Conversion Method

The signal summation conversion (SSC) method is used to refine the analysis of signals for which β (power spectrum analysis; see Section 4.2.1 and Figure 4.5) or α (detrended fluctuation analysis; see Section 4.2.2 and Figure 4.13) are near 1. This method is then used to refine fractal analysis near the fGn/fBm boundary. Taking the cumulative sum of a signal $x(t)$ converts an fGn to an fBm signal or an fBm to a summed fBm signal. Corrected scaled windowed variance analyses—that is, line-detrended scaled windowed variance analysis (ldSWV) or bridge-detrended scaled windowed variance analysis (bdSWV)—are then applied to the cumulate series to estimate the Hurst exponent H'. When $0 < H' \leq 1$ the signal is an fGn with $H = H'$. In contrast, when $1 < H'$, the signal is a fractional Brownian motion with $H = H' - 1$.

4.2.5 Dispersion Analysis

The dispersion analysis, originally introduced using relative dispersion of spatial data (Bassingthwaighte 1988), was later extended to the temporal domain (Bassingthwaighte and Raymond 1995). It is very similar to the original scaled windowed variance method but uses the standard deviation of the windows means (Figure 4.19). Specifically, a temporal signal $x(t)$ is divided into nonoverlapping windows of size τ. The mean $\bar{x}_i$ of each window is computed as:

$$\bar{x}_i = \frac{1}{\tau} \sum_{t=1}^{\tau} x(t) \tag{4.40}$$

The standard deviation of the local means is subsequently estimated as:

$$SD_{\bar{x}_i}(\tau) = \left[\frac{1}{n-1} \sum_{i=1}^{n} [\bar{x}_i - \bar{x}_\tau]^2 \right] \tag{4.41}$$

where n is the number of nonoverlapping windows of size τ and $\bar{x}_\tau$ the average of the n $\bar{x}_i$ values. This procedure is iterated for all possible window sizes, and for a fractal signal, $SD_{\bar{x}_i}(\tau)$ is related to τ following the power-law form:

$$SD_{\bar{x}_i}(\tau) \propto \tau^{H-1} \tag{4.42}$$

where H is the Hurst exponent. The dispersion analysis is applicable to fGn signals or to differentiated fBm signals.

4.2.6 Rescaled Range Analysis and the Hurst Dimension, D_H

4.2.6.1 Theory

Historically, the rescaled range (R/S) analysis is the first method developed for assessing H. R/S analysis was initiated by Hurst (1951) and Hurst et al. (1965) to describe the long-term dependence of water levels in river and reservoirs. Specifically, R/S analysis was developed to confront the question of how high the Aswan Dam had to be built so that it would contain the greatly varying levels of the Nile within a given temporal window of size τ (Figure 4.20). The rationale used to develop this method lies behind the three criteria of an ideal reservoir: (1) the outflow is uniform, (2) the water level is the same at the beginning and at the end of the observation window, and (3) the reservoir never overflows. Looking at retrospective records of water levels, $x(t)$, Hurst (1951) estimated the time series of the increase in water volume in the dam as the summed difference of inflow and outflow, $y(t)$. The range R of $y(t)$, $R = y_{\max}(t) - y_{\min}(t)$, then defines how high the dam should be built. Finally, dividing the range by the standard deviation of outflow fluctuations, $S(\tau)$, Hurst (1951) found that the ratio $R/S(\tau)$ showed a power-law

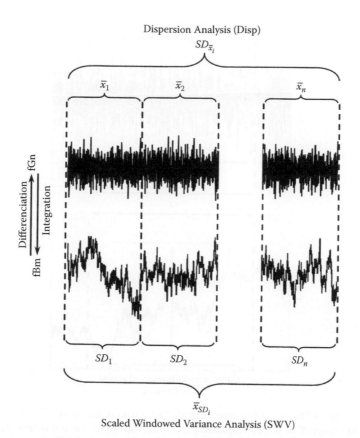

FIGURE 4.19 Principles of the scaled windowed variance (SWV) and dispersional (Disp) methods. SWV and Disp both use local statistical measures to derive the Hurst exponent H from fractional Brownian motion (fBm) and fractional Gaussian noise (fGn) signals, respectively. For a given interval size, these local statistical measures are the standard deviation SD_i for SWV and the mean $\bar{x}_i$ for Disp, and the related scale-dependent measures are the mean of local standard deviation ($\bar{x}_{SD_i}$) and the standard deviation of local means ($SD_{\bar{x}_i}$). (See Equations 4.37, 4.38, 4.41, and 4.42.)

relationship with the size of the observation window τ as $R/S(\tau) \propto \tau^H$, where H is the so-called Hurst exponent. This relation provides a sensitive method for revealing long-run correlations in random processes. It comes directly from the above and Figure 4.20 that R/S analysis was theoretically developed to work on fGn signals (or differentiated fBm signals) but provides irrelevant results for fBm signals.

R/S analysis can, however, more generally be applied to any regularly sampled temporal signal. Consider a discrete time series $x(t)$ and the interval, or window, of length τ. Within this window, one can define two quantities: $R(\tau)$, the range taken by the values of $x(t)$ in the interval τ, and $S(\tau)$, the standard deviation of the values of $x(t)$ within the window. $R(\tau)$ is measured with respect to a trend in the window, where the trend is estimated as the line connecting the first and the last points within the window. $R(\tau)$ is thus expressed by the following:

$$R(\tau) = \max_{1 \le t \le \tau} X(t, \tau) - \min_{1 \le t \le \tau} X(t, \tau) \qquad (4.43)$$

where $X(t, \tau)$ is the variable defined as $X(t, \tau) = \Sigma_{t=1}^{\tau}(x(t) - \langle x(t)\rangle_\tau)$. The mean over the time lag τ is substracted to remove a trend in the window when the expectation of $x(t)$ is not zero. $R(\tau)$ is then the

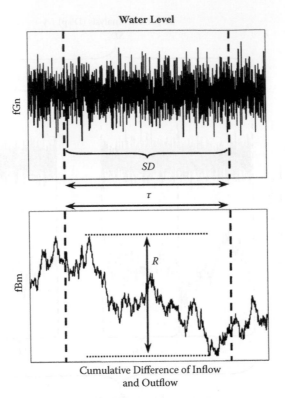

FIGURE 4.20 Principle of the rescaled range method. The standard deviation (*SD*) of a fractional Gaussian noise (fGn) and the range *R* of the corresponding fractional Brownian motion (fBm) are estimated over a window of size τ. The Hurst exponent *H* is subsequently estimated as the slope of the power-law relationship between the ratio between *R* and *SD* and the scale of observation τ; see Equation (4.46).

self-adjusted range. $S(\tau)$ is defined as:

$$S(\tau) = \left[\frac{1}{\tau} \sum_{t=1}^{\tau} \left(x(t) - \langle x(t) \rangle_{\tau} \right)^2 \right]^{1/2} \tag{4.44}$$

A reliable measurement of $S(\tau)$ requires data with a constant sampling interval, because the expected difference between successive values of $x(t)$ is a function of the distance separating them. Here, $S(\tau)$ is used to standardize the range $R(\tau)$ to allow comparisons of different data sets; if $S(\tau)$ is not used, the range $R(\tau)$ can be calculated on data sets that have a nonconstant sampling interval. A discussion of the importance of the division by the standard deviation $S(\tau)$ to obtain a statistical quantity of extreme robustness can be found in Mandelbrot and Wallis (1969).

The R/S statistics is subsequently defined as follows:

$$R/S(\tau) = \langle R(\tau)/S(\tau) \rangle \tag{4.45}$$

where $R/S(\tau)$ is the self-rescaled self-adjusted range. The basis of the method is that, because of self-affinity, one expects the range taken by the values of $x(t)$ in a window of length τ to be proportional to the window length to a power equal to the so-called Hurst exponent *H* following:

$$R/S(\tau) = k\tau^H \tag{4.46}$$

where k is a constant. In particular, Feller (1971) proved that the asymptotic behavior for any independent random process with finite variance is given by:

$$R/S(\tau) = k\tau^{1/2} \tag{4.47}$$

and are most of the time referred to as "Brownian" processes.

However, many (if not most) processes in nature are not independent random processes but show significant long-term correlations. In this case, the asymptotic scaling law is modified and $R/S(\tau)$ is asymptotically given by a power law, τ^H. The corresponding exponent H is referred to as the Hurst exponent, and can be conveniently used to characterize the long-range dependence of a random variable. A persistent behavior (that is, an increase in the value of the random variable is expected to be followed by another increase) is characterized by $0.5 < H < 1$. An antipersistent behavior (that is, an increase in the value of the random variable is expected to be followed by a *decrease*) is characterized by $0 < H < 0.5$. Many data on natural phenomena that show persistent behavior can be found in the literature, including fluctuations of argon concentration (Bejar et al. 1995), Rhine and Nile flush (Mandelbrot and Wallis 1969), and genome organization (Almirantis and Provata 1999). To contrast with the aforementioned Brownian processes, Mandelbrot (1983) introduced the concept of "fractional Brownian motion" to describe random processes characterized by Hurst exponents H such as $H \neq 0.5$.

In practice, for a given window length τ, one subdivides the input series in a number of intervals of length τ, measures $R(\tau)$ and $S(\tau)$ in each interval, and calculates $R/S(\tau)$ as the average ratio $\langle R(\tau)/S(\tau) \rangle$, as in Equation (4.45). This process is iterated for a number of window lengths, and the logarithms of $R/S(\tau)$ are plotted vs. the logarithm of τ. If the trace is self-affine, this plot follows a straight line whose slope equals the Hurst exponent H. The fractal dimension of the trace can then be calculated following:

$$D_H = 2 - H \tag{4.48}$$

where H is the Hurst exponent and D_H the Hurst fractal dimension.

Finally, although the R/S analysis described above has been illustrated in the framework of a temporal random process $x(t)$, one must note that it can be equivalently applied to any random process, recorded in time or in space.

4.2.6.2   Example: R/S Analysis and River Flushing Rates

R/S analysis was illustrated using the daily flushing rate of the Seine River (France) from 1993 to 2001 (Figure 4.21A). The resulting log-log plot of $R/S(\tau)$ vs. τ does not follow a straight line over the whole range of available τ, but instead two scaling regions are separated by a clear break for $\tau = 63$ days (Figure 4.21B). The related Hurst exponents are $H = 0.67$ and $H = 0.43$ for $\tau < 63$ and $\tau > 63$ days, respectively. These exponents reflect the presence of both persistence and antipersistence in the Seine water-flow statistics. The corresponding Hurst fractal dimensions are $D_H = 1.33$ and $D_H = 1.57$. This observation clearly diverges from the water-flow statistics of the Rhine (1808–1966), which is characterized by its Brownian properties ($H \approx 0.55$ and $D_H = 1.45$), and also from the water-level statistics of the Nile (622 to 1469), which show a high degree of persistence, and $H \approx 0.51$ ($D_H = 1.09$) (Mandelbrot and Wallis 1969).

4.2.7   Autocorrelation Analysis

Autocorrelation (AC) functions are widely used in time-series analysis to describe to what extent the value of a given event, $x(t)$, of a time series depends on its past values h lag apart, that is, $x(t - h)$; see, for example, Legendre and Legendre (2003). Autocorrelation functions have, however, rarely been directly used to estimate the Hurst exponent H and the related fractal dimension.

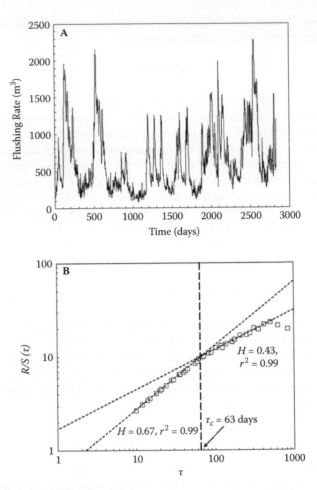

FIGURE 4.21 The Hurst dimension, illustrated using a time series of daily flushing rate of the Seine River (France) from 1993 to 2001 (A) that clearly show two scaling regimes (B), with $D_H = 1.33$ and $D_H = 1.57$ for scales lower and higher than 63 days, respectively.

The h-lagged autocorrelation coefficient, $r(h)$, defines how strongly the local value of the signal $x(t)$ depends on the one h lag before, that is, $x(t - h)$. The h-lagged autocorrelation coefficient is bounded between –1 and 1. Specifically, a positive correlation indicates that the trends of deviation of $x(t)$ and $x(t - h)$ relative to the mean of the signal are in the same direction. In contrast, negative values of $r(h)$ reveal opposite trends of deviation of $x(t)$ and $x(t - h)$ relative to the mean of the signal, that is, anticorrelation. The autocorrelation coefficient, $r(h)$, for lags h ($h = 0, 1, \ldots, n$) is defined as:

$$r_h = \left[\frac{1}{N-h-1} \sum_{i=h+1}^{N} (x(t)-\bar{x})(x(t-h)-\bar{x}) \right] \bigg/ \left[\frac{1}{N-1} \sum_{t=1}^{N} (x(t)-\bar{x})^2 \right] \tag{4.49}$$

Using the classical definition of the correlation coefficient (see Feller 1968), van Beek et al. (1989) derived that in the special case $h = 1$ the nearest-neighbor correlation between values of an fGn signal was expressible directly by the correlation coefficient r:

$$r = 2^\rho - 1 \tag{4.50}$$

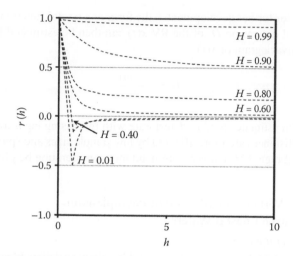

FIGURE 4.22 The autocorrelation function $r(h)$ of simulated fractional Gaussian noise (fGn) signals with different Hurst exponents H shown as a function of the lag, h. The correlation is trivially maximum at $r(0) = 1$, and the closer H is to 1, the slower n decays (that is, the longer the memory of the process). Long-term dependence (or long-memory correlation) occurs for $0.5 < H < 1$, $H = 0.5$ when there is no correlation, and when $H < 0.5$ the fGn signal shows anticorrelation.

where the exponent ρ relates to the Hurst coefficient of the fBm signal whose increments yield the fGn signal in question as $\rho = 2H_{fGn} - 1$. Equation (4.50) has further been extended to correlation between nonadjacent neighbors through the autocorrelation function of an fGn signal with $0 \leq H \leq 1$ (Bassingthwaighte and Beyer 1991) as:

$$r_h = \frac{1}{2}(|h+1|^{2H} - 2|h|^{2H} + |h-1|^{2H}) \tag{4.51}$$

The closer H is to 1, the slower $r(h)$ decays (Figure 4.22), that is, the longer the memory of the process. More generally, long-memory correlation (or long-term dependence) is for $0.5 < H < 1$, $H = 0.5$ when there is no correlation in the fGn signal, and when $H = 0.5$ the fGn signal shows anticorrelation. Note that Equation (4.51) is applicable to fGn or differenced fBm signals for both discrete and continuous values of h (Bassingthwaighte and Beyer 1991).

4.2.8 SEMIVARIOGRAM ANALYSIS

4.2.8.1 Theory

Semivariogram (SV) analysis is based on geostatistics and regionalized variables (RV) theory (Matheron 1971; Journel and Huijbregts 1978), and is applicable to stationary signals. RVs are continuous variables whose variations are too complex to be described by traditional mathematical functions (Phillips 1985). Patterns of variation in RVs can then be expressed by their semivariance $\gamma(h)$ defined as:

$$\gamma(h) = \frac{1}{2N(h)}\sum_{t=1}^{N(h)}[x(t) - x(t+h)]^2 \tag{4.52}$$

where $x(t + h)$ is the value of the dependent variable $x(t)$ at a point separated from point t by distance, or lag h, and $N(h)$ is the number of pairs of data points separated by the lag h. The semivariogram is the plot of $\gamma(h)$ as a function of h. The semivariance has, under certain conditions (see, for example,

Berry and Lewis 1980 for further developments), the form of a fractal function that scales with h^{4-2D} at the origin; the fractal dimension D_v of the RV $x(t)$ can then be estimated from the slope m of a log-log plot of the semivariogram of $x(t)$:

$$D_v = \frac{(4-m)}{2} \qquad (4.53)$$

Because semivariogram estimates tend to deteriorate with increasing lag h for finite-length sample series (that is, greater distances are more affected by low sample sizes and spurious properties of the data) (Journel and Huijbregts 1978), it is recommend that consideration be given to h values greater than $N(h)/3$ to $N(h)/2$.

4.2.8.2 Case Study: Vertical Distribution of Phytoplankton in Tidally Mixed Coastal Waters

4.2.8.2.1 Ecological Framework

As discussed in Section 4.2.1.3, marine systems exhibit intimate relationships between physical and biological processes (Legendre and Demers 1984; Mackas et al. 1985), as shown by the coupling between the distribution of phytoplankton populations and the structure of their physical environment over a wide range of spatial and temporal scales (Haury et al. 1978). Specifically, in tidally mixed coastal waters such as the eastern English Channel (Figure 4.8), the dissipation of tidal energy is regarded as responsible for the vertical homogenization of the shallow (50 m maximum depth) inshore and offshore water masses. However, recent investigations have shown that the vertical distribution of phytoplankton biomass, regarded as vertically homogenized by vertical mixing—and then characterized by a mean concentration and its associated variability (that is, the variance S^2)—should also be regarded as being vertically structured in terms of fractal dimension (Seuront and Lagadeuc 1998). Moreover, this fractal structure appears to be both space and time dependent, in relation respectively with the inshore–offshore hydrological gradient and the tidal advective processes (Seuront and Lagadeuc 1998). However, these results, based on the analysis of the data recorded along an inshore–offshore transect and characterized by an extreme intricacy of space-time scales and processes and by severe limitations in terms of sampling temporal resolution, led to a lack of generality concerning the processes responsible for the observed structure for both inshore and offshore locations.

Herein, the goal of this section is to provide a precise quantification of the vertical structure of phytoplankton distribution at the scale of the high-low tidal cycles and at the scale of neap-spring tidal cycles for both inshore and offshore waters of the eastern English Channel in order to specify and generalize preliminary results by Seuront and Lagadeuc (1998).

4.2.8.2.2 Experimental Procedures and Data Analysis

The data set studied in this paper consists of hourly measurements of physical parameters (temperature, salinity, and light transmission) and *in vivo* fluorescence (an index of phytoplankton biomass) taken from the surface to bottom with an SBE 25 Sealogger CTD and a Sea Tech fluorometer during seven sampling experiments (numbered from S1 to S7) conducted between 1993 and April 1997 in different tidal conditions both in offshore and inshore waters of the eastern English Channel (Figure 4.8; Table 4.4). Current speed and direction were recorded with an Aanderaa current meter every 5 minutes at different depths (Table 4.4). Water samples were collected from each sampled depth at 2-hour intervals for data sets S1 to S4 and at 1-hour intervals for data sets S5 to S7, and chlorophyll *a* concentrations (1-liter filtered frozen samples, extracted with 90% acetone, assayed in a spectrophotometer and the chlorophyll *a* concentration calculated following Strickland and Parson, 1972) were estimated for each sampled depth.

TABLE 4.4

Characteristics of the Seven Sampling Experiments Considered in the Present Study

	Sampling Date	Tidal Conditions	Sampling Site	Chl. a $(\mu g. \, l^{-1})$	N
S1	04/29/93—05/01/93	NT	OW	1.50	36
S2	03/20/94—03/21/94	NT	IW	1.50	36
S3	09/07/94—09/08/94	ST	IW	7.50	24
S4	02/04/96—04/04/96	ST	IW	13.80	47
S5	04/6/1996	ST	IW	15.00	15
S6	04/7/1996	ST	OW	3.00	13
S7	06/21/1998	ST	IW	8.40	24

Notes: NT: neap tide; ST: spring tide; OW: offshore waters; IW: inshore waters. Chl. *a*: chlorophyll a concentration; *N*: number of vertical profiles.

4.2.8.2.3   Results

Fractal dimensions D_v were estimated for *in vivo* fluorescence, temperature, and salinity, which exhibited a scaling behavior over the whole range of studied scales, for the whole data set (Figure 4.23). Their linearity over the whole range of spatial scales illustrates spatial dependence, suggesting that the same process, or at least similar processes, can be regarded as the source of physical and biological patterns, whatever the sampling locations or the hydrodynamical conditions. However, although the mean fractal dimensions of temperature, salinity, and *in vivo* fluorescence estimated for the whole data set respectively as 1.52 ± 0.02 ($\bar{x} \pm SD$ SD), 1.53 ± 0.02, and 1.63 ± 0.14 were significantly different (Kruskal-Wallis test, $p < 0.05$), the temperature and salinity fractal dimensions were not significantly different (Dunn test, $p > 0.05$; Siegel and Castellan 1988). At the scale of the whole sampling experiment, the vertical distribution of phytoplankton cells then cannot be regarded as being wholly driven by vertical mixing. Finally, as previously shown by Seuront and Lagadeuc (1998), it must be added that light transmission did not exhibit even a partial scaling behavior (that is, its variability is independent of scale), and therefore could not have been subjected to fractal analysis.

The mean empirical estimates of the fractal dimensions D_v of temperature, salinity, and *in vivo* fluorescence estimated for each sampling experiment led to further results (Table 4.5). There were no significant differences between salinity and temperature fractal dimensions between sampling experiments (Kruskal-Wallis test, $p > 0.05$). On the contrary, *in vivo* fluorescence fractal dimensions were significantly different ($p < 0.05$) and exhibited very specific behaviors. Fluorescence fractal dimensions were consistently significantly lower for inshore than for offshore locations (Wilcoxon-Mann-Whitney U-test, $p < 0.05$), with values ranging from 1.54 ± 0.12 to 1.82 ± 0.07, respectively. Moreover, correlation analysis demonstrated that fluorescence fractal dimensions were significantly correlated ($p < 0.05$) with current direction for each sampling experiment for both inshore and offshore waters (Table 4.6), except for sampling experiment S1 and S2, characterized by their very low chlorophyll *a* concentrations (cf. Table 4.4). There were no significant correlations between fluorescence fractal dimension and current speed, or between fluorescence fractal dimension and phytoplankton biomass at the scale of the high-low tidal cycles. In contrast, at the scale of the neap-spring tidal cycles, fluorescence fractal dimension exhibited significant ($p < 0.05$) positive correlations with current speed for both inshore and offshore waters (Figure 4.24). *In vivo* fluorescence fractal dimensions increased with hydrodynamical conditions. There was also a significant correlation ($p < 0.05$) between mean fluorescence fractal dimensions and mean chlorophyll *a* concentrations (Figure 4.25), suggesting a density-dependent control of the vertical fractal structure of phytoplankton biomass distribution.

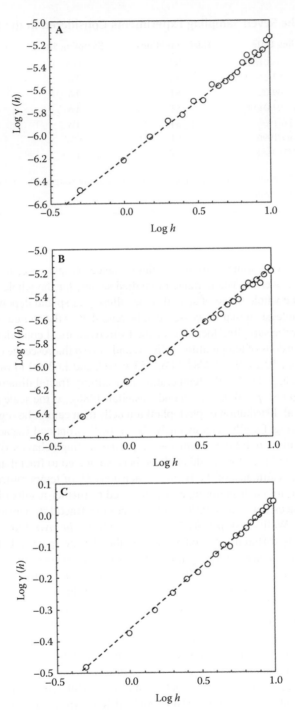

FIGURE 4.23 Double logarithmic semivariograms of temperature (A), salinity (B) and *in vivo* fluorescence (C) for sampling experiment S1, shown together with their best fitting line. The fractal dimension D_v is estimated as $D_v = (4 - m)/2$, where m is the slope of the log-log plot of the empirical semivariance $\gamma(h)$ vs. the lag h, in a log-log plot. Here, $m = 1.01$ ($r^2 = 0.99$), 0.99 ($r^2 = 0.97$), and 0.4 ($r^2 = 0.99$) for temperature, salinity, and fluorescence, respectively.

TABLE 4.5

Mean Fractal Dimension of Temperature, Salinity, and *In Vivo* Fluorescence Vertical Profiles for the Seven Sampling Experiments

	Temperature	Salinity	*In Vivo* Fluorescence
S1	1.50 (0.05)	1.52 (0.04)	1.82 (0.07)
S2	1.52 (0.06)	1.53 (0.05)	1.63 (0.09)
S3	1.54 (0.03)	1.50 (0.04)	1.57 (0.09)
S4	1.52 (0.04)	1.54 (0.06)	1.46 (0.12)
S5	1.53 (0.05)	1.53 (0.04)	1.54 (0.10)
S6	1.49 (0.03)	1.52 (0.05)	1.82 (0.06)
S7	1.53 (0.04)	1.55 (0.06)	1.56 (0.10)

Note: The numbers in parentheses are the standard deviations.

4.2.8.2.4   Ecological Interpretation

The fractal dimensions D_v estimated over the whole range of available spatial scales suggest that the scales of spatial dependence are very similar for *in vivo* fluorescence, salinity, and temperature, indicating similar sources of physical and biological patterns. However, the differences shown between fractal dimensions of temperature, salinity, and fluorescence fractal dimensions suggest that the vertical distribution of phytoplankton cells is very specific and cannot be regarded as being passively advected by mixing processes, even when chlorophyll *a* concentrations are very low, as it is the case for sampling experiments S1 and S2 (cf. Table 4.5). Thus, as shown by the correlation analysis, fractal dimensions of temperature and salinity are tidally and geographically independent, as opposed to fluorescence fractal dimensions, which are (1) significantly higher in offshore locations, (2) dependent on the current direction at the scale of the high-low tidal cycle, (3) dependent on the current speed at the scale of the neap-spring tidal cycles, and (4) dependent on phytoplankton concentration at the biological annual cycle. The vertical structure of the phytoplankton biomass is then more homogeneous, or less structured, in offshore locations and during flood tide, but also when hydrodynamical conditions are high and phytoplankton concentrations are low, showing that the structure of the vertical distribution of phytoplankton biomass is determined by different processes following the implied temporal scales.

TABLE 4.6

Spearman's Rank Correlation Coefficients between *in vivo* Fluorescence Fractal Dimensions *D* and Current Direction ($D/C_{Direction}$), Current Speed (D/C_{Speed}), and Mean Chlorophyll *a* Concentration (*D*/Chl. *a*)

	$D/C_{Direction}$	D/C_{Speed}	*D*/Chl. *a*
S1	0.22	0.22	0.10
S2	0.19	−0.10	0.20
S3	0.76 **	−0.33	0.01
S4	0.79 **	−0.14	−0.19
S5	0.95 **	−0.12	0.36
S6	0.80 **	−0.20	−0.37
S7	0.89 **	−0.13	0.28

** 1% Significance level.

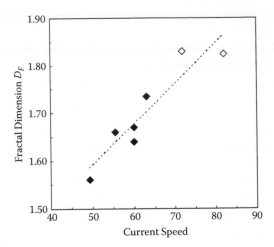

FIGURE 4.24 Relationship between the fractal dimensions D_v estimated for *in vivo* fluorescence vertical profiles and the speed of the tidal current (m s^{-1}) for inshore (black dots) and offshore waters (open dots) of the eastern English Channel. The best linear fit between D_v and current speed (dotted line) is significant at the 5% confidence level.

These results confirm and generalize previous studies conducted in the same environment (Seuront and Lagadeuc 1998). In particular, with fluorescence fractal dimensions higher for offshore than for inshore waters, these differences could be associated to a purely density-dependent effect (Seuront and Lagadeuc 1998), to a qualitative effect relative to the specific composition of phytoplankton assemblages (Truffier et al. 1997; Peta et al. 1998), or to a combination of the two previous hypothesized phenomenologies. Whatever that may be, an increase in fluorescence fractal dimensions should have been expected during flood tide because of the offshore water advection associated with the semidiurnal M2 tidal component, instead of the significant decrease generally observed. At the scale of the high-low tidal cycles, it can then be suggested that the observed differential tidal structure of phytoplankton distribution could be associated with the differential mixing occurring during a tidal cycle between water masses qualitatively and quantitatively different in terms of

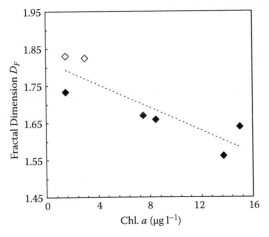

FIGURE 4.25 Relationship between the fractal dimensions D_v estimated for *in vivo* fluorescence vertical profiles and chlorophyll *a* concentration for inshore (black dots) and offshore waters (open dots) of the eastern English Channel. The best linear fit between D_v and chlorophyll *a* concentration (dotted line) is significant at the 5% confidence level.

phytoplankton populations that could then be regarded as a secondary source of heterogeneity. On the other hand, at the scale of the neap-spring tide cycles, phytoplankton distributions appear controlled by hydrodynamical conditions; high hydrodynamical conditions lead to more homogeneous distributions characterized by high fractal dimensions. Finally, comparisons between mean fractal dimensions of fluorescence and mean chlorophyll a concentrations for each sampling experiment confirm the density-dependent control of phytoplankton structure proposed by Seuront and Lagadeuc (1998) as a potential explanation of the different fractal dimensions observed for the vertical distribution of phytoplankton biomass between inshore and offshore waters. In particular, this means that the heterogeneity of phytoplankton is all the more high as its density is high, and the observed density-dependence could be a consequence of the aggregation processes of phytoplankton cells, mainly driven by phytoplankton density and hydrodynamism (see, for example, Kiørboe 1997).

4.2.9 WAVELET ANALYSIS

This method is only briefly described here, as the interested reader can refer to Dremin et al. (2004) and Fisher et al. (2004) for an introduction on the subject, and to Arnéodo et al. (1995), Jones et al. (1996), and Simonsen et al. (1998) for details on different wavelet methods to estimate the Hurst exponent H. Briefly put, a wavelet is a waveform of limited duration with an average value of zero. Although Fourier analysis breaks up a signal into cosine wave components of various frequencies, wavelet analysis decomposes a signal into shifted and stretched versions of the original wavelet. Wavelets are then parametrized by a scale parameter ($a > 0$) and a translation parameter ($-\infty < b < \infty$) that are incorporated into one single function $\psi(x)$ as (for example, Simonsen et al. 1998):

$$\psi_{a,b}(x) = \psi\left(\frac{x-b}{a}\right)$$ (4.54)

The wavelet transform of a signal $x(t)$ is then defined as:

$$W_x(a,b) = \frac{1}{\sqrt{a}} \int_{-\infty}^{\infty} \psi_{a,b}^*(t) x(t) dt$$ (4.55)

where $\psi_{a,b}^*(t)$ is the complex conjugate of $\psi_{a,b}(t)$. It comes from Equation (4.1) that any self-affine signal $x(t)$ scales with the time scale τ as:

$$x(t)_\tau \propto \tau^H$$ (4.56)

Incorporating Equation (4.56) into Equation (4.55) leads to:

$$W_x(a) \propto a^w$$ (4.57)

where $w = \frac{1}{2} + H$, and $W_x(a) = \langle W_x(a, b) \rangle_b$ is the average of $W_x(a, b)$ over all location parameters b. This method, referred to as the average wavelet component (AWC) (Simonsen et al. 1998) has been proven to be very useful in time-series analysis (Simonsen 2003) and basically consists in finding a representative wavelet amplitude for a given scale a. The exponent H is then estimated from the slope of $W_x(a)$ versus a on a log-log scale. The exponent H obtained from the time series of the abundance of the calanoid copepod *Temora longicornis* (Figure 4.18B) using power spectrum analysis (Figure 4.26A) and the AWC method (Figure 4.26B) are statistically undistinguishable. Note that this method can be applied to fBm signals or to cumulatively summed fGn signals.

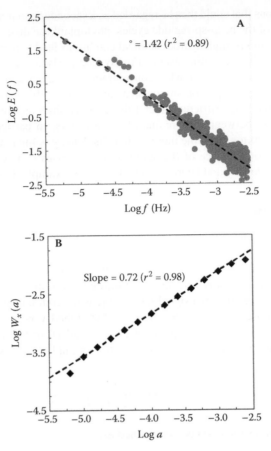

FIGURE 4.26 Hurst exponents estimated using power spectrum analysis (A) and the average wavelet component (AWC) (B). The power spectrum Hurst exponent is estimated from Equation (4.13) as $(\beta - 1)/2 = 0.21$ (Section 4.2.1.1); and the wavelet Hurst exponent from Equation (4.57) as $w - 1/2 = 0.22$.

It is finally stressed that wavelet-based spectra might be a preferable approach because wavelets are localized, in contrast to the infinite sine waves used in Fourier analysis, and can thus be directly applied to data that are anisotropic and nonstationary, resulting in one-dimensional series that have both inherent directionality and trend (Malamud and Turcotte 1999).

4.2.10 Assessment of Self-Affine Methods

4.2.10.1 Comparing Self-Affine Methods

Despite the dichotomy between fGn and fBm signals (see Section 4.1.5) and the related analysis techniques to be chosen to estimate the Hurst exponent H (Figure 4.27 and Figure 4.28), the fractal measures reported in this chapter relate to each other in a simple manner as shown in Table 4.7. However, inconsistencies between fractal measures can be found in the literature. For instance, a recent paper by Sims et al. (2008) that identified a consistent power law in the foraging behaviors of a range of marine predators reported mean values of $\alpha = 1.08$ and $\beta = 0.8$ for the diving time series of the five species they considered. However, from Table 4.7, it readily comes that $\beta = 2\alpha - 1$, which is not the case here. Similarly, for the same krill abundance time series, they contradictorily report $\alpha = 0.9$ and $\beta = 0.3$. A successful and meaningful fractal analysis requires that

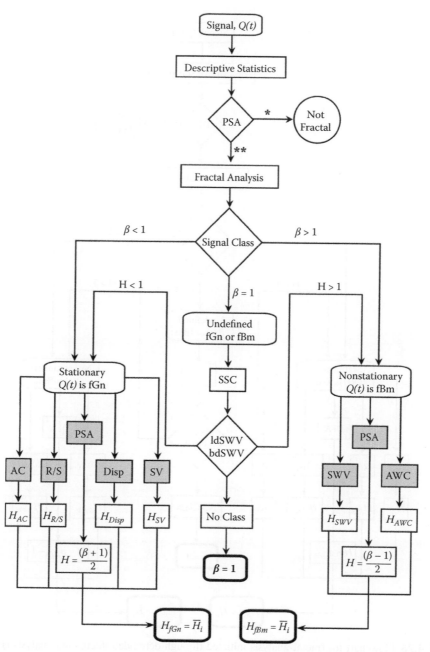

FIGURE 4.27 Flowchart for fractal analysis initiated through power spectrum analysis (PSA). First, power spectrum analysis is used to assess the existence of a power-law behavior, in which case the class to which the data set belongs (fGn or fBm) is identified using the value of the PSA exponent β. The signal is then analyzed with methods appropriate for fGn or fBm, and the average of method-specific H values is considered as the most reliable estimate of the Hurst exponent of the original signal. (*: a signal is considered as nonfractal in the absence of power-law behavior and when the power-law behavior spans less than 1 decade; **: a signal is considered as fractal in the presence of a power-law behavior spanning more than one decade, ideally more than two decades; AC: autocorrelation analysis; R/S: rescaled range analysis; Disp: dispersional analysis; SV: semivariogram analysis; SWV: scaled windowed variance analysis; AWC: average wavelet analysis; SSC: signal summation conversion method; ldSWV: line-detrended scaled windowed analysis; bdSWV: bridge-detrended scaled windowed analysis.)

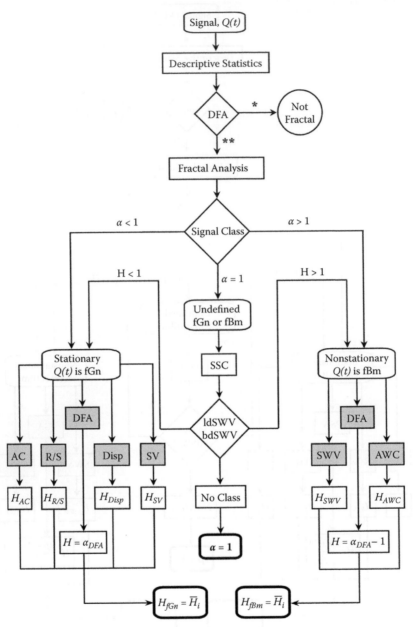

FIGURE 4.28 Flowchart for fractal analysis initiated through detrended fluctuation analysis (DFA). First, detrended fluctuation analysis is used to assess the existence of a power-law behavior, in which case the class to which the data set belongs (fGn or fBm) is identified using the value of the DFA exponent α. The signal is then analyzed with methods appropriate for fGn or fBm, and the average of method-specific H values is considered as the most reliable estimate of the Hurst exponent of the original signal. (*: a signal is considered as nonfractal in the absence of power-law behavior and when the power-law behaviour spans less than 1 decade; **: a signal is considered as fractal in the presence of a power-law behavior spanning more than one decade, ideally more than two decades; AC: autocorrelation analysis; R/S: rescaled range analysis; Disp: dispersional analysis; SV: semivariogram analysis; SWV: scaled windowed variance analysis; AWC: average wavelet analysis; SSC: signal summation convertson method; ldSWV: line-detrended scaled windowed analysis; bdSWV: bridge-detrended scaled windowed analysis.)

TABLE 4.7

Correspondence between the Characteristic Exponents Derived through Methods for Self-Affine Fractals and the Hurst Exponents of fBm, H_{fBm}, fGn, and H_{fGn}

Technique	Exponent	Hurst Exponent	
		H_{fBm}	H_{fGn}
Power spectrum analysis	β	$2H_{fBm}+1$	$2H_{fGn}-1$
Detrended fluctuation analysis	α	$H_{fBm}+1$	H_{fGn}
Scaled windowed variance analysis	H	H_{fBm}	—
Dispersional analysis	H	—	H_{fGn}
Rescaled range analysis	H	—	H_{fGn}
Autocorrelation analysis	ρ	—	$2H_{fGn}-1$
Semivariogram analysis	m	—	$2H_{fGn}$
Average wavelet component analysis	w	$1/2 + H_{fBm}$	—

fractal measures returned by different methods be consistent with each other. In addition, due to the dichotomy between fGn and fBm signals that underlie the choice of an appropriate analysis (Figure 4.27 and Figure 4.28), a reliable and meaningful fractal analysis will be achieved through the following steps:

1. Identify the class of the signal using either power spectrum analysis (Section 4.2.1, Figure 4.5) or detrended fluctuation analysis (Section 4.2.2, Figure 4.13).
2. Analyze the signal with different methods, and take the average of the resulting H values as the most reliable estimate of the Hurst exponent (Figure 4.27 and Figure 4.28).
3. Systematically report estimates of the Hurst exponent H along with the following:
   a. The class of the signal as H_{fGn} and H_{fBm}.
   b. The numerical method used to estimate H_{fGn} and H_{fBm}.

4.2.10.2 From Self-Affinity to Intermittent Self-Affinity

It is, however, implicit from the content in this chapter that all of the above methods return relevant results if the signals to be analyzed belong to fGn or fBm signal class. Instead, many natural phenomena are characterized by their deviation from "Gaussianity" and are characterized by local, sharp fluctuations over a wide range of low-density values (Figure 4.29A). In the context of those intermittent signals, the linear relationship assumed between, for example, the spectral exponent β and the Hurst exponent H—that is, $H_{fBm} = (\beta - 1)/2$ (see Equation 4.13) and $H_{fGn} = (\beta + 1)/2$ (see Equation 4.14)—does not hold anymore. Because the power spectrum signatures of a fractional Brownian motion and intermittent fractional Brownian motion do not significantly differ (Figure 4.29B), the identification of the class to which a data set belongs requires the dichotomy between fractional Brownian motion (fBm) and intermittent fractional Brownian motion (ifBm). This can be achieved using the qth-order structure functions that are an empirical generalization to high orders of moments in physical space of the power spectrum. The structure functions are defined as

$$\langle |\Delta Q(\tau)|^q \rangle \propto \tau^{\varsigma(q)} \tag{4.58}$$

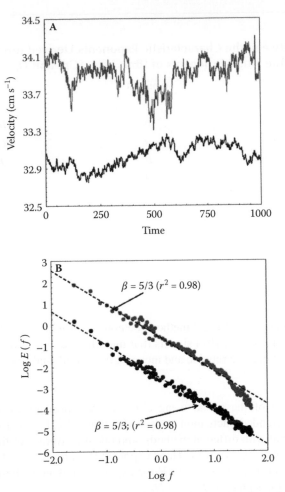

FIGURE 4.29 Time series of grid-generated turbulent velocity recorded by hot-wire velocimetry at 100 Hz in a circular flume (Seuront et al. 2004) and a synthetic fractional Brownian motion time series with the same spectral properties as the empirical one (A). Although the empirical time series is clearly more intermittent than the synthetic one, their power spectra are very similar, showing a power-law behavior with $\beta = 5/3$ over more than three decades (B). The roll-off observed for the empirical power spectrum at high frequencies is related to the electronic limitations of the instrument.

where $\Delta Q(\tau)$ is the fluctuations of a scalar signal at scale τ and "$\langle . \rangle$" means statistical average; that is, $\langle |\Delta Q(\tau)|^q \rangle$ is the statistical moments of the fluctuations $\Delta Q(\tau) = Q(t + \tau) - Q(t)$. Here, instead of considering discrete statistical moments as the mean ($q = 1$), variance ($q = 2$), skewness ($q = 3$), and kurtosis ($q = 4$), Equation (4.58) gives the scale-invariant structure functions' exponent $\zeta(q)$, which continuously characterizes all the statistics of the signal. The first moment $\zeta(1)$ gives the "intermittent Hurst exponent" and defines the scaling of the average fluctuation; $\zeta(1) = 0$ for scale-independent signals. The second moment is linked to the power spectrum exponent β as $\beta = 1 + \zeta(2)$. For fractional Brownian motion signals, $\zeta(q)$ is linear (Figure 4.30). In contrast, for intermittent signals, this function is nonlinear and convex (Figure 4.30). Further theoretical and practical developments of structure function can be found in Chapter 8. Specifically, Equations (4.13) and (4.14) would lead to a systematic underestimation of the Hurst exponent H_{fBm} for intermittent fractional Brownian motion (ifBm; $\beta > 1$) and a systematic overestimation

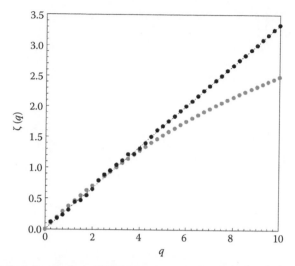

FIGURE 4.30 Structure function analysis of the empirical and simulated time series shown on Figure 4.29. The exponent $\zeta(q)$ is nonlinear and convex for the empirical (intermittent) time series (gray dots), while $\zeta(q)$ is linear for the simulated fractional Brownian motion (black dots) and fits the theoretical expectation $\zeta(q) = qH$, where $H = \zeta(1)$.

of H_{fGn} for intermittent fractional Gaussian noise (ifGn; $\beta < 1$) (see Figure 4.31). A meaningful fractal analysis thus requires first the dichotomy between intermittent and nonintermittent signals. Nonintermittent signals can be analyzed following the step-by-step approach proposed in Section 4.2.10.1. Intermittent signals have to be analyzed using appropriate techniques, as described in detail in Chapter 8.

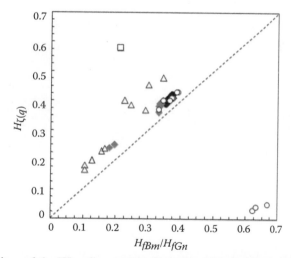

FIGURE 4.31 Comparison of the "Hurst" exponents H obtained for the time series of temperature (black diamonds), salinity (gray diamonds), *in vivo* fluorescence (open dots), dissolved inorganic nutrients (open triangles), and zooplankton abundance (open square) using methods developed for the analysis of fBm and fGn signals (H_{fBm}/H_{fGm}); and using structure function analysis, $H_{\zeta(q)}$. (Temperature, salinity, and fluorescence data are from Seuront et al., 1996a, 1999; Seuront, 1999, 2005b. The dissolved inorganic nutrient data are from Seuront et al., 2002; zooplankton data are from Seuront and Lagadeuc, 2001.)

FIGURE 4.30 Structure function analysis of the empirical and simulated time series shown in Figure 4.29. These percent $C(\tau)$ is nonlinear and convex for the empirical (intermittent) time series (gray dots), while $C(\tau)$ is linear for the simulated (fractional) Brownian motion (black dots) and has the theoretical expected slope $C(\tau) = \zeta(1)$ where $\zeta = \zeta(1)$.

of H_{ap} for intermittent (fractional Gaussian noise (fGn; $β = 1$) (see Figure 4.31). A meaningful fractal analysis thus requires first the dichotomy between intermittent and nonintermittent signals. Nonintermittent signals can be analyzed following the step-by-step approach proposed in Section 4.3.10.1. Intermittent signals have to be analyzed using appropriate techniques, as described in detail in Chapter 5.

FIGURE 4.31 Comparison of the Hurst exponent H obtained for the time series of temperature (blue diamonds), relative humidity (red triangles) or relative vapor stress measured from plant turbulence (open triangles), and fractal relation about lawn appearance or intermittency developed for the analysis of fluctuated H on the $β$-relationship (that form into the classes of $H = H(β)$, and color combinations (see later) of fluctuated time series and fluctuation error data. Mean data are in (1998a, 1998b Stanton, 1997, 1998). For intermittent turbulence error data from Stanton et al., 2002, employers of data (the Hurst function and Logatchev, 2004).

5 Frequency Distribution Dimensions

A common procedure for looking at the level of organization of any data set is to study the probability density function (PDF) or the cumulative density function (CDF). In particular, cumulative hypergeometric frequency distributions have been found in many areas of the natural sciences (see, for example, Laherrere and Sornette 1998 for a review) and imply a wide range of values with many small values and few large values. This chapter focuses on several aspects of the cumulative frequency distribution of self-similar and self-affine patterns. This includes theoretical investigations of the correspondence between cumulative distribution functions and probability density functions (Section 5.1) and the descriptions of the frequency distributions of intensities, areas, and volumes (Sections 5.2, 5.3, and 5.4). Special attention is finally given to rank-frequency distributions, from their original development in linguistics and their link to information theory and entropy to their effectiveness as a simple and direct diagnostic tool for ecologists to assess ecosystem complexity and their applicability to the analysis of symbolic sequences (Section 5.5).

5.1 CUMULATIVE DISTRIBUTION FUNCTIONS AND PROBABILITY DENSITY FUNCTIONS

5.1.1 THEORY

The Pareto law was originally introduced in economics to describe the number of people whose personal incomes exceeded a given value (Pareto 1896). More generally, the Pareto law of any random variable X is described in terms of the cumulative density function (CDF):

$$P[X \geq x] \propto x^{-\phi} \tag{5.1}$$

where x is a threshold value, and ϕ is the slope of a log-log plot of $P[X \geq x]$ vs. x. Note that Equation (5.1) can be equivalently rewritten in terms of the PDF as (Faloutsos et al. 1999):

$$P[X = x] \propto x^{-\mu} \tag{5.2}$$

where μ ($\mu = \phi + 1$) is the slope of a log-log plot of $P[X = x]$ vs. x. Note that Pareto's law has also been used to describe self-organized criticality (SOC) in a range of natural phenomena (see Section 6.3).

5.1.2 CASE STUDY: MOTION BEHAVIOR OF THE INTERTIDAL GASTROPOD *LITTORINA LITTOREA*

5.1.2.1 The Study Organism

The common periwinkle, *Littorina littorea* (Linnaeus 1758), is among the most abundant herbivorous gastropod molluscs of the Western and Northern European coasts. *L. littorea* is probably the best known of its family, as it has been collected and eaten for centuries. *L. littorea* was introduced to North America from Europe in the mid-1800s to Nova Scotia either through ballast waters or for food (Bertness 1999). Since their arrival, they have managed to outcompete most local species to become the dominant herbivore in the rocky intertidal zone from New England to Chesapeake Bay

FIGURE 5.1 The intertidal gastropod *Littorina littorea* in its typical rocky environment while sealed to a dry rock. (Modified from Seuront et al., 2007.)

(Bertness 1999). Their spread was limited to most of the East Coast of North America, which is compatible with the temperature ranges that exist in Europe (Levinton 2001).

Littorina littorea normally grow to about 2 to 3 centimeters in length (Bertness 1999) and have an average life span of 5 to 10 years (Buczaki 2002). The general morphology differs from region to region, but they have a dark grey or black conical shell (Figure 5.1a) with spiral ridges that evolves toward a smooth surface with age. *L. littorea* is widely distributed on most rocky shores from the upper shore into the sublittoral, except in the most exposed areas. It can also be found in sandy and muddy habitats such as estuaries and mud flats, and is fairly tolerant of brackish water.

Like its land relative, the snail, *L. littorea* move on a muscular, ciliated foot, secreting a film of slimy mucus on which they can slide and move (Brusca and Brusca 1990). They forage primarily underwater or during cool, low tides, or when ocean spray moistens the rocks (Figure 5.1b). The pedal retractor muscle shortens and lengthens the foot so that the organism can have the option of hiding in its shell for periods of time. When not walking or exposed to the sun for a long time, the *L. littorea* often seeks shelter in a shaded crevice and seals the gap between its shell and the rock with mucus to avoid desiccation or being swept away by currents and breaking waves (Lerman 1986).

Periwinkles graze on a wide range of food items from nonsiliceous microalgae over diatoms to leathery and coralline algae (Steneck and Watling 1982). *L. littorea*, however, graze preferentially on periphyton (that is, a complex matrix of microalgae, cyanobacteria, heterotrophic bacteria, and detritus attached to submerged surfaces such as rocks) and ephemeral macroalgae such as *Enteromorpha* sp. (Figure 5.1c). This species is so voracious that at high densities (up to 600 to 1000 per square meter; Bertness 1999), it will consume all ephemeral algae, similar to the destructive feeding patterns of sea urchins (Bertness 1999). *L. littorea* are important grazers in intertidal ecosystems and often control the dominant algae (Lubchenco 1983).

5.1.2.2 Experimental Procedures and Data Analysis

The motion behavior of *L. littorea* was investigated on a rocky platform typical of the rocky habitats found along the French coast of the eastern English Channel. This platform ranges over the whole intertidal zone, bounded between the upper and lower limits reached by the tidal flow at high and low tide, respectively. The platform was topographically homogeneous, dominated by bare rocks partially covered by the common barnacle *Balanus balanoides* and with a few cracks and crevices occupied by the blue mussel *Mytilus edulis*. Three sites were chosen for their decreasing immersion time during high tide at 50 m (site A), 70 m (site B), and 130 m (site C) from the lower limit of the low tide and were all submersed at high tide. At each of the three locations, 30 specimens of *L. littorea* were captured, individually measured, and marked with numbered plastic tags (2 mm × 3 mm) fixed to the dorsal part of the shell with inert glue and released from a single point. The shell sizes of the three groups of individuals were significantly different (Kruskal-Wallis H-test, $p < 0.01$) with 15.27 ± 0.15 mm at site A, 14.68 ± 0.17 mm at site B, and 13.69 ± 0.18 mm at site C. After the release of individuals on 20 March 2006, the three release areas were searched on 14 successive daylight low tides. The direction of a sighted *L. littorea* from the release stake was measured with a compass and distance (nearest centimeter) with a tape, and the apparent distances traveled from one low tide to the next were estimated. Each site thus provided 420 measurements of distance traveled. No movements were observed at low tide during the distance measurements.

5.1.2.3 Results

The successive displacements of *L. littorea* are consistently characterized by a very intermittent behavior, with a few localized large displacements over a wide range of small displacements (Figure 5.2). The corresponding displacements ranged from 1 to 1558 cm in site A, 1 to 715 cm in site B, and 1 and 1084 cm in site C. These intermittent distributions result in a significantly non-normal distribution ($p < 0.01$) with elevated positive skewness g_1, $g_1 = 6.1$, 3.4, and 5.1 in sites A, B, and C, respectively. None of the distributions were statistically different from the others (Kruskal-Wallis H-test, $p > 0.05$).

To further quantify the property of the extreme displacements leading to the observed positively skewed distributions, Equation (5.1) has been rewritten as:

$$P[l_d = l] \propto l^{-\mu} \qquad (5.3)$$

where l_d is the displacement length, l a threshold value, and μ ($1 < \mu \le 3$) characterizes the power-law behavior of the tail of the distribution. Equation (5.3) corresponds to a family of distributions defined according to the values of μ. These distributions mean that extremely long movements occurred more often than would be expected if the forager exhibited movement lengths with a normal distribution. For $\mu \ge 3$, the distribution is Gaussian (which according to the central-limit theorem has a finite variance) and the motion is equivalent to Brownian motion walks. For $2 \le \mu < 3$, the scaling is superdiffusive (Shlesinger et al. 1996), while the value $\mu = 2$ indicates that the scaling becomes quadratic in time and corresponds to the lower extreme of superdiffusive processes, that is, Lévy flight (Shlesinger et al. 1996). In contrast, values $\mu \le 2$ do not correspond to probability distribution that can be normalized (Shlesinger et al. 1996). The smaller μ is, the more

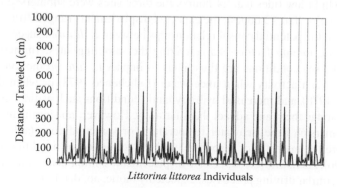

FIGURE 5.2 Intermittent in the successive displacements in *Littorina littorea* motion behavior observed at site A, located 50 m away from the lower limit of low tide. Each dotted line separates the 14 successive records of each of the 30 tracked individuals.

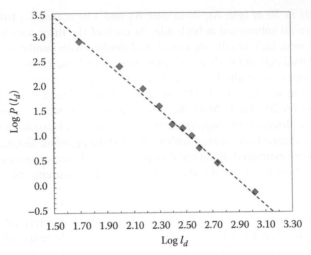

FIGURE 5.3 Log-log plot of the probability distribution function (PDF) of *Littorina littorea* successive displacements investigated at site B located 70 m away from the lower limit of low tide. The dotted lines correspond to the best linear fit of $P(l_d)$ vs. l_d, leading to estimate the exponent μ as $\mu = 2.43 \pm 0.05$. (Modified from Seuront et al., 2007.)

intermittent the distribution is. When $\mu \leq 3$, the variance of the process diverges, and when $\mu \leq 2$ the mean is not defined. The parameters μ were estimated for sites A, B, and C as the slope of $P(l_d)$ vs. l_d in log-log plots (Figure 5.3). The values of μ estimated from *L. littorea* displacements were all significantly smaller than 2 ($p < 0.01$), and significantly different from each other ($p < 0.01$) with $\mu = 2.22 \pm 0.02$ for site A, $\mu = 2.43 \pm 0.05$ for site B, and $\mu = 2.67 \pm 0.04$ for site C. The scaling then falls into the category of superdiffusive processes, that is, $2 \leq \mu$.

5.1.2.4 Ecological Interpretation

The increase in the parameter μ with increasing distances from the lower limit of the low tide suggests that the intermittency of movement patterns increases with increasing immersion time. This is consistent with reported increase in periwinkle foraging activity when underwater at high tide (Lubchenco 1978). In addition, the maximum displacements observed here (1558, 858, and 1084 cm for sites A, B, and C, respectively) are consistent with the movement speed found in the literature for *L. littorea*, ranging from 3 to 5 cm mn⁻¹ (Erlandson and Kostylev 1995). Considering that between two successive daylight low tides (ca. 24 hours), the three sites were submersed between 30% and 50% of the time (that is, 8 to 12 hours), the expected displacements ranged from 1440 to 3600 cm. We also investigated the potential causal relationships between *L. littorea* successive displacements and the main abiotic forcing factors characterizing the sampling sites. Immersion time, seawater temperature, and sea conditions are thus likely to impact *L. littorea* motion behavior as these organisms increase their activity when underwater (Lubchenco 1978) or when the temperature is rising (Lubchenco 1978), and increased hydrodynamic conditions such as tidal currents and breaking waves may dislodge them and advect them far away. However, no significant correlations were observed with immersion time, seawater temperature, or sea conditions (approximated by wind speed, as winds are the main factor responsible for the formation of breaking waves in this area) (Seuront et al. 2006). As *L. littorea* successive displacements exhibit heavy-tailed distributions, it is hypothesized that similar driving processes, expectedly biotic, are driving the dynamics of distance traveled by *L. littorea*.

By simulating a limiting generalized searcher-target model (for example, predator–prey, mating partner, pollinator-flower, parasite-host), recent theoretical results indicate that Lévy walks confer a significant advantage over the usual Gaussian (that is, Brownian) motion for increasing encounter rates when the searcher is larger or moves rapidly relative to the target, and when the target density is low (Viswanathan et al. 2001, 2002; Bartumeus et al. 2003). The heavy-tailed distributions observed here (with $\mu = 2.22$, 2.43 and 2.67) significantly diverge from Lévy walks where $\mu = 2$. According to optimal foraging theory (Stephens and Krebs 1986), evolution through natural selection should favor flexible behavior, leading to different optimum searching strategies (that is, searching statistics) under different conditions. Our results then suggest that the biotic conditions encountered at each of our three sites might have been different, leading to different (heavy-tailed) distributions for the most extreme displacements. Note that power-law behaviors have also been widely described from pause or flight duration in flies (Cole 1995), albatrosses (Viswanathan et al. 1999), rats (Kafetsopoulos et al. 1997), gilts (Harnos et al. 2000), and zoo-plankton (Bartumeus et al. 2003), and subsequently interpreted as a Lévy flight signature. A Lévy flight, however, is diagnosed by a power law in the probability density function (or cumulative density function) of flight amplitudes (see, for example, Shlesinger et al. 1996). This is fundamentally different from a power law in the probability density function of pause duration and flight duration. Specifically, a power law in the probability density function of flight duration is equivalent to a power law in the probability density function of flight length only if the considered organisms are moving at a constant velocity, which is very unlikely considering the intrinsic intermittent nature of animal locomotion (see, for example, Kramer and McLaughlin 2001; Seuront et al. 2004c, 2004d, 2007); see also Figure 8.1B and Figure 8.2A,B.

Although an inverse square probability density distribution $P(l_d) \propto l_d^{-2}$ of step lengths l_d leads to an optimal random strategy for organisms searching for randomly located objects that can be revisited any number of times (Viswanathan et al. 2001; Stephens and Krebs 1986), we are not aware of any attempt to investigate this issue when prey items are heterogeneously distributed as previously reported for the sampling site (Seuront and Spilmont 2002; Seuront and Leterme 2006). Although this is not an easy task, future work should concentrate on getting simultaneous measurements of predator motion behavior and prey concentration and distribution. As the main biotic factors driving organism motion behavior are the presence/absence, abundance, and distribution of prey items, predators, and mates, further investigations on the interplay between motion behavior statistics and the qualitative and quantitative nature of the biotic environments are essential to gain new insights into the origin of heavy-tailed distributions in biological systems.

5.2 THE PATCH-INTENSITY DIMENSION, D_{pi}

Equation (5.1) has also been independently used to describe the space-time dynamics of self-affine processes (see Chapter 4) that build up stress and then release the stress in intermittent pulses, such as earthquakes (Olami et al. 1992; Correig et al. 1997), landscape formation (Somfai et al. 1994a, 1994b), avalanches (Noever 1993), volcanic eruption (Diodati et al. 1991), and sediment deposition in the ocean (Rothman et al. 1994); the Gutember-Richter law of geophysics states that the number of earthquakes N with energy E greater that a given threshold E_0 scales with E_0 (Feder and Feder 1991). Equation (5.1) can then be rethought and adapted to a mosaic landscape/seascape composed of patches of different intensities as:

$$N(C \geq c) = kc^{-D_c} \tag{5.4}$$

where k is a constant, N is the number of patches of concentration C greater than c, and D_{pi} is the related fractal dimension (Figure 5.4). Equation (5.4) has recently been successfully used to characterize the distribution of microscale microphytobenthos biomass distribution on an intertidal sandy flat, leading to a patch-intensity dimension $D_{pi} = 5.31$ (Seuront and Spilmont 2002). Evidence for such distributions in ecological sciences is still scarce but nevertheless includes a

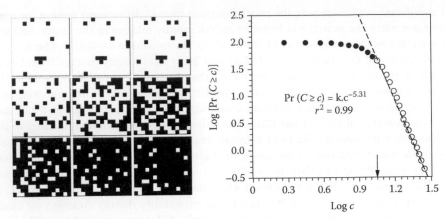

FIGURE 5.4 Patch-intensity dimension. From a two-dimensional pattern (here, microphytobenthos chlorophyll a concentration C; left panel), one can define a series of increasing thresholds (in white, the areas where the biomass exceeds a critical concentrations c such as $C \geq c$ with white areas indicating locations where $C \geq c$). The slope of the linear behavior of a log-log plot of the probability of exceeding a given threshold as a function of the threshold is an estimate of the patch-intensity dimension (right panel). (Adapted from Seuront and Spilmont, 2002.) (See Section 3.2.4.2 for a description of the data and their analysis using the mass dimension method, see Section 3.2.4.)

wide spectrum of ecological fields ranging from tree-fall gap formation in tropical rainforests (Manrubia and Solé 1996) and bird population dynamics (Keitt and Marquet 1996) to models of ecosystem (Bak et al. 1989) and evolution (Bak and Sneppen 1993; Paczuski et al. 1995). In particular, data sets related to the introduced Hawaiian avifauna support a scenario in which island communities build up to a critical number of species, above which avalanches of extinction occur (Keitt and Marquet 1996). The avalanches of extinction observed in the fossil record (Raup 1982) may then be indicative of a self-organized critical state, as suggested from simple coevolutionary models (Bak and Sneppen 1993; Kauffman and Johnsen 1991; Flyvbjerg et al. 1993). In the case of coevolving species, one may note that exact analytical solutions have been given and demonstrate that extinction cascades following the distributions given in Equation (5.4) can emerge spontaneously in simple models of coevolution (Flyvbjerg et al. 1993; de Boer et al. 1994). More generally, these results suggest that ecological communities are not characterized by a well-defined equilibrium but rather by a detailed balance that is minimally stable to perturbations such that the introduction of species can trigger extinction cascades. In particular, an interesting feature of Equation (5.4) is its applicability to self-similar patterns and also to self-affine processes; its connections with self-organized criticality and multifractals are respectively discussed in Section 6.3 and Chapter 8.

Equation (5.4) can be generalized to the analysis of surface structure previously described in a self-similar fashion using the transect dimension described in Section 3.2.9.1. If one considers the cumulative frequencies $f_i(x)$ and $f_j(x)$ of observing values of the descriptor $X(i, j)$ greater than a threshold value x, the patch-intensity dimensions of the ith and jth longitudinal and latitudinal transects are given, following Equation (5.4), as:

$$N(C \geq c_i) = kc^{-D_{pi_i}}$$
$$N(C \geq c_j) = kc^{-D_{pi_j}}$$
(5.5)

where $i = 1,..., n_i$, $i = 1,..., n_j$, k is a constant, and D_{pi_i} and D_{pi_j} are the patch-intensity dimensions of the ith and jth longitudinal and latitudinal transects, respectively.

The mean fractal dimensions of the latitudinal and longitudinal transects are subsequently given as:

$$\bar{D}_{p_i} = \frac{1}{n_i} \sum_{i=1}^{n_i} D_{p_i}$$

$$\bar{D}_{p_j} = \frac{1}{n_j} \sum_{j=1}^{n_j} D_{p_j}$$

(5.6)

The one-dimensional patch-intensity dimension $\bar{D}_{p_{i,j}}$ is then defined as:

$$\bar{D}_{p_{ij}} = \frac{1}{2}(\bar{D}_{p_i} + \bar{D}_{p_j})$$

(5.7)

and the two-dimensional patch-intensity dimension of the surface finally comes as:

$$D_{p_{ij}} = 1 + \bar{D}_{p_i}$$

(5.8)

As stressed in Section 3.2.9, note that Equation (5.6) and consequently Equation (5.7) and Equation (5.8) are relevant only when the isotropy condition is fully satisfied (see Section 7.2.2) and then requires an appropriate statistical test of homogeneity between the n_i and n_j dimensions D_{t_i} and D_{t_j} (see Zar 1996).

5.3 THE KORCAK DIMENSION, D_K

Looking for structure in the complex distribution of areas of islands in the Aegean Sea, Korcak (1938) used the conventional cumulative frequency distribution and empirically found a relationship between the number of islands of area greater than the threshold area a:

$$N(A \geq a) = ka^{-K}$$

(5.9)

where K is referred to as the Korcak exponent of patchiness ($0 \leq B \leq 1$) and k is a constant. It was later demonstrated (after Mandelbrot's discovery of fractal geometry) that the Korcak exponent of patchiness K and the fractal dimension D_K were related as:

$$K = \frac{1}{D_E} D_K$$

(5.10)

where D_E is the dimension of the embedding Euclidean space; see Mandelbrot (1977, 1983), and Hastings and Sugihara (1993) for mathematical proof. An examination of the data of the whole world yields $K = 0.65$, thus $D_K = 1.30$; see Equation (5.10). More local estimates using restricted regions range from $K = 0.50$ (that is, $D_K = 1.00$) for Africa (typically an enormous island with smaller islands whose sizes decrease rapidly) up to $K = 0.75$ (that is, $D_K = 1.50$) for Indonesia and North America (where the predominance of the largest islands is less overwhelming; Mandelbrot 1975). The Korcak exponent K measures the number of small patches relative to the number of larger patches, with smaller values of K corresponding to fewer small patches. Distributions with high Korcak exponents of patchiness (that is, low Korcak dimension D_K) are thus patchier (more small patches) than distributions characterized by smaller values of K (that is, larger values of D_K). Hastings et al. (1982) used this method to measure patchiness in vegetative ecosystems, while

Sugihara and May (1990a) discussed its general relevance for a variety of different systems, ranging from remote sensing to patch dynamics of bryozoans and coral colonies. In particular, they showed the relationship between the Korcak dimension D_K and the exponent H characterizing fractional Brownian motions (Sugihara and May 1990a) (see also Chapter 4):

$$D_K = 2 - H \tag{5.11}$$

where H characterized the degree of persistence (or autocorrelation) of a pattern: For $H < 0.5$ and $H > 0.5$, a pattern is respectively negatively and positively autocorrelated, while $H = 0.5$ corresponds to the Brownian (that is, random) case (see Chapter 4 for further details). Sugihara and May (1990a) additionally stated that increased persistence (more memory in the process) should correspond to smoother boundaries and patches with larger and more uniform areas, whereas reduced persistence will correspond to more complex and highly fragmented landscapes dominated by many small areas.

5.4 FRAGMENTATION AND MASS-SIZE DIMENSIONS, D_{fr} AND D_{ms}

The Rosin law, widely used to describe the distribution of particle size in soils and other geological material (Turcotte 1992), can be regarded as the volumetric analogy to the Korcak patchiness exponent (Equation 5.9) and is defined as:

$$N(R \geq r) = kR^{-D_{fr}} \tag{5.12}$$

where k is a constant, $N(R \geq r)$ is the number of particles whose radius R is greater than a threshold radius r, and D_{fr} is referred to as the fragmentation dimension, and condenses the information about the scale dependence of the number-size distribution of soil aggregates (Turcotte 1986, 1989; Perfect and Kay 1991; Perfect et al. 1992; Rasiah et al. 1992). A soil with a high fragmentation dimension is then more fragmented and dominated by small particles, while $D_{fr} = 0$ is indicative of a homogeneous soil where all particles are of equal diameter.

However, counting the total number of aggregates of a given size is not always possible, especially for smallest sizes. This inconvenience can be avoided by inferring the number-size distribution from the mass-size distribution function $M(x < X)$ of the cumulative mass of aggregates of characteristic size less than X as (Perfect et al. 1992; Rasiah et al. 1993; Kozac et al. 1996):

$$M(x < X) = kX^{D_{ms}} \tag{5.13}$$

where k is a constant, x is the aggregate size, and D_{ms} is the so-called mass-size dimension. The relationship between the fragmentation dimension D_f and the mass-size dimension D_{ms} is given by (Tyler and Wheatcraft 1992):

$$D_{ms} = 3 - D_f \tag{5.14}$$

The fractal structure of soil aggregates has also been investigated following (Perfect et al. 1994):

$$N(X \geq x_*) = kx_*^{-D_{fr}} \tag{5.15}$$

where x_* is a normalized measure of aggregate size (for example, sieve aperture divided by aperture of the largest sieve), $N(X \geq x_*)$ is the cumulative number of fragments with normalized length $X \geq x_*$ and k is the number of fragments not passing the largest sieve. Note that Equation (5.15) is strictly equivalent to the size distribution of fragments resulting from a fractal reduction of a Euclidean initiator (Mandelbrot 1983; Turcotte 1986) and to the Rosin law; see Equation (5.12).

5.5 RANK-FREQUENCY DIMENSION, D_{rf}

5.5.1 ZIPF'S LAW, HUMAN COMMUNICATION, AND THE PRINCIPLE OF LEAST EFFORT

One of the most surprising instances of power laws is probably Zipf's law, named after the Harvard linguistic professor G. K. Zipf (1902–1950), which is the observation of frequency of occurrence of any event as a function of the rank r, when the rank is determined by the above frequency of occurrence (that is, from n events, the most and least frequent ones will then have ranks $r = 1$ and $r = n$, respectively). More specifically, the Zipf law states that the frequency f_r of the rth largest occurrence of the event is inversely proportional to its rank r as:

$$f_r = \frac{f_1}{r} \tag{5.16}$$

where f_1 is the frequency of the most frequent event in the distribution. Zipf's law emerges from almost all languages' letters and words as an approximate slope of −1 in log-log plots of f_r vs. r, a result Zipf (1949) stated due to the "Principle of Least Effort" in communication systems, representing a balance between the repetition desired by the listener and the diversity desired by the transmitter.

Zipf's law is based on what Zipf (1949) termed the "Principle of Least Effort" in which he proposes that human speech and language are structured optimally as a result of two opposing forces: unification and diversification. If a repertoire is too unified or repetitive, a message is represented by only a few signals, and therefore less communication complexity is conveyed. Alternatively, if a repertoire is too diverse or randomly distributed, the same message can be overrepresented by a multitude of signals and, again, less communication is conveyed. These two opposite forces result in a balance between unification and diversification. Zipf's Principle of Least Effort can be statistically represented by regressing the log of the frequency of occurrence of some event (that is, letters, characters, words, morphemes, phonemes) as a function of the rank, where the rank is determined by the above-mentioned frequency of occurrence. The balance between unification and diversification leads to a power-law function with a slope close to unity. Zipf subsequently showed that a multitude of diverse human languages (for example, English words, Nootka varimorphs and morphemes, Plains Cree holophrases, Dakota words, German stem forms, Chinese characters, Gothic root morphemes and words, Aelfric's Old English morphemes and words, English writers from Old English to Present, Old and Middle High German and Yiddish sources, and Norwegian writings), whether letters, written words, phonemes, or spoken words, followed this principle and the predicted slope of ca. −1.00. This balance has also been found in the study of manuscripts of unknown origin such as the Voynich manuscript* (see, for example, Landini 2001; Schinner 2007) and optimizes the amount of potential communication that can be carried through a channel from speaker to receiver or from writer to reader. The structure of the system is thus neither too repetitive (the extreme would be one signal for all messages) nor too diverse (the extreme would be a new signal for each message, and in practice a randomly distributed repertoire would represent the highest degree of diversity). A system exhibiting such a balance (that is, a −1.00 slope in a log-log plot of frequency of occurrence vs. rank) can be thought to have a high potential capacity for transferring information, and as such has a high communication capacity. Note, however, that it only has the potential to carry a high degree of communication, because Zipf's statistic only examines the structural complexity of the repertoire, not how the composition is internally organized within the repertoire.

---

* The Voynich manuscript is a 16th-century manuscript written in an unknown language and alphabet.

5.5.2 Zipf's Law, Information, and Entropy

Information theory (Shannon 1948, 1951; Shannon and Weaver 1949), although originally illustrated using statistically significant samples of human language, generally provides quantitative tools to assess and compare communication systems across species. Specifically, this theory has been applied to a wide range of animal communicative processes or sequential behavior (for example, MacKay 1972; Slater 1973; Bradbury and Vehrencamp 1998). These include aggressive displays of hermit crabs (Hazlett and Bossert 1965), aggressive communication in shrimp (Dingle 1969), intermale grasshopper communication (Steinberg and Conant 1974), dragonfly larvae communication (Rowe and Harvey 1985), social communication of macaque (Altmann 1965), waggle dance of honeybees (Haldane and Spurway 1954), chemical paths of fire ants (Wilson 1962), structure of songs in cardinals and wood pewees (Chatfield and Lemon 1970), vocal recognition in Mexican free-tailed bats (Beecher 1989), and bottlenose dolphin whistles (McCowan et al. 1999). In contrast, only few investigations have assessed animal behavior using Zipf's law (Hailman et al. 1985, 1987; Hailman and Ficken 1986; Ficken et al. 1994; Hailman 1994). The Zipf law and Shannon entropy* are conceptually and mathematically related but nevertheless subtly differ. Zipf's law measures the potential capacity for information transfer at the repertoire level by examining the "optimal" amount of diversity and redundancy necessary for communication transfer across a "noisy" channel (that is, all complex audio signals will require some redundancy). In comparison, Shannon entropies were originally developed to measure channel capacity, and the first-order entropy differs from Zipf's statistic as Zipf does not specifically recognize language as a noisy channel.

Both the similarity and difference between Zipf's law and Shannon entropy prompted Mandelbrot (1953) to analyze the question of how the value of the Zipf exponent α relates to the Shannon entropy† H_0 (Equation 5.20; see also Box 5.1) for an information source following a Zipf's distribution.

The main problem here is that the maximum rank is intrinsically controlled by the length of the data set—that is, the vocabulary or repertoire size, or the number of species. In theory, the maximum rank can grow to infinity. In practice, however, the maximum rank is limited to a finite value

---

BOX 5.1 THERMODYNAMIC ENTROPY

In scientific fields such as information theory, mathematics, and physics, entropy is generally considered as a measure of the disorder of a system. More specifically, in thermodynamics (the branch of physics dealing with the conversion of heat energy into different forms of energy—for example, mechanical or chemical), entropy, S, is a measure of the amount of energy in a system that cannot be used to do work. Entropy can also be seen as a measure of the uniformity of the distribution of energy. Central to the concept of entropy is the second law of thermodynamics, which states that "the entropy of an isolated system which is not in equilibrium will tend to increase over time, approaching a maximum value at equilibrium."

Ice melting illustrated in Figure 5.B1.1 is the archetypical example of entropy increase through time. First, consider the glass containing ice blocks as our system. As time elapses, the heat energy from the surrounding room will be continuously transferred to the system. Ice will then continuously melt until it reaches the liquid state, and the liquid will then keep receiving heat energy until it reaches thermal equilibrium with the room. Through this process, energy has become more dispersed and spread out in the system at equilibrium than when the glass was only containing ice.

---

* Here, Shannon entropy specifically refers to the first-order Shannon entropy; Shannon higher-order entropies provide a more complex examination of communicative repertoires and are discussed and illustrated in Section 5.5.6.

† Note that entropy is defined here as a measure of the informational degree of organization and is not directly related to the thermodynamic property used in physics; see also Box 5.1.

FIGURE 5.B1.1 Ice melting as an example of entropy increase over time.

A striking property of entropy, the arrow of time, coined in the seminar work of the Nobel laureate Ilia Prigogine, is implicit in the second law of thermodynamics. As the entropy of an isolated system naturally tends to increase over time, it cannot decrease. This gives to thermodynamic processes a temporal directionality and irreversibility, especially clear from the case of ice melting (Figure 5.B1.1).

due to finite sample size. This limitation, referred to as "rank truncation," is at the core of the link between α and H_0 (Figure 5.5):

- The entropy of a Zipf's distribution cannot be defined for $\alpha \geq -1$ because the sum of all probabilities is an infinite series $\sum_{r=1}^{\infty} p(r)$, where $p(r) = cr^{\alpha}$, that diverges (Figure 5.5). In contrast, for the region $\alpha < -1$, where $\alpha \approx -1$, H_0 changes sharply with α. H_0 is then much

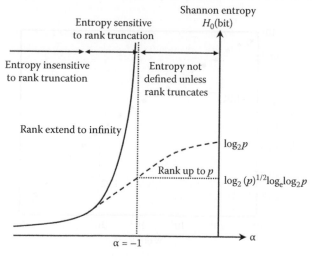

FIGURE 5.5 Shannon entropy H_0 shown as a function of the Zipf's exponent α. The solid curve is the theoretical case where the rank r as no upper bound, that is, $r \to \infty$. The dashed curve is the "rank truncation" case that is practically encountered. (Modified from Mandelbrot, 1953.)

more sensitive than α to a change in the source properties. As a consequence, two information sources with very different values of H_0 will have similar values of α. α is thus a
poor parameter to characterize the communicative properties of information sources.

- Rank truncation distorts entropy estimates in the region $\alpha \approx -1$, which is also the region of
  greatest interest for information sources close to the Zipf law (see Equation 5.16). Considering
  the rank r of the least frequent word, Mandelbrot (1953) identified two cases where H estimates are erroneously finite; that is, $H_0 = \log_2 r$ when $\alpha = 0$ (the case where words are uniformly distributed) and $H_0 = \log_2 \sqrt{r} + \log_e \log_2 r$ when $\alpha = 1$.

As a consequence, while both Zipf's law and Shannon entropy directly relate to information theory,
this suggests that the Zipf distribution parameter α is a less reliable estimate and a less reliable representation of the source properties than Shannon entropy.

5.5.3 FROM THE ZIPF LAW TO THE GENERALIZED ZIPF LAW

Ferrer i Cancho and Solé (2001) described a double law for Zipf; that is, the Zipfian curve given by
Equation (5.16) is best described by two functions (Figure 5.6). This suggests the existence of two
regimes in English and questions the generality of Equation (5.16), as clear deviations from the expected
$\alpha \approx 1$ have been documented. This is the case for language-affecting diseases such as schizophrenia
(Ferrer i Cancho 2005a) and also certain types of words such as English nouns and verbs (Ferrer i
Cancho 2005b, 2005c). Studies on multiauthor collections of texts showed two distinct regimes for the
most frequent words (the core lexicon) and for the less frequent words (the peripheral lexicon; Ferrer
i Cancho and Solé 2001). Shakespearean works also exhibit the shape of a peripheral lexicon, which
leads to the controversial statement that it can be a case of multiauthorship (Michell 1999).

In many natural phenomena, large events are scarce but small ones quite common. For example,
there are few large earthquakes and avalanches but many small ones. There are a few words, such as
"the," "of" and "to" that occur very frequently, but many that occur rarely, such as "Zipf." The Zipf
law (Equation 5.16) can thus be a generalized Zipf's law and subsequently rewritten as:

$$f_r = \frac{f_1}{r^\alpha} \tag{5.17}$$

where the log-log plot can be linear with any slope α.

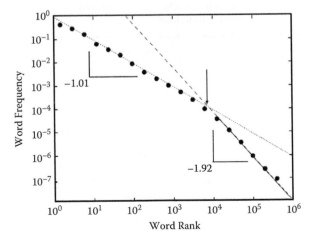

FIGURE 5.6 Word frequency as a function of rank for English language, showing two distinct regimes scaling with $\alpha \approx 1$ (dotted line) and $\alpha \approx 2$ (dashed line). The arrow indicates the cutoff rank between the two
regimes identified as dashed and dotted lines. (Modified from Ferrer i Cancho and Solé, 2001.)

This family of power laws has been successfully applied to a wide variety of problems related to species competition (Lotka 1926) ($\alpha = 2$), linguistics and social dynamics (Zipf 1949) ($\alpha = 1$), and species diversity (Frontier 1985, 1994); see also Table 5.1. In particular, Mandelbrot (1977, 1983) demonstrates that the fractal dimensions of such power laws are given by:

$$D_{rf} = 1/\alpha \tag{5.18}$$

The original law, Equation (5.16), was further modified by Mandelbrot (1953) as:

$$f_r = f_0 (r + \phi)^{-\alpha} \tag{5.19}$$

Equation (5.19) has proven to be extremely useful to describe living communities in both aquatic and terrestrial ecosystems (Frontier 1985, 1994). Thus, in Equation (5.19) f_r must be thought of as the frequency of the rth species after ranking the species in decreasing order of their frequency. The two parameters α and ϕ are conditioning the species diversity and the evenness of a given community, where the diversity H is given by (Shannon and Weaver 1963):

$$H = -\sum_{i=1}^{N} f_i \log_2 f_i \tag{5.20}$$

and the evenness J by (Pielou 1966):

$$J = \frac{H}{\log_2 N} \tag{5.21}$$

where f_i is the relative frequency of the species i and N is the number of species. It can be easily seen from Equations (5.20) and (5.21) that for the same number of species, the diversity is high when species have equivalent probability (high evenness) and low when a weak number of species is frequent and others are scarce (low evenness). Strictly speaking, it is implicit from Equation (5.19) that α depends on the average probability of a species; all the prerequisite conditions necessary for the development of this species have thus been fulfilled. ϕ depends on the average number of alternatives per category of previous conditions, hence the potential diversity of the environment. More specifically, a low value of α means a slow decrease in the species abundance (that is, a more even distribution of individuals among species), and a high value of α means a rapid decrease of species abundance (that is, a more heterogeneous distribution). The former and the latter give less and more vertical rank-frequency distributions (RFDs), hence high and low evenness and diversity. On the other hand, a positive value of ϕ (Figure 5.7) results in a greater evenness among the most frequent species than a higher diversity index. Alternatively, a negative ϕ (Figure 5.7) describes a community marked by the dominance of a few (even one) species and provides a low diversity index and a low evenness. In summary, both ϕ and α act upon the diversity and evenness respectively through the niche diversity (that is, the number of alternatives in each type of previous environmental condition) and through the predictability of the ecosystem (probability of the appearance of a species when its environmental conditions are satisfied; Frontier 1985). Despite appealing and meaningful properties, Equation (5.19) has seldom been used in ecology. For instance, Margalef (1957) was the first to fit the Mandelbrot distribution to Mediterranean tintinnids with $\phi = 0.84$ and $\alpha = 0.45$, while Frontier and Bour (1976) and Frontier (1977) respectively estimated $\alpha = 1$ and $\alpha = 2$ for chaetognaths and pteropods. Changes in the shape of the RFDs characterize temporal changes in the community structure (Frontier 1985). More specifically, a linear-concave curve (or S-shaped curve) indicates the dominance of one or two species that have fast growth and reproduction rates

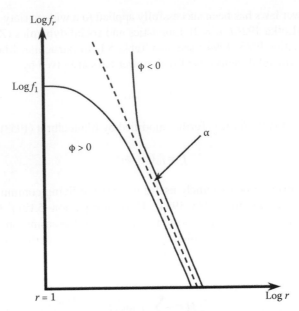

FIGURE 5.7 Schematic illustration of the expected shape of a log-log plot of the rank-frequency diagram. The dashed line is the best fit of the linear part of the rank-frequency diagram in a log-log plot, and its slope α provides an estimate of the rank-frequency dimension as $D_{rf} = 1/\alpha$. Negative and positive values of the parameter ϕ lead to different shapes of the rank-frequency diagram, and describe communities marked respectively by the dominance of a few (even one) species (low diversity and low evenness) and a greater evenness among the most frequent species, than a higher diversity index.

in a low-species-richness assemblage (that is, stage 1, pioneer community). In contrast, a more convex shape among the first-ranked species indicates a more even distribution among species (that is, stage 2, mature community), and a linear RFD is observed at the end of an ecological succession when the first-ranked species becomes more dominant and the species richness is lower (that is, stage 3, senescent community). After a disturbance,* few species can quickly develop (that is, "r strategists" and "opportunists") so the RFD appears coarsely rectilinear with successive steps (stage 1', intermediate stage between stages 1 and 2) (Frontier 1985; Legendre and Legendre 2003).

5.5.4 GENERALIZED RANK-FREQUENCY DIAGRAM FOR ECOLOGISTS

Equations (5.16) and (5.17) can be more generally written as:

$$X_r \propto r^{-\alpha} \tag{5.22}$$

where X_r is the value taken by any random variable relative to its rank r, and $\alpha = 1$ and $\alpha \neq 1$ for the Zipf's and the generalized Zipf's law, respectively. The concept related to X_r is very general and refers without distinction to frequency, length, surface, volume, mass, or concentration. Discrete processes such as linguistics, species assemblages, and genetic structures would nevertheless still require frequency computations, and thus refer to Equations (5.16) and (5.17). Alternatively, Equation (5.22) can be thought of as a more practical alternative that can be directly applied to any continuous process. The relevance of Equation (5.22) to describe and interpret ecological patterns is extensively discussed and illustrated in Section 5.5.5 hereafter. Note that the appeal of both Zipf's law (Equations 5.17, 5.19,

---

* Here, the concept of disturbance is very general and includes, for example, seasonal overturn, massive enrichment events (such as upwelling or eutrophication), substrate destruction (such as fire or flood), and human disturbance (such as pollution).

and 5.22) and Pareto's law (Equation 5.11) lies in the fact that they do not require any assumptions about the distribution of the data set or the regularity of the sampling interval, and are easy to implement. Zipf and Pareto laws have then been widely used in areas such as human demographics, linguistics, genomics, and physics, but surprisingly seldom in terrestrial and marine ecology (see Table 5.1).

5.5.5 Practical Applications of Rank-Frequency Diagrams for Ecologists

5.5.5.1 Zipf's Law as a Diagnostic Tool to Assess Ecosystem Complexity

Before illustrating the applicability of Zipf's law to original data sets of centimeter-scale, two-dimensional spatial distributions of bacterioplankton, phytoplankton, and microphytobenthos (Section 5.5.5.2), the characteristic shapes expected for Zipf's law when a distribution of interest is driven by (1) pure randomness, (2) power-law behavior, (3) power-law behavior contaminated by internal and external noise, and (4) competing power laws are investigated.

5.5.5.1.1 Zipf's Law of Random Processes

Figure 5.8 shows the characteristic signatures of five simulated random processes (that is, white noise) with 100, 500, 1,000, 5,000, and 10,000 data points in linear and logarithmic plots of Zipf distributions. In linear plots (Figure 5.8A), the Zipf's law for random noise appears as linear. On log-log plots (Figure 5.8B), the simulated random noises do not produce any power-law behavior as

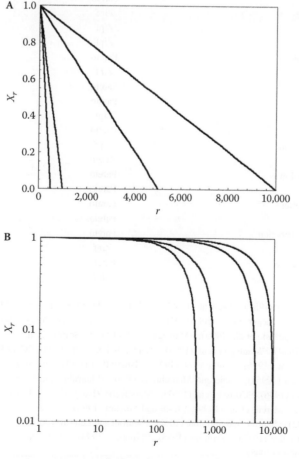

FIGURE 5.8 Linear and log-log plots of random processes with 100, 500, 1000, 50,000, and 100,000 data points (from left to right). (Modified from Seuront and Mitchell, 2008.)

TABLE 5.1
Nonexhaustive Review of the Systems Studied Using Pareto or Zipf Laws in Physical, Biological, and Ecological Sciences

System	Pareto/Zipf Law	Reference
X-ray intensity from solar flares	Pareto	1
Ecosystem model dynamics	Pareto	2
Sand pile dynamics	Pareto	3
Volcanic acoustic emission	Pareto	4
Earthquake dynamics	Pareto	5–6
Granular pile dynamics	Pareto	7
Himalayan avalanches	Pareto	8
Intensity of "starquakes"	Pareto	9
Evolution model dynamics	Pareto	10–11
Noncoding DNA sequences	Zipf	12
Landscape formation	Pareto	13–14
Sediment deposition in the ocean	Pareto	15
Coding/noncoding DNA sequences	Zipf	16
Word frequencies	Zipf	17
Word frequencies	Zipf	18
Formation of river networks	Pareto	19–20
Rice pile dynamics	Pareto	21
Noncoding DNA sequences	Zipf	22
Percolation process	Zipf	23
Linguistics	Zipf	24
Tropical rain forest dynamics	Pareto	25–27
City formation	Zipf	28
Bird population dynamics	Pareto	29
Aftershock series	Pareto	30
City distribution	Zipf	31
Discrete logistic systems	Pareto	32
Procaryotic protein expression	Zipf	33
Ion channels	Pareto	34
Dynamics of atmospheric flows	Pareto	35
U.S. firm sizes	Pareto	36
Distribution of city populations	Pareto	37
Economics	Pareto	38
Microphytobenthos 2D distribution	Pareto	39–40
Marine species diversity	Zipf	41–45
Size spectra in aquatic ecology	Pareto	46
Phytoplankton distribution	Zipf	47–48

Sources: [1]McHardy and Czerny (1987); [2]Bak et al. (1989); [3]Held et al. (1990); [4]Diodati et al. (1991); [5]Feder and Feder (1991); [6]Olami et al. (1992); [7]Jaeger and Nagel (1992); [8]Noever (1993); [9]Garcia-Pelayo and Morley (1993); [10]Bak and Sneppen (1993); [11]Paczuski et al. (1995); [12]Mantegna et al. (1994); [13]Somfai et al. (1994a); [14]Somfai et al. (1994b); [15]Rothman et al. (1994); [16]Mantegna et al. (1995); [17]Kanter and Kessler (1995); [18]Czirók et al. (1995); [19]Rigon et al. (1994); [20]Rinaldo et al. (1996); [21]Frette et al. (1996); [22]Israeloff et al. (1996); [23]Watanabe (1996); [24]Perline (1996); [25]Solé and Manrubia (1995a); [26]Solé and Manrubia (1995b); [27]Manrubia and Solé (1996); [28]Makse et al. (1995); [29]Keitt and Marquet (1996); [30]Correig et al. (1997); [31]Marsili and Zhang (1998); [32]Biham et al. (1998); [33]Ramsden and Vohradsky (1998); [34]Mercik et al. (1999); [35]Joshi and Selvam (1999); [36]Axtell (2001); [37]Malacarne et al. (2002); [38]Burda et al. (2002); [39]Seuront and Spilmont (2002); [40]Seuront and Leterme (2006); [41]Margalef (1957); [42]Frontier and Bour (1976); [43]Frontier (1977); [44]Frontier (1985); [45]Frontier (1994); [46]Vidondo et al. (1997); [47]Mitchell (2004); [48]Mitchell and Seuront (2008).

expected from Equation (5.22) but instead produce a continuous roll-off from a horizontal line (that is, $\alpha \rightarrow 0$) to a vertical line (that is, $\alpha \rightarrow \infty$). This is representative of the fact that no value is more likely to be more common than any other value.

The previous observations can be extended to fractional Brownian motions (fBm) (Figure 5.9A and Figure 5.9B). Because fBm have the desirable property of exhibiting antipersistent (an increase in the value of the random variable is expected to be followed by a decrease) and persistent (an increase in the value of the random variable is expected to be followed by another increase) behavior, they explore a certain range of values before moving off more or less gradually to another range of values. These properties lead to a weaker version of randomness in the Zipf framework (Figure 5.9C,D,E,F). For antipersistent fBm (Figure 5.9A), the Zipf plots do not exhibit any clear linear behavior (Figure 5.9C), mainly because of the upward and downward roll-off observed for low and high rank values, respectively. This is, however, simply the result of an undersampling of

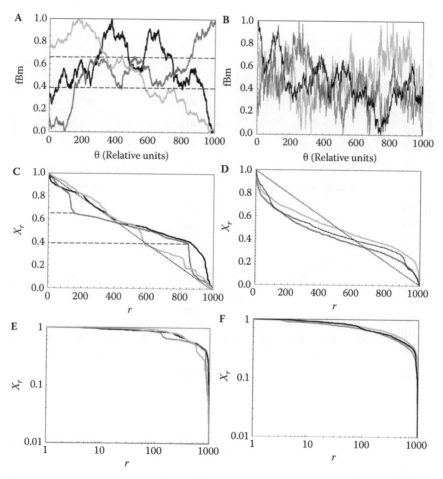

FIGURE 5.9 Antipersistent (A) and persistent (B) fractional Brownian motions (fBm), shown together with their characteristic signatures in linear (C, D, E) and log-log (D, E, F) plots. In antipersistent and persistent processes, an increase in the value of a random variable is expected to be followed by a decrease and an increase, respectively. The resulting Zipf plot exhibits different deviations from randomness. The dashed lines in (B, C, D) indicate a range of values explored by the fBM before moving off more or less gradually to another range of values. The same colors have been used for the different fBm (A, B) and their related Zipf plots (C, D, E, F); the darker colors characterize the more antipersistent/persistent fBm. In both cases, the symbol θ represents space or time in case of time series or transect studies, respectively. (Modified from Seuront and Mitchell, 2008.)

the highest and lowest values that can be regarded as an implicit consequence of antipersistence. The distributions are characterized by a weak evenness for high and low values, the distribution being dominated by a few (ultimately one) high and low values. On the other hand, for persistent fBm (Figure 5.9B), the step shape of the Zipf plot (Figure 5.9D,F) reflects the property of persistent processes to visit one particular range of values and then to change to another range sharply. This step function becomes clearer when the fBm exhibit more persistence (Figure 5.9D,F). The main difference between antipersistent and persistent Zipf plots then relies on the quantity of values taken by the fBm between transitions, which will be more gradual in the antipersistent case and thus contain more points than in the persistent case. Because the scale expansion related to log-log plots may hide, at least partially, the specific structural features of Zipf plots when compared to noise (see Figure 5.9C,D,E,F), the use of both linear and logarithmic plots is recommended. Finally, it is stressed that any step in Zipf plots indicates structural discontinuities within the data set.

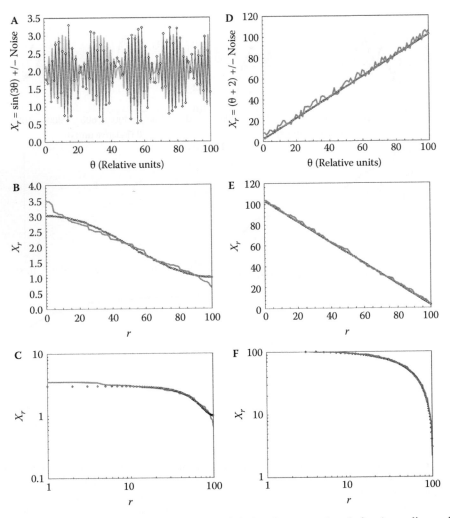

FIGURE 5.10 Simulated monotonic and periodic trends before (gray curve) and after (open diamonds) being contaminated by observational white noise (A, D), their subsequent Zipf signatures in linear (B, C), and log-log (D, E) plots. In both cases, the symbol θ represents space or time in case of time series or transect studies, respectively. (Modified from Seuront and Mitchell, 2008.)

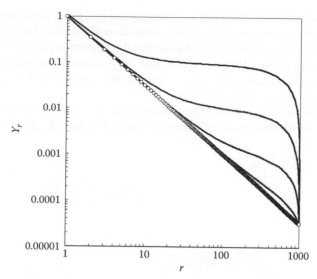

FIGURE 5.11 Log-log signature of a power law (dots) with different percentage of additive noise (0.01%, 0.1%, 1%, and 10%, from bottom to top). (Modified from Seuront and Mitchell, 2008.)

5.5.5.1.2 Zipf's Law of Deterministic Processes

Deterministic patterns and processes are well known in time-series analysis and referred to as monotonic and periodic trends. Gradients and sine waves are archetypical examples. Periodicity is common in both terrestrial and marine ecology. Here, we simulate an increasing linear trend and a sine wave trend. Both of them have been subsequently contaminated by observational white noise (Figure 5.10A,D). The Zipf plots of the increasing trends exhibit the characteristic signature of white noise (Figure 5.10B,C). In contrast, the Zipf plots of the sine waves exhibit distinct features (Figure 5.10E,F). The sine wave has the characteristic Zipf shape of a pattern oversampled for its higher and lower values, while the noisy sine wave converges toward a random Zipf signature.

5.5.5.1.3 Zipf's Law of Pure Power Laws

Patterns and processes characterized by a power law function (for example, Equation 5.22) will appear as a straight line in log-log plots (Figure 5.11). However, this theoretical case is rare in nature, and a more realistic series where power laws may be hidden by a wide range of contaminating processes is investigated hereafter.

5.5.5.1.4 Zipf's Law of Contaminated Power Laws

Before focusing on the processes susceptible to modify the characteristic exponents of Zipf power laws, we will consider the potential effects of external and internal noise on the extent of the power laws. In the first approach, the variability of a given descriptor is driven by "new" events, which represent exogenous variables (exogenous in the sense that they are not a part of an internal mechanism that drives the descriptor fluctuations), for instance, the motion of dinoflagellate cells induced by vertical turbulent eddy diffusivity. On the other hand, internal noise refers to the existence of an engine within the cells (that is, endogenous) that generates motion by mechanisms of feedback of the motion of the cells upon themselves.

5.5.5.1.4.1 Zipf's Law of Power Laws Contaminated by External (White) Noise

If varying amounts of noise are added to the power function $X_r \propto r^{-\alpha}$ as:

$$Y_r = (r^{-\alpha} + \varepsilon) \tag{5.23}$$

where ε is a white noise whose amplitude is defined as being a given percentage of the maximum value of X_r, then the noise causes a rightward departure from the straight line at a rank proportional to the amount of noise added (Figure 5.11). Measuring the point of departure from a power law for a variety of noise levels (here, 0.01%, 0.1%, 1%, and 10%) recovers the original function for Zipf plots. Such a graphical approach could be very valuable to estimate the extent to which noise contaminates or contributes to the measured signal.

Consider two situations where a simulated power-law function ($X_r \propto r^{-\alpha}$, with $\alpha = 0.18$) is mixed with a random noise ε_1, vertically offset so as not to overlap, that is, $\varepsilon_2 \in [\min X_r, \chi_1]$ and $\varepsilon_2 \in [\chi_2, \min X_r]$, as

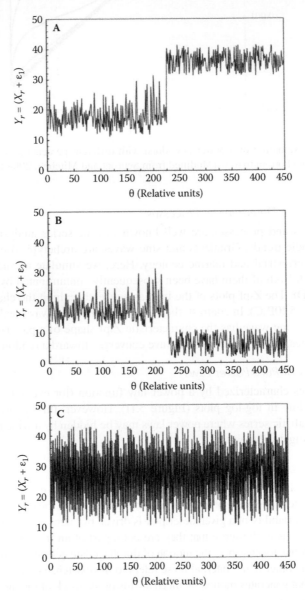

FIGURE 5.12 Power-law distribution $X_r \propto r^{-0.18}$ combined with a white noise distribution ε_i as $Y_r = X_r + \varepsilon_i$, as a caricature of two populations separated by a sharp gradient (A, B), or mixed (C). The two populations, characterized by a power law X_r and a random distribution ε_i, have been considered as fully separated (A, B), with $\varepsilon_i = \varepsilon_1 (\varepsilon_1 \in [\max X_r, \chi_1])$ and $\varepsilon_i = \varepsilon_2 (\varepsilon_2 \in [\chi_2, \max X_r])$, and fully mixed, with $\varepsilon_i = \varepsilon_1$ (C). The symbol θ represents space or time in case of time series or transect studies, respectively. (Modified from Seuront and Mitchell, 2008.)

$Y_r = X_r + \varepsilon_i$. This could illustrate the expected outcome of a transect crossing a boundary separating two distinct structural entities or a vertical profile crossing a strong thermocline (Figure 5.12A,B). In both cases, the subsequent Zipf plots exhibit a clear step function indicative of a structural discontinuity (Figure 5.13A,B) between the characteristic behaviors expected in cases of randomness and power law. However, while we used the same power law in both cases, the exponents and the goodness of the power-law fits are different (Figure 5.13).

This result could lead to misinterpretation of Zipf plots. The widely acknowledged assumption that any range of values with the same extent (for example 10 to 100, or 10,000 to 10,090) on the x axis produces the same range of values of the y axis is no longer valid in the nonlinear framework of power laws. Thus, different ranges of rank, r, values 225 to 450 (Figure 5.12A) and 1 to 450 (Figure 5.12B), return different ranges on the y axis, and thus different laws. As a consequence, to conduct Zipf analyses successfully and for the results to be meaningful, we recommend separate analyses of the different ranges of values identified in a preliminary global analysis as being separated by a step function. Figure 5.14 thus illustrates how the simulated

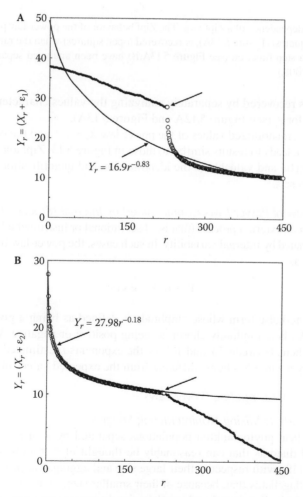

FIGURE 5.13 Zipf plots of the two theoretical situations illustrated in Figure 5.12A (A) and Figure 5.12B (B). Note that while the same power laws have been used in both situations, the original power law $X_r \propto r^{-0.18}$ is recovered only when $X_r > \varepsilon_i$ (b); when $X_r < \varepsilon_i$ the power law fit to the power law population is not significant (A). The circles indicate the step function behavior of the Zipf plot that should be regarded as being indicative of structural changes within the data set. (Modified from Seuront and Mitchell, 2008.)

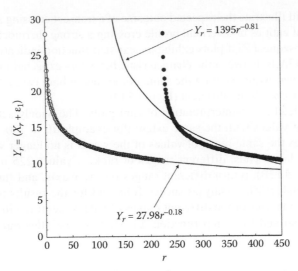

FIGURE 5.14 Density dependence of a Zipf plot. The Zipf behavior of the power-law population X_r character-ized by $X_r > \varepsilon_i$ (black squares) (Figure 5.13A), is recovered (open squares) when the range of values identified as being separated by a step function (see Figure 5.13A,B) have been analyzed separately. (Modified from Seuront and Mitchell, 2008.)

power law $X_r \propto r^{-\alpha}$ is recovered by separately analyzing the values characterized by ranks rang-ing from 225 to 450 for ε_1 (see Figure 5.12A and Figure 5.13A).

The combination of randomized values of the power law $X_r \propto r^{-\alpha}$ and the nonoverlapping noise, ε_i, (Figure 5.12A,B,C), leads to results similar to those in Figure 5.13. Zipf analysis is then revealed to be extremely powerful and valuable in the identification and quantification of hidden structural properties of any data sets.

5.5.5.1.4.2 Zipf's Law of Power Laws Contaminated by Internal (Process) Noise

Instead of considering an external process (that is, observational or instrumental noise), the power law itself can be contaminated by internal variability. In such cases, the power-law function $X_r \propto r^{-\alpha}$ (with $\alpha = 0.18$) is rewritten as

$$Y_r = (r \pm r \times \varepsilon)^{-\alpha} \tag{5.24}$$

where ε is still a white-noise term whose amplitude is defined as being a given percentage of the maximum value of X_r, here randomly chosen as being positive or negative. Whatever the amount of noise considered (here, between 5% and 100%), the exponents α estimated from Equation (5.24) cannot be statistically regarded as being different from the expected values of 0.18 ($p < 0.01$).

5.5.5.1.5 Zipf's Law of Competing Power Laws

5.5.5.1.5.1 Case Study 1: Mixing Noninteracting Species

Consider two theoretical phytoplankton populations separated by a sharp hydrological gradient. One is composed of diatoms that can reasonably be thought of as following a power-law form, $X_r \propto r^{-\alpha}$ (here, $\alpha = 0.18$), with respect to their large size and aggregative properties. The other one is composed of dinoflagellates that, because of their smaller size, high concentration, and motility are more homogeneously distributed and are then simply represented here as a background con-centration k_i. The resulting pattern can then be thought of as the combination $Y_r = X_r + k_i$. It is emphasized here that any change in the background concentration k_i does not affect the power-law nature of the original data set X_r. However, the exponent α' of the resulting power laws $Y_r \propto r^{-\alpha'}$ decreases with increasing values of k_i. The addition of an increasing background concentration

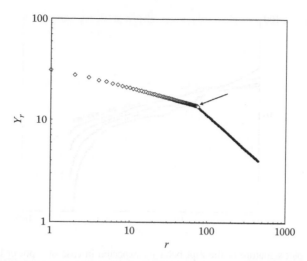

FIGURE 5.15 Log-log plot signature of the Zipf behavior resulting from mixing two theoretical populations characterized by two distinct power laws and overlapping ranges of concentrations. The range of values corresponding to the overlapping of the two power laws presents an intermediate power-law behavior with a characteristic exponent α' defined as $\alpha_1 < \alpha' < \alpha_2$ and $\alpha' = k\alpha_1 + (1 - k)\alpha_2$. (Modified from Seuront and Mitchell, 2008.)

thus smoothed out the differences between ranks. The observation of such a decrease in empirical power-law exponents from field data sampled at the same point before and after the disruption of a hydrological gradient, or at different period of the seasonal cycle, would strongly indicate a structural change in the relative organization of the studied biological communities.

Now consider a situation where two spatially separated phytoplankton populations are mixed—for example, two monospecific diatom populations—characterized by overlapping ranges of concentrations and distinct power-law forms, $X_{1r} \in [2.99, 13.79]$ and $X_{1r} \propto r^{-\alpha_1}$ with $\alpha_1 = 0.18$, and $X_{2r} \in [3.19, 31.37]$ and $X_{2r} \propto r^{-\alpha_2}$ with $\alpha_1 = 0.24$, respectively. Evenly mixing these two populations without considering any interactions will result in the Zipf structures shown in Figure 5.15. The range of values corresponding to the overlapping of the two power laws presents an intermediate power-law behavior with a characteristic exponent $\alpha' = 0.196$ (Figure 5.15). More generally, the values of α' are implicitly bounded between $\alpha_1 < \alpha' < \alpha_2$, where α_1 and α_2 are the Zipf exponents of the original power laws and depend on the proportion of values from each original power law, following $\alpha' = k\alpha_1 + (1 - k)\alpha_2$. Finally, as stated above, a separate analysis of the values greater than the critical concentration (13.79) associated with the step function shown in Figure 5.15 is necessary to recover the original exponents $\alpha_2 = 0.24$.

5.5.5.1.5.2 Case Study 2: Mixing Interacting Species

Here, we consider one of the previous phytoplankton populations whose concentration X_r is characterized by a power-law form $X_r \propto r^{-\alpha}$, with $\alpha = 0.24$. We will now investigate the effects of processes capable of locally decreasing (that is, mortality related to inter- and intraspecific competition, or grazing) or increasing (phytoplankton growth or coagulation processes) phytoplankton concentration on the Zipf signature of the population $X_r \propto r^{-0.24}$. Note that while the following examples are based on the interactions between phytoplankton and zooplankton organisms, this does not hamper the generality of the results, as the same approach can be used to describe the interactions between terrestrial plants and grazers.

Decrease in local phytoplankton concentration. First, under the assumption of evenly distributed grazers, the grazing impact of copepods can be estimated as a percentage or a Michaelis-Menten function of the local phytoplankton concentration. Assuming the ingestion of phytoplankton cells by copepods is a percentage of a random function of food availability, the resulting food

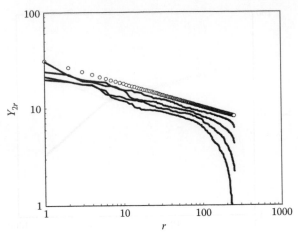

FIGURE 5.16 Log-log plot signature of the Zipf behavior expected in case of a power law X_r (open dots) competing with a random mortality component ($Y_{2r} = X_r - \varepsilon X_r$), $\varepsilon = 0.05, 0.25, 0.50,$ and 0.75 (from top to bottom). (Modified from Seuront and Mitchell, 2008.)

distributions can be described by

$$Y_{1r} = X_r - kX_r \tag{5.25}$$

and

$$Y_{2r} = X_r - \varepsilon X_r \tag{5.26}$$

where k is a constant, $0 \le k \le 1$, and ε is a random-noise process, that is, $\varepsilon \in [0, 1]$. For increasing values of k, the function Y_{1r} is simply shifted downward on a log-log Zipf plot (not shown), indicating a decrease in the background concentration of the population. A similar trend can be identified in the variable Y_{2r} for an increasing amount of noise, but with a characteristic "noise roll-off" for low rank values (Figure 5.16).

Alternatively, following laboratory data on the feeding of copepods suggesting that ingestion rate can be fairly represented by a Michaelis-Menten function (see, for example, Mullin et al. 1975), Equations (5.25) and (5.26) are modified as:

$$Y_{3r} = X_r - I_{max}X_r/(k_s - X_r) \tag{5.27}$$

where I_{max} is the maximum ingestion rate, k_s is the half-saturation constant for feeding, and X_r the concentration of food. Figure 5.17 shows the Zipf structure of the resulting phytoplankton concentration Y_{3r} for different values of the half-saturation constant k_s and the maximum ingestion rate I_{max}. It clearly appears that the effect of grazing is mainly perceptible for low values of Y_{3r}, a direct consequence of the convex form of the Michaelis-Menten function (see Equation 5.27), and leads to a significant divergence from a power law when I_{max} is high and k_s is low (compare Figure 5.17A,B,C).

However, the two previous approaches are implicitly based on the hypothesis of a homogeneous phytoplankton distribution, which is now recognized as an oversimplified hypothesis (Seuront et al. 1996a, 1999; Waters and Mitchell 2002; Waters et al. 2003) and did not take into account potential behavior adaptation of grazers to varying food concentrations (Tiselius 1992). If one considers that the remote sensing abilities (Doall et al. 1998) of copepods can lead to aggregation of grazers in areas of high phytoplankton concentrations as investigated both empirically and numerically (Tiselius 1992; Saiz et al. 1993; Leising and Franks 2000; Seuront et al. 2001), Equation (5.26) can be modified as:

$$Y_{4r} = X_r - 10^{(X_r/k)} \tag{5.28}$$

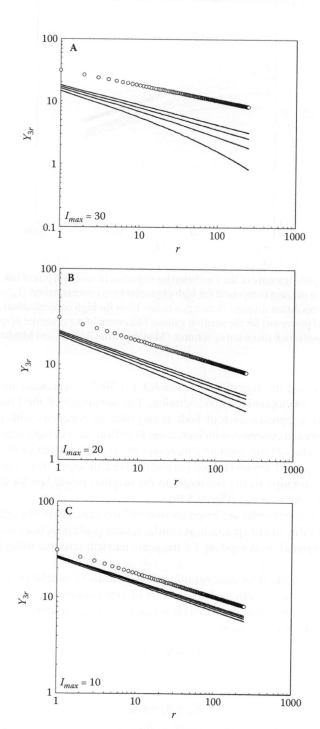

FIGURE 5.17 Log-log plot signature of the Zipf behavior expected in case of a power law X_r (open dots) competing with a Michaelis-Menten grazing component ($Y_{3r} = X_r - I_{max} X_r/(k_s + X_r)$). For a given maximum ingestion rate I_{max}, the effect is stronger for high values of the half-saturation constant k_s. (Modified from Seuront and Mitchell, 2008.)

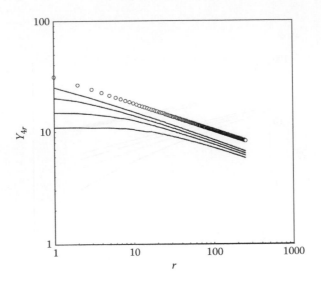

FIGURE 5.18 Log-log plot signature of the Zipf behavior expected in case of a power law X_r (open dots) competing with a preferential grazing component for high phytoplankton concentrations ($Y_{4r} = X_r - 10(X_r/k_s)$). The grazed phytoplankton population diverges from a power-law form for high concentrations, but asymptotically converges to the original power law for the smallest values. The extent of the observed divergence is controlled by increasing grazing pressure k (from top to bottom). (Modified from Seuront and Mitchell, 2008.)

where k is a constant and the ingestion function $I(X_r) = 10^{(X_r/k)}$ represents an increased predation impact on higher phytoplankton concentration. The advantage of the function $I(X_r)$ is that it can be regarded as a representation of both aggregation of copepods with constant ingestion rates and evenly distributed copepods with increasing ingestion rates in high-density phytoplankton patches. Decreasing values of the constant k increases the grazing impact on high-density patches (Equation 5.28). The grazed phytoplankton population then diverges from a power-law form for high values of Y_{4r} but asymptotically converges to the original power law for the smallest values of Y_{4r}; that is, $Y_{4r} \propto r^{-\alpha}$ for $r \to r_{\min}$ (Figure 5.18).

Although the previous examples are based on zooplankton grazing on phytoplankton, we nevertheless stress the generality of our approach, as similar results could have been obtained considering two phytoplankton populations competing for the same nutrient resource using Michaelis-Menten or Droop functions.

Increase in local phytoplankton concentration. For the sake of simplicity, consider that phytoplankton growth (in response to physical coagulation or nutrient uptake) could be represented as a percentage or a random function of the actual phytoplankton concentration X_r. Equations (5.25) and (5.26) are then respectively rewritten as:

$$Y_{5r} = X_r + kX_r \tag{5.29}$$

and

$$Y_{6r} = X_r + \varepsilon X_r \tag{5.30}$$

where k is a constant, $0 \le k \le 1$, and ε is a random noise process, that is, $\varepsilon \in [0, 1]$. For increasing values of k, the function Y_{5r} is, in full agreement with what has been concluded from Equation (5.25), shifted upward on a log-log Zipf plot (not shown), indicating an increase in the background concentration of the population. Using different values of k in Equation (5.29) has no effect on the shape of the related Zipf plots and exponents α' ($Y_{5r} \propto r^{-\alpha'}$), which remain equal to the original

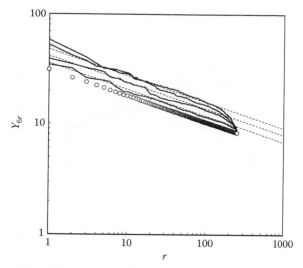

FIGURE 5.19 Log-log plot signature of the Zipf behavior expected in case of a power law X_r (open dots) competing with a random growth component ($Y_{6r} = X_r - \varepsilon X_r$), where $\varepsilon = 0.25$, 0.50, 0.75, and 1.00 (from bottom to top). The arrows indicate the minimum of a range of Y_{6r} values locally diverging from a power law because of successively increasing random increments. The dashed lines indicate slope of the power-law behavior of the initial values X_r. (Modified from Seuront and Mitchell, 2008.)

power law, that is, $Y_{5r} \propto r^{-0.24}$. Slightly different conclusions can be drawn from the behavior of the Zipf plots of function Y_{6r} (Figure 5.19). First, increases in the amount of noise ε (ranging from 25% to 100%) lead to a vertical offset of function Y_{6r} when compared to the original power law. The resulting functions exhibit the characteristic downward roll-off signature related to randomness and might also locally show increasing trends that are intrinsically caused by the random component in Equation (5.30). They could be misleading, especially when they occur over the highest rank range and must not be related to breakpoints indicative of structural discontinuities that would erroneously lead to a separate analysis of different subsections of the original data set. Finally, even if the exponents α' fluctuate around the original value, they are never significantly different ($p < 0.05$).

Increase vs. decrease in local phytoplankton concentration. Because the two previous situations are unlikely to be found individually in the ocean but should also occur concomitantly, combining Equations (5.25) and (5.29) with Equations (5.26) and (5.30) leads to:

$$Y_{7r} = X_r + (k_1 - k_2) X_r \tag{5.31}$$

and

$$Y_{8r} = X_r + (\varepsilon_1 - \varepsilon_2) X_r \tag{5.32}$$

where k_1 and k_2 are constants ($0 \leq k_1 \leq 1$, and $0 \leq k_2 \leq 1$), and ε_1 and ε_2 are random-noise processes; that is, $\varepsilon_1 \in [0, 1]$ and $\varepsilon_2 \in [0, 1]$. The resulting functions Y_{5r} and Y_{6r} exhibit intermediate behaviors between what have been observed from Equations (5.25) and (5.29), and Equations (5.26) and (5.30). For $k_1 = k_2$, the original power law, $Y_{7r} \propto r^{-0.24}$, is recovered; the growth component compensates for the death component. In contrast, when $k_1 < k_2$ and $k_1 > k_2$, the resulting function Y_{7r} is shifted downward and upward on a log-log plot as previously observed from Equations (5.25) and (5.29). Although the overall structure is preserved, the latter and the former cases lead to decreases and increases in the background concentration of the population. The Zipf plot of function Y_{8r}, shown in

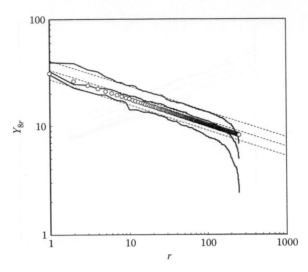

FIGURE 5.20 Log-log plot signature of the Zipf behavior expected in case of a power law X_r (open dots) competing with random growth and mortality components ($Y_{8r} = X_r + (\varepsilon_1 - \varepsilon_2) X_r$), where the random processes ε_1 and ε_2 have been chosen as $\varepsilon_1 = 0.75$ and $\varepsilon_1 = 0.25$, $\varepsilon_1 = 0.50$ and $\varepsilon_1 = 0.25$, and and $\varepsilon_2 = 0.75$ (from bottom to top). The dashed lines indicate the power-law behavior of the initial values X_r. (Modified from Seuront and Mitchell, 2008.)

a log-log plot, exhibits the different characteristic features previously identified: a power-law behavior not significantly different from the original one (that is, $\alpha' \approx \alpha = 0.24$) followed by a roll-off toward low Y_{8r} values (Figure 5.20). As stated above, successive positive random fluctuations might lead to local increasing trends slightly diverging from a power law (gray arrow) but should not be associated with a step function.

Next, consider a situation where the positive and negative fluctuations are purely randomly driven as:

$$Y_{9r} = X_r \pm kX_r \tag{5.33}$$

and

$$Y_{10r} = X_r \pm \varepsilon X_r \tag{5.34}$$

where k is a constant ($0 \leq k \leq 1$) and ε is a random-noise process, that is, $\varepsilon \in [0, 1]$, whose amplitude is defined as being a given percentage of the maximum value of X_r; k and ε are randomly chosen as being positive or negative. The resulting Zipf signatures of the functions Y_{9r} and Y_{10r} are shown in Figure 5.21 as log-log plots. The positive and negative components of Equation (5.33) clearly appear as separated by a step function (Figure 5.21A). The positive components lead to power laws that are not significantly different from the original. A separate analysis of the range of values separated from the power laws by step functions (arrows; Figure 5.21A) did not show any power-law behavior. Alternatively, the effects of Equation (5.34) on the initial power-law behavior are the characteristic downward roll-off signature related to randomness and the fluctuations around a power-law behavior that are not significantly different from the original (Figure 5.21B). To ensure the relevance of Zipf analysis, as introduced by Equations (5.17) and (5.22), the few data points diverging upward, or flattening, toward the first-rank values must not be included in the regression analysis aimed at estimating α. Indeed, the former case describes a distribution marked by the dominance of a few (ultimately one) "hotspots" that is likely to be chronically undersampled (Seuront et al. 1999), while the latter case refers to a distribution that has been systematically oversampled. These issues have

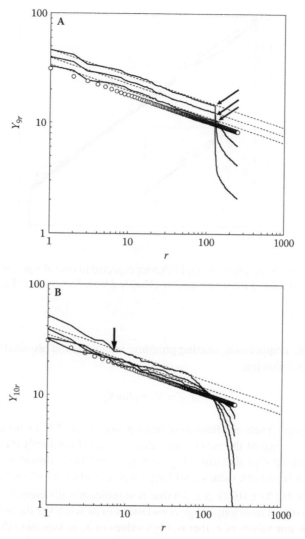

FIGURE 5.21 Log-log plot signature of the Zipf behavior expected in case of a power law X_r (open dots) competing with combined constant random and growth components ($Y_{9r} = X_r \pm kX_r$; (A) and combined random growth and mortality components ($Y_{9r} = X_r \pm \varepsilon X_r$; (B), where the constant k and the noise ε have been chosen as $k = 0.25$, 0.50, and 0.75, and $\varepsilon = 0.25$, 0.50, 0.75, and 100 (from bottom to top). The arrows indicate a step function (A) and the beginning of a local departure from a pure power law due to successively increasing random increments (B). The dashed lines indicate the power-law behavior of the initial values X_r. (Modified from Seuront and Mitchell, 2008.)

been specifically detailed elsewhere in the framework of information theory (Mandelbrot 1953) and led to the modified version of the generalized Zipf law presented above; see Equation (5.19).

It is now known that the distributions of nutrients, phytoplankton, and zooplankton exhibit different levels of persistence (Tsuda 1995; Seuront et al. 1996a, 1996b, 1999, 2002; Seuront and Lagadeuc 2001). In addition, the interplay between the biotic properties of individuals and populations and abiotic processes produce space-time structures characterized by long-range correlation (that is, persistence) (Kendall et al. 2000). We consider, finally, a situation where the local concentration of a phytoplankton population initially driven by a power law ($X_r \propto r^{-0.24}$) could be influenced by a fractional Brownian motion resulting from the combined effects of local biological (nutrient uptake,

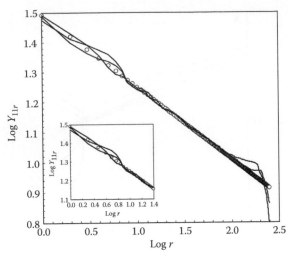

FIGURE 5.22 Log-log plot signature of the Zipf behavior expected in case of a power law X_r (open dots) competing with a persistent fractional Brownian motion. (Modified from Seuront and Mitchell, 2008.)

inter- and intraspecific competition, grazing pressure, infection) and physical (advection, diffusion, turbulence) processes following

$$Y_{11r} = X_r \pm fBmX_r \tag{5.35}$$

where fBm is a persistent fractional Brownian motion (see Figure 5.9b) whose amplitude is defined as being a given percentage of the maximum value of X_r, and randomly chosen as being positive or negative. The resulting Zipf signature (Figure 5.22) exhibits the downward roll-off characteristic of randomness for high rank values, and long-range correlations around a power-law behavior ($Y_{11r} \propto r^{-\alpha'}$, with $\alpha' = 0.240 \pm 0.005$; $\bar{x} \pm$SD) that is significantly different from the original power law ($X_r \propto r^{-0.24}$, $p > 0.05$). These long-range correlations exist whatever the values of r but are more clearly visible for the low values of r, that is, high values of X_r in Equation (5.35).

5.5.5.1.6 On the Relevance of Zipf's Law to Diagnose Ecosystem Complexity

The previous sections illustrate the potential for the seldom-used Zipf's law to be a powerful tool in the analysis and the classification of marine and terrestrial ecosystems in the presence of randomness, monotonic and periodic trends, internal and external noise, and both biotic and abiotic forcings. Specifically, Zipf analysis can be directly and easily applied to any data set without intensive computational, mathematical, or statistical analysis, and with a minimum amount of calculation. It can be conducted in a few minutes with most standard software packages, even for a data set of several thousand data points. The results of Zipf analysis should not, however, be used without a preliminary visual inspection of the data (an absolute prerequisite in data analysis that is often neglected, especially by undergraduate students), as they could erroneously be used as a direct index of patchiness. For instance, a distribution characterized by a patch of 10 high-density data points, 10 randomly or regularly spaced hotspots, or 10 ranked hotspots will return exactly the same Zipf shape. This issue has also been raised in the framework of power spectrum analysis in marine ecology (Franks 2005).

The one-to-one correspondence between Zipf and Pareto distributions analytically derived here (see Box 5.1) could be regarded as a way to reconcile previous and future works using one or the other technique. The strength of the Zipf framework also has the desirable properties of not requiring any assumptions about the statistical distribution, regularity of sampling intervals, and stationarity of the data set that are sometimes absolute prerequisites to some statistical data analysis techniques.

Finally, the Zipf framework can be conveniently used as a tool to identify and classify structures in marine ecosystems and also to infer the underlying processes that generate the observed patterns. The characteristic shapes introduced above, and most importantly their potential changes in time and space, make it possible to hypothesize the origin and the ecological implications of such modifications, as well as providing useful insights on what further analysis to conduct and how to design sampling schemes. For instance, a transect study providing a step function in a Zipf plot (see Figure 5.9d) will indicate different levels of organization within the same populations, and maybe different subpopulations, that would require separate analysis or additional sampling. In the specific case of phytoplankton distribution, it is easy to imagine that mixing or changing nutrient and/or zooplankton concentrations will alter distribution and intensity to the extent that the characteristic exponents α and β for a set of data will vary due to natural processes. Thus, a phytoplankton population exhibiting a single power-law behavior before wind stratification and investigated temporally from a fixed point might exhibit successive changes (see, for example, Figure 5.17 and Figure 5.18), and the identification and the classification of the Zipf shapes will then allow one to infer the nature of the observed changes. In turn, a study mainly focusing on phytoplankton distributions that results in transitions such as those shown in Figure 5.13 and Figure 5.15 in spring and autumn respectively may well be modified and adapted to investigate the potential differences in the grazer community. More generally, if phytoplankton properties such as growth or distribution follow a power law, then mortality processes such as grazing and lysis may well follow a similar but competing power law, as hypothesized and illustrated above. Thus, if such power-law behavior can be shown in phytoplankton, the removal of the first ranks (large values) could be interpreted as an indication of predation. However, the ubiquity of power laws is not an absolute requirement as many non-power-law processes could be involved in the modification of the pure power-law behavior as well as the removal of the last ranks (low values).

5.5.5.2 Case Study: Zipf Laws of Two-Dimensional Patterns

5.5.5.2.1 Ecological Framework

An extensive amount of work has still to be done to cover spatial gaps between remote sensing, which provides low-resolution data over large areas, and experimental approaches, which give information about local processes but are unable to provide continuous, spatially explicit data over large areas. This is a particularly salient issue considering the increasing awareness of the heterogeneous nature of the microscale distribution of nutrient, bacteria, phytoplankton, and microphytobenthos (Seymour et al. 2000, 2004, 2007, 2008; Seuront and Spilmont 2002; Seuront and Leterme 2006; Seuront et al. 2002; Waters and Mitchell 2002; Waters et al. 2003). Large-scale and small-scale patterns and processes could thus be reconciled by achieving a full understanding of how the effects of small-scale and microscale processes on the biology and the ecology of individual organisms propagate toward larger scales, for example, at the population level.

In this framework, the objective of the present case study is to demonstrate the applicability of the Zipf method described above (Section 5.5.5.1) to characterize two-dimensional patterns. This method, investigated theoretically by Seuront and Mitchell (2008) and applied to one-dimensional data sets (Mitchell and Seuront 2008), does not require any assumptions about the distribution of the data set or regular sampling intervals, and presents the desirable feature of being extremely easy to implement. To ensure the generality and the relevance of this work, original data sets of centimeter-scale spatial distributions of bacterioplankton, phytoplankton, and microphytobenthos are considered.

5.5.5.2.2 Experimental Procedures and Data Analysis

5.5.5.2.2.1 Centimeter-Scale (1.2 cm) Bacterioplankton Distribution

Sampling sites. The distribution of bacterial marine populations has been investigated from two coastal sites, Port Noarlunga and Port River, in the metropolitan area of Adelaide, South Australia. These sites have been chosen, first, because their bacterial concentrations have been shown to differ by one order of magnitude and exhibit different degrees of variability in their total abundance (Seymour et al. 2000, 2004) and, second, because they are characterized by two different hydrodynamic and hydrological regimes. Port Noarlunga is an oligotrophic environment, exposed to turbulence induced by waves breaking over a reef, while Port River is located in a eutrophic, sheltered estuary influenced by high levels of urban and industrial waste and intermittent flows of freshwater. Samples were taken from subsurface waters from the end of a coastal pier at Port Noarlunga (35°09′S, 138°28′E) on March 21, 2002, and from a floating pontoon platform at Port River (34°49′S, 138°30′E) on May 9, 2002.

Microscale sampling. The two-dimensional distribution of bacterial distributions has been investigated using a sampling device conceptually similar to the millimeter-scale resolution system extensively described elsewhere (Seymour et al. 2000). The system consists of a 10×10 array of 1 ml syringes, each separated by a distance of 1.2 cm and set to sample volumes of 100 μl. A messenger weight would release the sampling mechanism and 100 subsamples would simultaneously be taken across an area of 116 cm^2 (Seymour et al. 2004). Subsamples were subsequently transferred to 1 ml cryovials and immediately incubated with 2.5% paraformaldehyde for 20 minutes, before being quick-frozen in liquid nitrogen and subsequently stored at −80°C.

Enumeration of bacterioplankton. Prior to flow cytometric analysis, frozen samples were quick-thawed and transferred to 5 ml cytometry tubes. Samples were then stained with SYBR-I Green solution (1:10000 dilution; Molecular Probes), and incubated in the dark for 15 minutes (Marie et al. 1997, 1999). Fluorescent beads of 1 μm diameter (Molecular Probes) were added to samples in a final concentration of ca. 10^5 beads ml^{-1} (Gasol and del Giorgio 2000), and all measured cytometry parameters were normalized to bead concentration and fluorescence. After each cytometry session, working bead solutions were enumerated using epifluorescent microscopy to ensure consistency of the bead concentration (Gasol and del Giorgio 2000). Samples were analyzed using a Becton Dickinson FACScan flow cytometer, with phosphate buffered saline (PBS) solution employed as a sheath fluid. For each sample, forward-angle light scatter (FALS), right-angle light scatter (RALS), green (SYBR-I) fluorescence, red fluorescence, and orange fluorescence were acquired. Acquisition was run until at least 50 to 100 μl of the sample was analyzed at an approximate rate of 40 μl mn^{-1}. To avoid coincidence of particles, it was ensured that the rate of analysis was kept below 1000 events sec^{-1} by diluting samples with 0.2 μm filtered seawater collected from the study site at time of sampling when necessary (Gasol and del Giorgio 2000). Data were analyzed and bacterial populations were identified and enumerated using WinMDI (Scripps Research Institute) and CYTOWIN (Vaulot 1989) flow cytometry analysis software.

5.5.5.2.2.2 Centimeter-Scale (1.2 cm) Phytoplankton Distribution

Sampling sites. Sampling was conducted on 9 December 2003, from a floating pontoon platform, in the above described Port River estuary, Adelaide, South Australia (34°49′S, 138°30′E).

Microscale sampling. Two-dimensional samples were collected using the above-mentioned spring-loaded 10×10 syringe array sampler, set up to simultaneously collect 100 samples of 200 μl each. During sample collection, the array sampler was oriented vertically, 10 cm below the water surface with the syringe inlets facing upstream. Sampling consisted of four sets of 100 samples, collected in succession, at a time interval of 10 minutes, and referred to as P1, P2, P3, and P4 hereafter. At the completion of sampling, syringe contents were subsampled (150 μl), transferred to cryovials with 2% final concentration paraformaldehyde, frozen in liquid nitrogen, and subsequently stored at −80°C.

Enumeration of phytoplankton. Total phytoplankton cell concentrations were estimated using a FACScan flow cytometer (Becton Dickinson) at the Flinders Medical Centre of South Australia. The nozzle diameter of the flow cytometer was 70 μm, which was taken to be the maximum size of cells enumerated. Samples were quick-thawed and analyzed at a rate of approximately 20 μl min^{-1}, employing PBS as sheath fluid. For each sample, natural orange fluorescence (from phycoerythrin) and red fluorescence (from chlorophyll), together with FALS and RALS parameters, were recorded on three decade logarithmic scales, sorted in list mode, and analyzed using CYTOWIN custom-designed software (Vaulot 1989). All parameters were normalized to a known concentration of 1 μm fluorescent marker beads (Molecular Probes), which were added to the sample prior to analysis at a final concentration of ca. 10^5 beads ml^{-1}.

5.5.5.2.2.3   Centimeter-Scale (6.6 cm) Microphytobenthos Distribution

Sampling sites. The two study sites, located on the French coast of the eastern English Channel, were chosen because of their intrinsic sharp differences in terms of hydrodynamic exposure, sediment nature, and biotic properties.

The first study site, an intertidal flat of sand in Wimereux (50°45′896 N, 1°36′364 E), is typical of the hydrodynamically sandy beach habitats that dominate the littoral zone along the French coast of the eastern English Channel. Measurements were performed on a flat area located in the upper intertidal zone, without sharp topographical features such as ripple marks, high pinnacles, or deep surge channels. The substrate was homogeneous, medium-size sand (200 to 250 μm, modal size), typical of the surrounding sandy habitat. Because of the substrate homogeneity and the weak biomass, productivity and production of phyto- and zoobenthic organisms, the microphytobenthos biomass distribution is *a priori* expected to be rather homogeneous (Seuront and Spilmont 2002). In addition, due to the highly dynamic environment, microphytobenthos is resuspended, and surface concentrations at low tide are low.

The second study site is located in the Bay of Somme, at Le Crotoy (50°13′524 N, 1°36′506 E), which is the second-largest estuarine system in France, after the Seine estuary, and the largest sandy-muddy (72 km²) intertidal area on the French coasts of the eastern English Channel. The sampling site was chosen in a topographically homogeneous area, where the substrate grain size typically varied between 125 and 250 μm (modal size), and is characterized by higher phyto- and zoobenthos biomass, activity, and spatial heterogeneity when compared to the Wimereux site (Seuront and Leterme 2006). Because of the weak hydrodynamic conditions, the microphytobenthos biomass is only weakly influenced by resuspension processes, and surface concentrations at low tide are high.

Microscale sampling. All measurements were performed at low tide, on October 9 and 10, 2003, at the Wimereux and Bay of Somme study sites, respectively. The two-dimensional spatial distribution of microphytobenthos was investigated for scales smaller than 1 m², which is usually the finest grain considered in both landscape ecology (He et al. 1994) and intertidal benthic ecology (MacIntyre et al. 1996; Blanchard and Bourget 1999). A rigid 1 m² aluminum quadrant was used, and 225 sediment samples were collected every 6.67 cm using 1.9 cm² plastic cores. The cores were pushed into the sediment down to a depth of 1 cm, where the most of the active cells are concentrated, carefully removed, mixed with 5 ml of methanol, and then stored in an insulated, cool box, brought back to the laboratory, and stored in the dark at –20°C.

Chlorophyll content analysis. Five ml of methanol were added directly to the sampled sediment sections, and the chlorophyll content was assayed in a Turner 450 fluorometer previously calibrated with a pure chlorophyll *a* solution (*Anacystis nidulans* extract, Sigma Chemicals) after an extraction time of 4 hours. Chlorophyll *a* concentrations in the sediment sections were then converted in terms of Chl.*a* m^{-2}, taking into account the surface (1.9 cm²) of the sampling units.

5.5.5.2.3   Results

5.5.5.2.3.1   Centimeter-Scale Bacterioplankton Distribution

Zipf analyses clearly show that the two-dimensional bacterial distributions (Figure 5.23) are not uniformly distributed (Figure 5.24). The Zipf plots estimated for bacterial populations sampled at Port Noarlunga and Port River exhibit a linear behavior with $\alpha = 0.05$ ($r^2 = 0.98$) for bacterial concentrations ranging from 3.1×10^5 to 3.7×10^5 cell ml^{-1} (Figure 5.24A), and with $\alpha = 0.03$ ($r^2 = 0.98$) for bacterial concentrations ranging from 37.7×10^5 to 39.8×10^5 cell ml^{-1} (Figure 5.24B). Respectively, 33% and 28% of the values observed in Port Noarlunga and Port River are included in the identified power-law behaviors. While the Port Noarlunga Zipf plot exhibits a power-law behavior up to the highest bacterial concentrations (Figure 5.24A), note that the three highest concentrations (that is, the three first ranks) observed in Port River were not included in the regression analysis (Figure 5.24B). As discussed on the basis of Zipf analyses of simulated data (Seuront and Mitchell 2008), such a local increasing trend is intrinsically caused by random fluctuations and should be regarded as a source of contamination of the observed power law resulting in a distribution dominated by a few hotspots rather than a breakpoint indicative of structural discontinuities (see Section 5.5.5.1). A proper normalization further shows that for concentrations lower than 0.37×10^6 cell ml^{-1} and 3.98×10^6 cell ml^{-1} respectively at Port Noarlunga and Port River, the Zipf plots are extremely similar (Figure 5.25A), continuously diverging from a power-law behavior as a step function toward the lowest concentrations (Figure 5.24A,B and Figure 5.25A). Because a step function might be indicative of the presence of structural discontinuities within the distributions, we

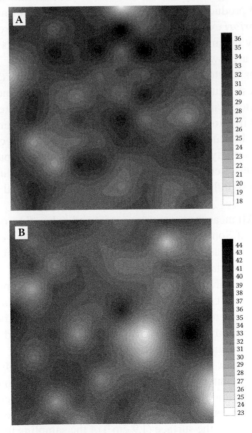

FIGURE 5.23 Two-dimensional distributions of the bacterioplankton abundance sampled in (A) Port Noarlunga ($\times10^4$ cell ml^{-1}) and (B) Port River ($\times10^5$ cell ml^{-1}) with a spatial resolution of 1.2 cm.

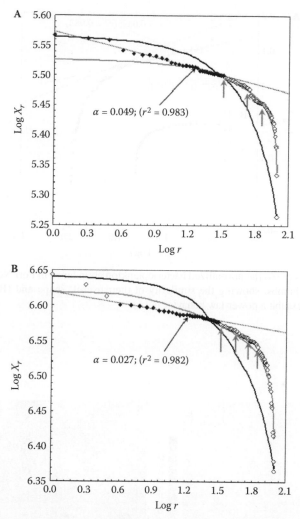

FIGURE 5.24 Zipf plots of the bacterioplankton abundance in Port Noarlunga (A) and Port River (B), shown in log-log plots. The black diamonds correspond to the range of abundance values exhibiting a power-law behavior, and used to estimate the exponent α as the slope of the linear fit maximizing the coefficient of determination and minimizing the total sum of the residuals in the regression (dotted lines). The continuous black and gray lines correspond to the Zipf plots obtained from 100 simulated uniform distributions with the same minimum and maximum values as the empirical ones and from 100 simulated normal distributions with the same mean and variance as the empirical ones, respectively. The gray arrows indicate the structural break points in the Zipf distributions.

performed separate Zipf analyses for the ranges of concentrations separated by identified break-points (gray arrows in Figure 5.24A,B). The resulting Zipf plots (Figure 5.25B), shown here for concentrations lower than 0.37×10^{-6} cell ml^{-1} at Port Noarlunga and 3.98×10^{-6} cell ml^{-1} at Port River, do not exhibit any power-law behavior, but instead produce a continuous roll-off from a horizontal line (that is, $\alpha \to 0$) to a vertical line (that is, $\alpha \to \infty$). This is representative of the fact that no value is more likely to be more common than any other one, a characteristic of uniformity.

5.5.5.2.3.2 Centimeter-Scale Phytoplankton Distribution

The 1.2-cm resolution, two-dimensional phytoplankton distributions obtained from the four "replicate" experiments conducted in this study (hereafter referred to as P1, P2, P3, and P4) exhibit specific

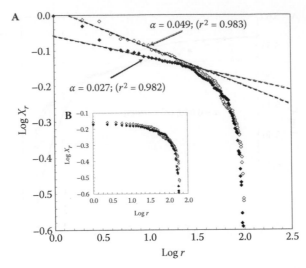

FIGURE 5.25 Zipf plots of (A) the normalized bacterioplankton abundance in Port Noarlunga (open rhombs) and Port River (black rhombs) showing the similarity of their overall shape and (B) the bacterioplankton abundance that did not exhibit a power-law behavior in (A).

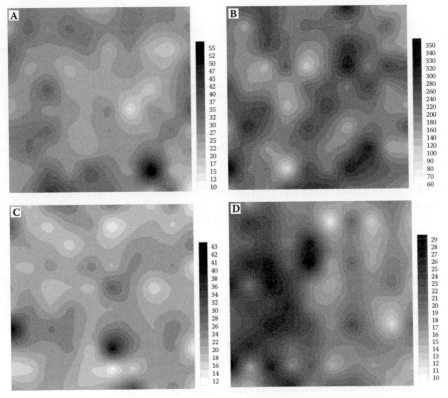

FIGURE 5.26 Replicate two-dimensional distributions of the phytoplankton abundance ($\times 10^3$ cell ml^{-1}) sampled in Port River with a spatial resolution of 1.2 cm.

features such as localized gradients, hotspots, and "coldspots" (Figure 5.26) that are not compatible with a homogeneous or a normal distribution. As previously observed for bacterioplankton distributions, the Zipf analysis of two-dimensional phytoplankton patterns shows that phytoplankton cells are not uniformly distributed ($p < 0.01$) and exhibit two different types of organization (Figure 5.27). P2 and P4 thus exhibit a linear behavior starting from the highest values, while distributions P1 and P3 present local increasing trends characterizing a distribution dominated by a few hotspots. Phytoplankton patterns thus exhibit a power-law behavior for cell concentrations ranging from 38.2 $\times 10^3$ to 31.8 $\times 10^3$ cell ml^{-1} with $\alpha = 0.12$ ($r^2 = 0.99$) for P1 (Figure 5.27A), 35.9 $\times 10^3$ to 24.3 $\times 10^3$ cell ml^{-1} with $\alpha = 0.13$ ($r^2 = 0.99$) for P2 (Figure 5.27B), 26.7 $\times 10^3$ to 22.8 $\times 10^3$ cell ml^{-1} with $\alpha = 0.09$ ($r^2 = 0.97$) for P3 (Figure 5.27C), and 29.8 $\times 10^3$ to 25.6 $\times 10^3$ cell ml^{-1} with $\alpha = 0.06$ ($r^2 = 0.97$) for P4 (Figure 5.27D). The percentage of values contributing to the power laws are 20% for P1, 23% for P2, 36% for P3, and 11% for P4. As stated above, separate analyses were performed for the ranges of concentrations separated by breakpoints (arrows in Figure 5.27). Except in the case of the distribution P4 that shows a power law for concentrations ranging from 21.4 $\times 10^3$ to 25.1 $\times 10^3$ cell ml^{-1} with $\alpha = 0.05$ ($r^2 = 0.97$), no power laws were observed. The corresponding Zipf plots instead exhibit a continuous roll-off from a horizontal line (that is, $\alpha \to 0$) to a vertical line (that is, $\alpha \to \infty$), representative of uniformity (Seuront and Mitchell 2008).

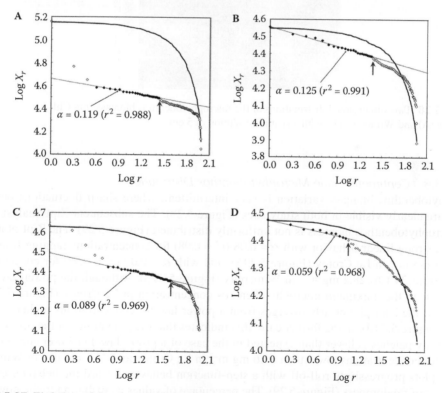

FIGURE 5.27 Zipf plots of the phytoplankton abundance sampled in Port River, shown in log-log plots. The black diamonds correspond to the range of abundance values exhibiting a power-law behavior, and used to estimate the exponent α as the slope of the linear fit maximizing the coefficient of determination and minimizing the total sum of the residuals in the regression (dotted lines). The continuous black lines correspond to the Zipf plots obtained from 100 simulated uniform distributions with the same minimum and maximum values as the empirical ones. The gray arrows indicate the structural breakpoints in the Zipf distributions.

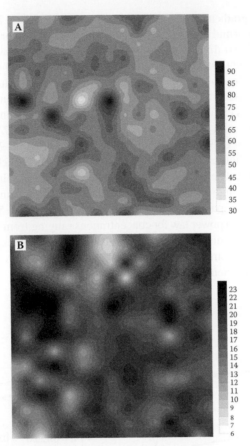

FIGURE 5.28 Two-dimensional distributions of the microphytobenthos biomass (mg Chl. *a* m^{-2}) sampled in Le Crotoy (A) and Wimereux (B) with a spatial resolution of 5 cm.

5.5.5.2.3.3 Centimeter-Scale Microphytobenthos Distribution

Microphytobenthos biomass variation is very intermittent, where sharp fluctuations occurring locally are clearly visible in both study sites (Figure 5.28). The subsequent Zipf analysis shows that microphytobenthos biomass is not uniformly distributed (Figure 5.29). The Zipf plots show instead a strong linear behavior with $\alpha = 0.08$ ($r^2 = 0.98$) for concentrations ranging from 82.60 to 113.98 mg m^{-2} in Le Crotoy (Figure 5.29A), and with $\alpha = 0.07$ ($r^2 = 0.98$) for concentrations ranging from 24.1 to 28.2 mg m^{-2} in Wimereux (Figure 5.29B). Although the power-law behavior expands to the maximum microphytobenthos concentration in Le Crotoy (Figure 5.29A), in Wimereux, the Zipf plot clearly diverges from a power law for concentrations higher than 28.2 mg m^{-2} (Figure 5.29B). In the former case, this indicates that the probability of the occurrence of high-density patches is lower than expected in the case of a power law. For lower concentrations (that is, for concentrations lower than 82.6 mg m^{-2} in Le Crotoy and 24.1 mg m^{-2} in Wimereux), the Zipf plots progressively roll-off with a step-function behavior toward the behavior expected in the case of randomness (Figure 5.29). The percentage of values contributing to the power laws are 26% and 29% in Le Crotoy and Wimereux, respectively.

In contrast to what has been observed for bacterioplankton distributions (Figure 5.24 and Figure 5.25) and phytoplankton distributions (Figure 5.27), the continuous roll-offs toward the lowest concentrations are clearly different for microphytobenthos (Figure 5.29). This could be indicative of differential driving processes competing with the pure power-law behavior observed for

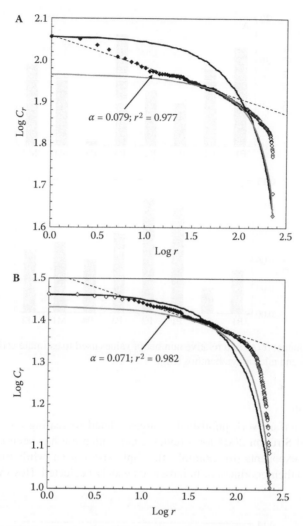

FIGURE 5.29 Zipf plots of the microphytobenthos biomass sampled in Le Crotoy (A) and Wimereux (B), shown in log-log plots. The black diamonds correspond to the range of abundance values exhibiting a power-law behavior, and used to estimate the exponent α as the slope of the linear fit maximizing the coefficient of determination and minimizing the total sum of the residuals in the regression (dotted lines). The continuous black and gray lines correspond to the Zipf plot obtained from 100 simulated uniform distributions with the same minimum and maximum values as the empirical ones and from 100 simulated normal distributions with the same mean and variance as the empirical ones, respectively.

higher concentrations. In particular, the more progressive roll-off observed in Le Crotoy when compared to Wimereux (Figure 5.29) could be related to the differences in grazing pressure observed in these two study sites or to the ease of moving in different grain size or organic matter contents. The grazing pressure is expected to be low in Wimereux where the meiobenthic biomass is negligible (Seuront and Spilmont 2002; Seuront and Leterme 2006). In contrast, considering the elevated abundance of the deposit-feeding amphipod *Corophium* sp., estimated as 800 ± 100 ind. m^{-2}, during the sampling experiment in Le Crotoy (Seuront, unpublished data; Figure 3.17A), the grazing pressure on microphytobenthos population should be high. This is fully congruent with the theoretical roll-off expected in the case of a grazing process driven by a Michaelis-Menten function (see Figure 5.17A).

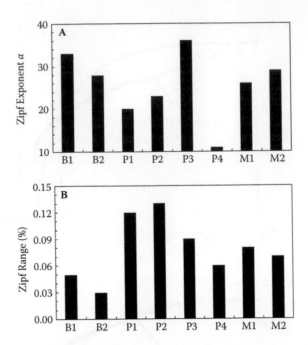

FIGURE 5.30 Zipf exponent α (A) and relative number of values used to estimate α (B) for bacterioplankton (B_i), phytoplankton (P_i) and microphytobenthos (M_i).

5.5.5.2.4 Discussion

In the extensive amount of work published in areas related to scaling and power laws (see, for example, Seuront and Strutton 2004 for a review), most attention has been given to the values of the so-called scaling exponents (for example, the Zipf exponent α), while the range of scales and values contributing to their goodness of fit have been widely neglected. However, because the value

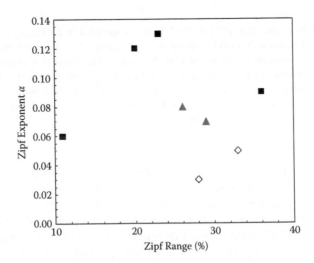

FIGURE 5.31 *Zipf topological mandala*: Zipf exponent α, shown as a function of the relative number of values used to estimate α, for bacterioplankton (gray triangles), phytoplankton (black squares), and microphytobenthos (open diamonds).

of any scaling exponent may intrinsically be scale dependent (Seuront et al. 1999) or density-dependent (Seuront and Mitchell 2008; Mitchell and Seuront 2008), their values are likely to change depending on the range of scales or values at which they are estimated. Consequently, without relevant information relative to their scaling range, there is no way to interpret and to compare Zipf exponents without leading to potentially spurious conclusions. Here, the Zipf exponents and the percentage of values that contribute to their goodness of fit (that is, their scaling range) range respectively from 0.03 to 0.13 and from 11% to 33% (Figure 5.30) and are significantly negatively correlated ($p < 0.01$). While Zipf analysis should be applied to a wider range of datasets, it is suggested that the combined knowledge of the Zipf exponents and their scaling range may represent the first step toward a seascape Zipf typology. The concept of "Zipf typological mandala" (Figure 5.31) is then introduced as a potential tool to classify the structure of marine and terrestrial ecosystems.

The fact that bacterioplankton, phytoplankton, and microphytobenthos abundance greater (or smaller) than a given threshold have a specific slope in a Zipf plot indicates that there is something unique about this set of values. Due to the implicit link between Pareto's and Zipf's laws (Box 5.2), then to self-organized criticality (see Section 6.3), the range of abundance values characterized by a Zipf power law can be considered as being in a critical state.

BOX 5.2   FROM ZIPF TO PARETO LAWS

Zipf and Pareto laws have often been described as separate power laws (Faloutsos et al. 1999), having been compared in a paper demonstrating that Zipf's law for the rank statistics is strictly equivalent to a power-law distribution of frequencies (Troll and Graben 1998). This comparison is unfortunately based on complicated mathematical analyses and does not provide any link between the Zipf and Pareto exponents α and ϕ. Such a comparison is nevertheless a crucial prerequisite step to reconciling and comparing results that could be obtained using one of these two methods. We demonstrate in a simple manner that Zipf and Pareto laws are strictly equivalent, and subsequently provide a one-to-one correspondence between the exponents α, ϕ, and μ.

Practically, Equation (5.22):

$$X_r \propto r^{-\alpha} \tag{5.B2.1}$$

shows that there are kr variables X_r (where k is a constant) greater than or equal to $r^{-\alpha}$. This leads us to rewrite the cumulative distribution function (CDF) of Pareto's law:

$$P[X_r \geq x] \propto x^{-\phi} \tag{5.B2.2}$$

as

$$P[X \geq kr^{-\alpha}] \propto r \tag{5.B2.3}$$

and

$$P[X > X_r] \propto X_r^{-1/\alpha} \tag{5.B2.4}$$

From Equations (5.B2.1), (5.B2.2), and (5.B2.4), the relationship between the exponents α, ϕ, and μ is given by

$$\begin{cases} \alpha = \dfrac{1}{\phi} \\[2ex] \mu = 1 + \dfrac{1}{\alpha} \end{cases} \tag{5.B2.5}$$

where μ comes from Equation (5.2):

$$P[X = x] \propto x^{-\mu} \tag{5.B2.6}$$

As a consequence, the Zipf and Pareto laws can be regarded as equivalent. Specifically, the x axis of the Zipf law is conceptually identical to the y axis of the Pareto law, Equation (5.B2.3) and Equation (5.B2.4). The use of one or the other distribution is simply a matter of convenience.

Alternatively, below and above these values (identified as structural breakpoints in the log-log Zipf plots), the system is in a subcritical state. It could be thought, in analogy with the sand pile, that a patch builds up from a cluster in which more and more cells are added, and eventually gets so that the patch gets bigger than some critical size, at which point it is split or spread. It is nevertheless still difficult to provide a clear phenomenological explanation of the processes involved in the generation of the observed critical distributions. However, it is likely that a simple combination of short-term and long-term cooperative and antagonistic processes (for example, predation, inter- and intraspecific competition, growth, death, reproduction), together with the intrinsically intermittent properties of the surrounding environment (Pascual et al. 1995; Seuront et al. 1996a, 1996b, 1999, 2002; Lovejoy et al. 2001), generates intermittent, critical patterns and dynamics. However, the resolution of this issue, numerically investigated elsewhere (Bak et al. 1989; Solé et al. 1992; Manrubia and Solé 1996), is beyond the scope of the present work.

5.5.5.3 Distance between Zipf's Laws

Based on the observations that although the Zipf laws of two books may look very similar, the same words may have different frequency (and rank) in both books, Havlin (1995) introduced a "distance" to characterize the differences between the Zipf structures of groups of words contained in books. This concept can be generalized to compare the difference between the rank-frequency distributions of two groups of discrete elements as described by Equation (5.17)—for example, DNA base pair sequences from two different organisms, species compositions of two samples or locations, or vocalizations of two organisms. Specifically, consider the species observed in two distinct environments.[*] Let $r_A(S_i)$ and $r_B(S_i)$ be the ranks of species S_i in environments A and B, respectively. The distance $d_{AB}(S_i)$ between the ranks of S_i in the two environments is:

$$d_{AB}(S_i) = [(r_A(S_i) - r_B(S_i))^2]^{1/2} \tag{5.36}$$

---

[*] Here, "environment" is considered in the general sense and could also be thought of as "sample" or "location."

The distance between the two environments is then defined as the mean square root distance between the ranks of all common species as:

$$d_{AB} = \left[\frac{1}{N} (r_A(G_i) - r_B(G_i))^2 \right]^{1/2} = \left[\frac{1}{N} \sum_{i=1}^{N} d_{AB}^2(G_i) \right]^{1/2} \tag{5.37}$$

where N is the total number of common species, G_i, that appear in both environments. Equation (5.37) has been used to estimate the distance between each pair of books from a set of nine books, written by Herbert G. Wells (*Dr. Moreau, The Time Machine*, and *The War of the Worlds*), Jules Verne (*20,000 Leagues under the Sea, Around the World in 80 Days*, and *From the Earth to the Moon*), and Mark Twain (*The Adventures of Huckleberry Finn, The Adventures of Tom Sawyer*, and *What Is a Man? And Other Essays*). The mean distance between books written by different authors ($d = 21.8 \pm 2.8$) significantly differs from the distance between books written by the same author ($d = 16.1 \pm 1.3$), showing that each author has his own hierarchy of words (Havlin 1995). In ecology, Equation (5.36) and Equation (5.37) have only been applied to the characterization of microscale spatial heterogeneity in flow cytometrically defined populations of heterotrophic bacteria (Seymour et al. 2004; see their Figure 5B).

5.5.6 BEYOND ZIPF'S LAW AND ENTROPY

This section explores how techniques initially developed for the analysis of natural languages (Ebeling and Nicolis 1992; Ebeling and Pöschel 1995) and essentially applied to the analysis of coding and noncoding DNA sequences (Mantegna et al. 1994, 1995; Stanley et al. 1999), the complexity of time series of electroencephalograms (Graben et al. 2000), and the neuronal activity of sensory receptors (Steuer et al. 2001) can be more generally applied to the symbolic dynamics of ecological processes. The basic idea of symbolic dynamics is to represent a continuous time process (that is, the behavior of an organism) by a series of sequences labeled by a symbol, each of which corresponds to a state of the system (Alekseev and Yakobson 1981). For instance, the behavior of the ferret (*Mustela putorius furo*) can be decomposed into a series of activities, each identified by a letter. Similarly, behavioral states can be identified in the swimming behavior of the copepod *Centropages hamatus*, that is, slow swimming, fast swimming, sinking, and breaking. Subsequent questions of critical ecological relevance are then to assess the complexity of the behavioral repertoire of the organism in relation to, for example, interaction with conspecifics, humans, and abiotic forcings such as turbulence and pollutants. These different issues will be illustrated on the basis of the symbolic dynamics of both ferret and zooplankton hereafter.

5.5.6.1 *n*-Tuple Zipf's Law

5.5.6.1.1 Theory

As stated above, Zipf behavior (Equation 5.16) has been universally observed in analyses of natural and technical languages. Note that Zipf analysis can be performed on texts of unknown languages, with the only limitation being the ability to recognize the basic semantic unit: the word. Conventional Zipf analysis has, however, been criticized since Zipf scaling can emerge in a purely random symbolic sequence if one character is defined as a "word" delimiter (Mandelbrot 1983; Li 1992). Hence, while the observation of a power-law behavior in a conventional Zipf analysis is necessary in natural and formal languages, it is not sufficient to prove the existence of non-Markovian correlations in the analysis of symbolic sequences. Although Zipf analysis can be performed on texts of unknown languages, a critical limitation is then to be able to identify the basic semantic unit.

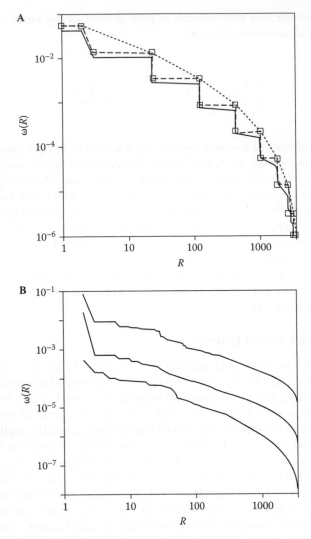

FIGURE 5.32 n-tuple Zipf plots of (A) a Markovian sequence with transition probability 0.80 (the steps are due to the fact that ω is determined by the number of consecutive digit pairs with both digits different, hence many words have the same frequency of occurrence); and (B) long-range correlation sequences generated from top to bottom by inverse Fourier transform, Lévy walk, and the expansion-modification system and exhibiting similar power-law behaviors with $\alpha = 0.80 \pm 0.02$ in the range $10 < R < 300$. (Modified from Czirók et al., 1995.)

In this context, n-tuple Zipf analysis has been introduced to analyze the complexity of symbolic sequences when the elementary semantic unit is not immediately recognizable. Symbolic sequences may not be composed of natural words but instead of strings of characters carrying information such as DNA. In the case of coding DNA, the words are the 64 3-tuples that code for the amino-acids, that is, AAA, AAT,..., GGG. In contrast, for noncoding DNA, the words are not known. In n-tuple analysis, the length n of a word is considered as a free parameter, and a "word" defined as an n-digit-long string of a sequence. Practically, an n-tuple analysis is carried out moving a window of length n along the sequence by shifting progressively by one character a window of length n. The number of occurrences of each n-tuple is then ranked ordered, and the relative occurrence $\omega(r)$

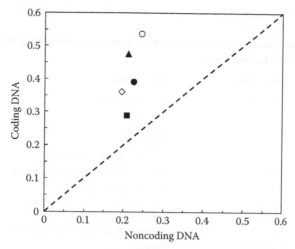

FIGURE 5.33 Exponents α returned by n-tuple Zipf analyses of noncoding and coding DNA of mammals (black square), the free-living nematode *Caenorhabditis elegans* roundworm living in temperate soils (open dot), other invertebrates (black triangle), yeast chromosome III (black dot), and eukaryotic viruses (open diamond). (Data from Mantegna et al., 1994, Table 1.)

plotted against rank r. For a sequence of length L, there are $L - n + 1$ words.* The set of possible words is finite; for example, for binary sequences (0 or 1) there are $N = 2^n$ different n-tuples and for the DNA alphabet of four characters (A,C,G,T), there are $N = 4^n$ different n-tuples. More generally, an alphabet with λ letters will be characterized by $N = \lambda^n$ different n-tuples. The n-tuple analysis of simulated Markovian process and long-range correlation respectively resulted in a stepwise decay and a power-law behavior of the log-log plots of $\omega(r)$ vs. r (Czirók et al. 1995; Figure 5.32). In contrast, if each character is considered as an independent variable, all possible n-tuples tend to have the same frequency ($1/N$), so the Zipf plot is horizontal, that is, $\alpha = 0$.

n-tuple Zipf analysis has previously been successfully applied to noncoding and coding DNA sequences (Mantegna et al. 1994, 1995), which exhibited significant differences in the exponents α_n (Figure 5.33). The consistently higher values of α_n found for noncoding DNA suggest that noncoding sequences bear more resemblance to a natural language than coding sequences. The applicability of n-tuple Zipf analysis to "nonnatural languages" (here, symbolic sequences) has been confirmed through the analysis of a collection of articles taken from an encyclopedia comprising 500,000 letters that returned $\alpha_n = 0.57$. In contrast, a conventional analysis of the same text using the actual words led to $\alpha = 0.85$.

5.5.6.1.2 Case Study: On the Behavioral Activities of the Ferret (Mustela Putorius Furo)

The ferret (*Mustela putorius furo*) (Figure 5.34A) has become an increasingly popular pet animal, yet little is still known about their behavior. Two aspects of ferret behavior are considered here: noninteractive and interactive behaviors. Noninteractive behavior corresponds to the behavioral activities conducted by the animal in its enclosure without any external disturbances. In contrast, interactive behavior corresponds to the activities conducted by the animal outside its enclosure when stimulated by its owner with a familiar toy, that is, play behavior. Noninteractive and interactive behaviors were categorized in, respectively, 12 and 7 activities, and each of them was associated with a symbol (here, a letter; Table 5.2). Noninteractive behavioral activities (Table 5.2) were observed in a four-level enclosure, dimension 80 cm (length) × 50 cm (width) × 100 cm (height), and behavior was recorded with a digital camera (DV Sony DCR-PC120E) at a rate of 25 frame s⁻¹.

---

* Note that if the window is moved n character at a time, this results in different "reading frames," each of which contains words. In coding DNA, $n = 3$, and there are three distinct reading frames.

TABLE 5.2
Ethogram Used in the Assessment of Noninteractive and Interactive Behavioral Activities of the Ferret (*Mustela Putorius Furo*)

Main Pattern	Subpattern	Description	Symbol
Noninteractive behavior	Exploring	Walking around	A
		Sniffing	B
		Scratching	C
		Climbing	D
	Defecating	Animal defecates or urinates	E
	Rubbing	Animals rubs the ground or wall with face or neck	F
	Fur shaking	Animals shake water off of fur	G
	Grooming	Animal licks or rubs its pelt with forepaws and tongue	H
	Coughing	Animal coughs/sneezes	I
	Alerting	Animal actively observes surrounding	J
	Drinking	Drinking	K
	Eating	Eating	L
Interactive behavior (play behavior)	Locomotory play	Facing still	A
		Galloping forward (bouncing jerky gait)	B
		Galloping away	C
		Jumping on toy	D
	Rough and tumble play	Rolling over	E
		Pushing with paws	F
		Warding off with paw	G

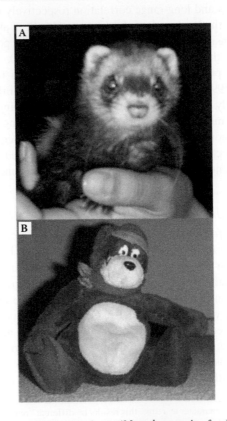

FIGURE 5.34 (A) Sasha, the 1-year-old female ferret (*Mustela putorius furo*) and (B) her favorite toy.

Interactive behavioral activities (Table 5.2) were stimulated by presenting a familiar toy to the test animal (Figure 5.34B) and recording the resulting behavioral responses with a digital camera (DV Sony DCR-PC120E) at a rate of 25 frame s^{-1} in a 2 m × 2 m enclosure. The n-tuple Zipf analyses were conducted on three data sets containing 5245, 6578, and 7027 successive symbols for noninteractive behavioral activities and on five data sets containing 7854, 8241, and 8759 successive symbols for interactive behavioral activities.

n-tuple Zipf analyses were conducted for all values of n in the range 3 to 6. Considering alphabets with 12 and 7 symbols of noninteractive and interactive behavioral activities, they will be characterized by $N = 12^n$ ($1728 \leq N \leq 2985984$) and $N = 7^n$ ($343 \leq N \leq 117649$) different words. The n-tuple Zipf analysis resulted in clear power-law behaviors for both behavioral activities (Figure 5.35). The resulting exponents α were not significantly different for different n-tuples for both behavioral activities, that is, for values of n from 3 to 6. The exponents α_n ($\alpha_n = 0.64 \pm 0.03$) estimated for interactive activities were, however, significantly higher than those returned by the n-tuple Zipf analysis of noninteractive behavioral activities, $\alpha_n = 0.47 \pm 0.02$. Note than in both cases, the exponent α_n is significantly larger than the value $\alpha_n = 0$ expected for a control sequence of random numbers.

5.5.6.2 *n*-Gram Entropy and *n*-Gram Redundancy

5.5.6.2.1 *n-Gram Entropy*

Most past research has focused almost exclusively on the use of Shannon's measure for information (Shannon 1948; Shannon and Weaver 1949). Shannon's entropies, however, examine the information content at increasingly complex levels of signaling organization. As an example from human speech, information content can be evaluated at the phonemic or letter level, the word level, and various levels of sentence organization. Each level can be represented by a series of increasing orders (for example, zero, first, second, and so on) of entropy. Entropy is defined here as a measure of the informational degree of organization and is not directly related to the thermodynamic property used in physics; see Box 5.1. Specifically, the Shannon n-gram entropies are defined as (Shannon 1948):

$$H_n = -\sum_{i=1}^{\lambda^n} p(E_i \ldots E_n) \log_2 p(E_i \ldots E_n) \tag{5.38}$$

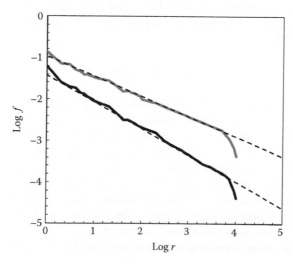

FIGURE 5.35 n-tuple Zipf analysis of noninteractive (gray) and interactive (black) behavioral activities in the ferret (*Mustela putorius furo*). For $n = 6$, both behavioral sequences exhibit a power-law behavior over nearly four decades, with $\alpha_6 = 0.64 \pm 0.03$ and $\alpha_6 = 0.47 \pm 0.02$ for interactive and noninteractive behaviors, respectively. No significant differences were found between α_n values for n in the range 3 to 6.

where $p(E_i \ldots E_n)$ is the probability of the n-tuple and $N = \lambda^n$ is the number of different n-tuples, with λ the number of letters (or symbols) in the alphabet. The zero-order entropy H_0:

$$H_0 = \log_2 N \tag{5.39}$$

is the number of bits of information required to represent a particular sample of different events N (for example, letters, words, phonemes, musical notes, or behavioral activities). The first-order entropy takes into account the probability of occurrence of each event as:

$$H_1 = -\sum_{i=1}^{\lambda} p(E_i) \log_2 p(E_i) \tag{5.40}$$

where $p(E_i)$ is the probability of event E_i. Note that Equation (5.40) is equivalent to the Shannon information index widely used in ecology to describe species diversity (Rosenzweig 1995); see also Equation (5.20). The second-order entropy introduces conditional probabilities into the structure of the stream of events (letters, words, phonemes, musical notes, behavioral activities):

$$H_2 = -\sum_{i=1,j=1}^{\lambda^2} p(E_i E_j) \log_2 p(E_i E_j) \tag{5.41}$$

where $p(E_i E_j)$ is the probability of event E_j given that event E_i has occurred. Similarly, the third-order entropy of an event E_k includes the conditional probability that both events E_i and E_j have occurred:

$$H_3 = -\sum_{i=1,j=1,k=1}^{\lambda^3} p(E_i E_j E_k) \log_2 p(E_i E_j E_k) \tag{5.42}$$

As n-gram entropies, H_n is a measure of uncertainty and gives the average amount of information contained in a word of length n; it comes from Equation (5.38) that the conditional entropies h_n (Ebeling and Nicolis 1991, 1992):

$$h_n = H_{n+1} - H_n \tag{5.43}$$

where $h_0 = H_1$ gives the average amount of information required to predict the $(n + 1)^{\text{th}}$ symbol when the preceding n symbols are known (Ebeling 1997). Note that h_n is decreasing as $H_{n+1} \leq H_n$. The slope of the conditional entropy h_n plotted against n provides information on the complexity of languages or, more generally, symbolic sequences. In the presence of long-range correlations, the entropies h_n decrease, hence predictability increases. In contrast, a truly random sequential system would show a slope of zero. As high-order conditional entropies drop from one order to the next, less statistical information and more organizational complexity are present. The sharper the decrease in h_n, the more sequential organization in the system. It is then suggested that "entropic slope" can provide a measure of organizational complexity that can be used to evaluate the importance of sequential order in the communication systems of different species.

5.5.6.2.2   n-Gram Redundancy

Redundancy is another common feature of languages. Letters or even entire words can be omitted or changed without the text becoming indecipherable; a text with typing errors does not become unintelligible. The concept of redundancy, which can be estimated from entropy, then also stems back to the seminal work of Shannon (1948). The n-gram redundancy present in a text or a symbolic sequence is defined as (Almagor 1985; Mantegna et al. 1994):

$$R_n = 1 - \frac{H_n}{kn} \qquad (5.44)$$

where $k = \log_2 \lambda$, with λ being the number of letters (or symbols) in the alphabet. The redundancy is a manifestation of the flexibility of the underlying language. Note that a sequence of random numbers will return values of $R_n = 0$ (Mantegna et al. 1994).

5.5.6.2.3 Case Study: Zooplankton Behavioral Response to Hydrocarbon Contamination

5.5.6.2.3.1 Ecological Framework
Massive crude oil spills such as *Torey Canyon* (1967), *Amoco Cadiz* (1978), *Ixtoc-I* (1979–1980), *Exxon Valdez* (1989), *Sea Empress* (1996), *Erika* (1999), and *Prestige* (2002) are a major source of polycyclic aromatic hydrocarbons (PAHs) in estuarine and coastal waters (Cachot et al. 2006). However, leakage from ships, petroleum transport, refining, and intentioned spills are more pernicious, but equally important, sources of PAHs in the ocean (Fernandes et al. 1997; Cachot et al. 2006), especially in coastal and shelf waters (Doval et al. 2006).

The effects of PAHs contamination on marine planktonic organisms have been studied extensively in the laboratory and in the field, and a variety of reactive changes have been found in relation to incidental oil spills for a range of plankton species, manifested as alterations of biomass, abundance, and ecophysiological effects. To date, the few studies regarding sublethal effects of hydrocarbons on copepods show, however, a very variable scenario depending on the chemicals used, their concentrations, and time of exposure, including anomalous metabolism (Samain et al. 1981), decreased or inhibited feeding (Barata et al. 2002), increased mortality (Gajbhiye et al. 1995), reduction in egg production (Ott et al. 1978), hatching rates (Cowles and Remillard 1983), and clutch size (Barata et al. 2005).

At low concentration and for short time exposure, hydrocarbons did not have any significant effect on feeding and egg production (Calbet et al. 2007). Decreases in egg production are observed, however, after long exposures to low hydrocarbon concentrations (Ott et al. 1978), indicating detrimental cumulative effects unidentifiable under short-term incubations. PAHs concentrations can reach dramatic concentrations. However, the "natural" concentrations of PAHs typically range between 1 and 100 µg l^{-1} (Doval et al. 2006). The ability to assess rapidly any increase in the background concentration of PAHs related to, for example, incidental, localized oil spills is then critical to anticipate their pernicious cumulative effects on copepod biology and ecology.

Since swimming and feeding are intertwined in most copepod species, any disruption of copepod swimming is predicted to have detrimental consequences to copepod biology and ecology. Despite the few attempts to use the swimming behavior of the freshwater cladoceran *Daphnia* sp. as an indicator of exposure to toxic chemicals (Piao et al. 2000; Shimizu et al. 2002), similar information for marine invertebrates is still very limited (Burlinson and Lawrence 2006). In particular, no attempts have been made to assess the effects of hydrocarbons on copepod behavior, despite an impressive body of literature devoted to their behavioral ecology; see, for example, Kiørboe (2008).

In this context, the potential for n-gram entropy and redundancy to detect behavioral changes relates to exposure to "natural" and dramatic concentrations of naphthalene, the most abundant hydrocarbon dissolved in oil-contaminated waters.

5.5.6.2.3.2 Toxicity Assay
The polycyclic aromatic hydrocarbon tested was naphthalene, as it has been widely used in toxicological assays involving copepods (Calbet et al. 2007) and is one of the most abundant hydrocarbons dissolved in oil-contaminated waters. Naphthalene (96% purity, Sigma-Aldrich, St. Louis) stock solutions

were prepared using acetone as a carrier (HPLC grade, 0.5 ml l⁻¹). Naphthalene stock solutions were transferred into the behavioral container and diluted with GF/C filtered (porosity 0.45 μm) *in situ* seawater at 50, 100, 500, 1,000, 2,500, 5,000, and 10,000 μg l⁻¹ to assess the effect of natural (1 to 100 μg l⁻¹) and extreme (up to 10,980 μg l⁻¹) PAHs concentrations encountered in the ocean. In addition to the toxicity experiments, two controls were considered with (0.5 ml l⁻¹ in GF/C filtered *in situ* seawater) and without (GF/C filtered *in situ* seawater) acetone to assess the potential effect of acetone on *C. hamatus* swimming behavior (Box 5.3).

BOX 5.3 NAPHTHALENE CONTAMINATION AND SYMBOLIC SEQUENCES OF *CENTROPAGES HAMATUS* SWIMMING BEHAVIOR

Copepods were collected in the coastal waters of the eastern English Channel using a WP2 net (200-μm mesh size) with closed end horizontally towed between 0 and 5 m. Specimens were gently diluted in 30-liter isotherm tanks using *in situ* seawater (*S* = 34 PSU) and transported to the laboratory, where *Centropages hamatus* adult females (Figure 5.B3.1) were sorted by pipette under a dissecting microscope and acclimated for 12 hours in a 2-liter beaker containing fresh *in situ* seawater vacuum filtered through Whatman GF/C glass-fiber filters (porosity 0.45 μm) prior to the behavioral experiments. All subsequent handling of animals was done at 18°C in a temperature-controlled room.

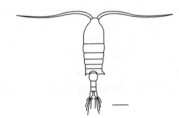

FIGURE 5.B3.1 Calanoid copepod *Centropages hamatus*; scale bar: 0.2 mm.

Behavioral experiments were conducted in a temperature controlled room (18°C), in aerated (100% air saturation) fresh *in situ* seawater (*S* = 34 PSU) in the dark and at night to avoid any behavioral artifact related to endogenous swimming rhythms. Prior to each experiment, 10 females (1.31 ± 0.05 mm cephalotorax length, mean ± SD) were randomly selected from the female stock, transferred into the experimental container (a cubical glass container, 15 × 15 × 15 cm) filled up with the test solutions, and allowed to acclimatize for 10 minutes (Seuront 2006). The free-swimming behavior of *C. hamatus* was recorded in three dimensions at a rate 25 frame s⁻¹ using two orthogonal, synchronized infrared digital cameras (DV Sony DCR-PC120E) facing the experimental container. To avoid any bias related to phototropism, the only light source was provided by six arrays of 72 infrared light emitting diodes (LEDs). For each test condition, 10 individual females were recorded swimming for 20 minutes, after which valid video clips were identified for analysis. Valid video clips consisted of pathways in which the animals were swimming freely, at least two body lengths away from any chamber's walls or the surface of the water.

Free-swimming *C. hamatus* patterns typically exhibit slow- and fast-swimming bouts separated by breaking events during which they remain motionless or sink with a horizontal orientation with tail pointed upward. Fast-swimming bouts were identified as movements longer than 1 body length within 0.08 second. These four behavioral sequences were associated to four letters, that is, S (slow swimming), F (fast swimming), M (motionless), and D (sinking) for each video frame.

A total of 270,000 swimming sequences were analyzed, corresponding to ca. 30,000 sequences for each experimental conditions, that is, 7 naphthalene solutions and 2 controls with (0.5 ml l^{-1} in GF/C filtered *in situ* seawater) and without (GF/C filtered *in situ* seawater) acetone to assess the potential effect of acetone on *C. hamatus* swimming behavior.

5.5.6.2.3.3 Results

Centropages hamatus adult females consistently swam in helical loops, and no differences were perceptible between uncontaminated control seawater (Figure 5.36A) and naphthalene-contaminated seawater (Figure 5.36B). More specifically, the slow-swimming speed (6.32 ± 0.04 mm s^{-1}), fast-swimming speed (27.3 ± 0.11 mm s^{-1}), sinking speed (1.52 ± 0.05 mm s^{-1}), and time spent motionless (1.52 ± 0.05 mm s^{-1}) were not significantly different for the control experiments with and without acetone ($p > 0.05$). No significant differences in swimming and sinking speeds were found between *C. hamatus* considered in uncontaminated and naphthalene-contaminated seawater ($p > 0.05$). In contrast, the time spent swimming exhibits a significant linear increase ($p < 0.01$) with naphthalene concentration from 66% to 81%. This indicates an increase in swimming activity under conditions of naphthalene contamination and is consistent with previous observations

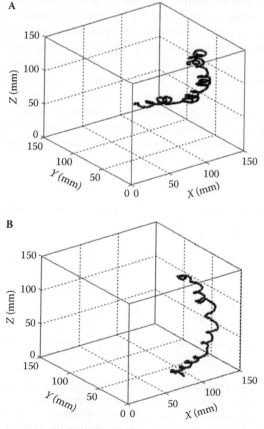

FIGURE 5.36 Swimming behavior of the calanoid copepod *Centropages hamatus* in uncontaminated seawater (A) and in naphthalene contaminated (1000 μg l^{-1}) seawater (B).

showing an increase in the foraging activity of *C. hamatus* exposed to microscale turbulence, a ubiquitous natural stressor of the marine environment. Note, however, that no increase in the occurrence of fast-swimming events, considered as an escape behavior, was recorded.

The n-gram entropies H_n (Equation 5.38) were used to estimate the conditional entropies h_n (Equation 5.43) and n-gram redundancies R_n (Equation 5.44) for n in the range 1 to 5 (Figure 5.37). The two controls with and without acetone did not exhibit any significant differences in the resulting behavioral sequence organization. The conditional entropies h_n (Figure 5.37A) consistently decrease with increasing sequence length, indicating correlations and long memory in *C. hamatus* behavioral activities. However, this decrease is much sharper for uncontaminated than contaminated seawater (Figure 5.37A). This shows that the complexity of the behavioral sequences of *C. hamatus* decreases with increasing naphthalene concentrations, thus the predictability of the related symbolic sequences decreases. In addition, the slopes of the function h_n vs. n found for naphthalene-contaminated seawater were not significantly different from the zero-slope expected for a truly random sequential system (that is, Markovian sequences) for n ranging from 1 to 4. Similarly, the n-gram redundancies R_n were consistently higher for uncontaminated seawater than for naphthalene treatments (Figure 5.37B). The swimming sequences of the copepod *C. hamatus* observed for uncontaminated seawater then possess a larger amount of structure than those observed for naphthalene-contaminated water, which are closer to random.

5.5.6.2.3.4 Discussion

The previous results, derived from methods originally designed to investigate the complexity of natural and artificial languages, are consistent with the existence of a structured "behavioral language" present in the sequential swimming behavior of the copepod *C. hamatus*. Both the existence of long-range correlation in this sequential behavior in uncontaminated seawater and disruption of the complexity of this language under conditions of naphthalene contamination are consistent with previous results showing that behavioral time series, though they often appear erratic, reveal $1/f$-like spectra (Quenette and Desportes 1992; Alados et al. 1996). Long-range correlation in biological systems is thus adaptive because it serves as an organizing principle for highly complex, nonlinear processes, and it avoids restricting the functional response of an organism to highly periodic behavior (Buldyrev et al. 1994). For example, $1/f$ temporal fluctuations are found in the heart rate of healthy individuals (Meesmann et al. 1993), respiratory intervals in animals (Kawahara et al. 1989), and neuronal discharges during sleep (Yamamoto et al. 1986). The time series of interbeat intervals in

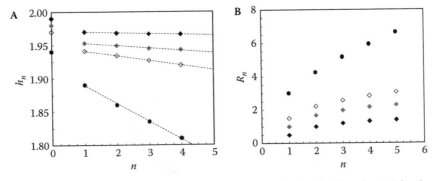

FIGURE 5.37 The conditional entropies h_n (A) and n-gram redundancies R_n (B) estimated for the sequential behavior of the copepod *Centropages hamatus* in uncontaminated seawater (black dots), and in seawater contaminated with increasing naphthalene concentrations, that is, 50 (open diamonds), 1000 (gray diamonds), and 10,000 μg l^{-1} (black diamonds). The dashed lines in (A) are the best linear fits of h_n vs. n.

healthy subjects have more complex fluctuations than patients with severe cardiac disease (Stanley et al. 1992; Buldyrev et al. 1994). Long-range correlations have been also observed in the stride interval of human gait (Hausdorff et al. 1995, 1997). This present analysis, conducted on marine invertebrates, then generalizes previous studies conducted on a range of vertebrates (from fish to primates) and shows that fractal dimension can detect impairments in the behavior sequences of individuals under stress, indicating its value in early stress assessment (Alados et al. 1996; Alludes and Huffman 2000).

6 Fractal-Related Concepts: Some Clarifications

6.1 FRACTALS AND DETERMINISTIC CHAOS

6.1.1 CHAOS THEORY

Chaos theory has been epitomized by the "butterfly effect" detailed by Lorenz (1963), and subsequently extensively investigated from popular descriptions (Gleick 1987; Kauffman 1995; Smith 2007) to more rigorous mathematical underpinning and specific algorithms (Baker and Gollub 1990; Peitgen et al. 1992; Turcotte 1992; Çambel 1993; Ott 1993, 2002; Hilborn 1994; Strogatz 1994; Kaplan and Glass 1995; Alligood et al. 1996; Williams 1997; Sprott 2003). In his attempt to simulate numerically a global weather system, Lorenz discovered that minute changes in initial conditions steered subsequent simulations toward radically different final states. This dependence on initial conditions is generally exhibited by systems containing multiple elements in nonlinear interactions, particularly when the system is forced or dissipative. A system is said to be forced when its internal dynamics are driven by externally supplied energy (for example, solar energy driving the global weather system). A system is considered dissipative when useful energy* is converted into a less useful form, most prominently through friction. In aquatic ecology, examples of dissipative systems are numerous, if not the rule. For instance, in the well-known turbulent energy cascade, the kinetic energy generated at a large scale by processes such as wind and tide are transferred without dissipation through the inertial subrange (that is, a hierarchy of eddies of decreasing size) to the viscous Kolmogorov scale where it is dissipated into heat. On the other hand, the solar energy first converted into chemical energy is subsequently transferred from one trophic level to the other one with considerable losses, responsible for the low energetic efficiency of both benthic and pelagic food chains.

Sensitive dependence on initial conditions is not only observed in complex systems but also in the simplest logistic equation model in population biology (May 1976). This equation describes the size of a self-reproducing population, P, at time $t + 1$ as a nonlinear function of the population at time t:

$$P_{t+1} = P_t(a - bP_t)$$ (6.1)

Considering the normalized population size $x_t = bN_t/a$, Equation (6.1) rewrites as:

$$x_{t+1} = ax_t(1 - x_t)$$ (6.2)

or equivalently as:

$$x_{t+1} = ax_t - ax_t^2$$ (6.3)

---

* Energy able to perform work.

where the driving parameter a is positive, thus ax_t represents a linear growth.* When x_t is small (that is, $x_t \rightarrow 0$), the nonlinear term ax_t^2 can be neglected, thus $x_{t+1} \approx ax_t$, and the growth of the population is indeed linear. It then comes that when $a > 1$, the population should increase. However, because of the recursive nature of Equation (6.3), ax_t^2 becomes significant as x_t increases; first an inflection point occurs and then a steady state (Figure 6.1A) when the two terms oppose one another because of their opposite signs. In contrast, when $a < 1$, the population will always decrease and eventually becomes extinct regardless of the size of the initial population (Figure 6.1B). More specifically, when $a > 1$ the value of parameter a determines whether a population stabilizes at a constant size, oscillates between a limited sequence of sizes, or behaves erratically in an unpredictable pattern (Figure 6.2).

The dichotomy between those different states can be established using bifurcation diagrams (Figure 6.3). A bifurcation diagram is a plot of the normalized population size x_t as a function of the values of the driving parameter a. This is illustrated in Figure 6.3 starting with an initial value of x_0 (here, $x_0 = 0.05$) and the driving parameter a in the range $2.95 < a < 4.00$. For low values of a (that is, $a < 3$), x_t (as t goes to infinity) eventually converges to a single number that represents the population of the species (Figure 6.3A). Now, when $a = 3$, x_t no longer converges but oscillates

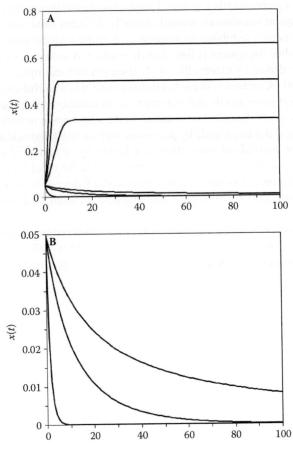

FIGURE 6.1 The logistic equation, $x_{t+1} = ax_t(1 - x_t)$, shown for values of ranging from (A) $a = 0.50, 0.95, 1.00,$ 1.50, 2.00, and 3.90 from bottom to top, and (B) from $a = 0.50, 0.95,$ and 1.00 from bottom to top. The initial value x_0 was set at $x_0 = 0.05$.

---

* The growth will be steady if a is constant, but we will see that a nonconstant a may lead to counterintuitive results.

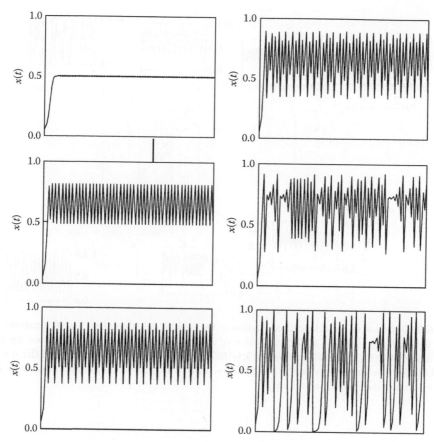

FIGURE 6.2 100 realizations of the logistic equation, $x_{t+1} = ax_t(1 - x_t)$, for values of a ranging from $a = 2.000000$, 3.261205, and 3.511687 from top to bottom (left), and from $a = 3.57494$, 3.683735, and 4.000000 from top to bottom (right). The initial value x_0 was set at $x_0 = 0.05$.

between two values (Figure 6.3B). This characteristic change in behavior is called a *bifurcation*. Turn up the driving parameter even further and x_t oscillates between four values (Figure 6.3C), and further goes through bifurcations of period 8, then 16, and then chaos (Figure 6.3E). When the value of the driving parameter a equals 3.57, P_t neither converges nor oscillates; its value becomes completely random (Figure 6.3D). For values of r larger than 3.57, the behavior is largely chaotic (Figure 6.3E,G). However, in the range $3.00 < a < 4.00$, there is a particular value of a where the sequence again oscillates with period of three ($a = 3.828427$) (Figure 6.3F). The bifurcations then begin again with period 6, 12, 24, and then back to chaos (Figure 6.3G). Within the chaotic regime are evident bands or windows containing few points that can result in attractors (Rasband 1990; see Section 6.1.2). Ultimately, for $a \geq 4$, values of $x_t > 1$ can be generated, and therefore $x_{t+1} < 0$; that is, the equation becomes unphysical (unbiological) beyond this point.

As stressed above, a bifurcation diagram contains regions with singular equilibrium populations for low values of a, bifurcating into an oscillating population as the parameter a increases and in turn deteriorating into a chaotic pattern as a reaches a critical value (Strogatz 1994). Within the chaotic region, however, smaller areas of stable periodicity are discernible (associated with certain minute ranges of the value of a), and these stable areas appear over and over on every possible scale of examination. Note that the repetition of this pattern across different scales (Figure 6.4) presents the intrinsic properties of self-similarity reviewed above, and has thus widely been identified as a fractal (Turcotte 1992; Strogatz 1994).

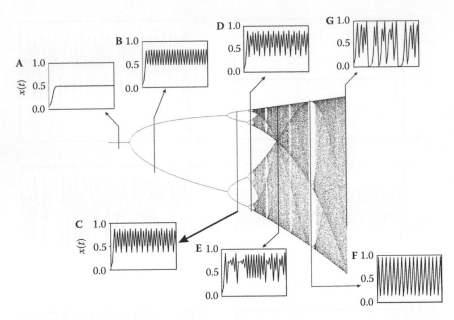

FIGURE 6.3 Depiction of the bifurcation diagram of the logistic equation after 10,000 iterations, x_t, of the recursive equation $x_{t+1} = ax_t(1 - x_t)$ for each value of a in relation to time series x_t obtained for increasing values of a; $a = 2.00$ (A), $a = 3.261205$ (B), $a = 3.511687$ (C), $a = 3.57494$ (D), $a = 3.683735$ (E), $a = 3.828427$ (F), $a = 4.00$ (G). The initiating value x_0 is set to $x_0 = 0.05$. (The bifurcation diagram was created with the freeware Fractint of the Stone Soup Group.)

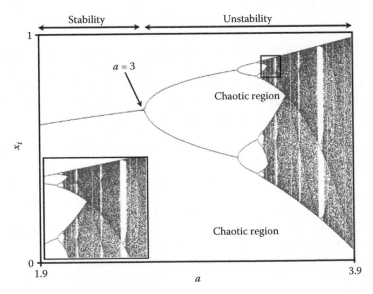

FIGURE 6.4 Depiction of the bifurcation diagram of the logistic equation after 10,000 iterations, x_t, of the recursive equation $x_{t+1} = ax_t(1 - x_t)$ for each value of a. The initiating value x_0 is set to $x_0 = 0.05$. Upon magnification of a chaotic region (box) self-similar areas appear (inset). (The bifurcation diagram was created with the freeware Fractint of the Stone Soup Group.)

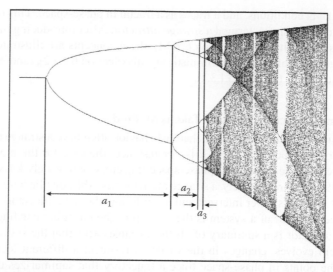

FIGURE 6.5 First set of three successive values of the driving parameter a of the logistic equation bifurcation diagram used to estimate the Feigenbaum number δ. (The bifurcation diagram was created with the freeware Fractint of the Stone Soup Group.)

6.1.2 FEIGENBAUM UNIVERSAL NUMBERS

As stressed above, the intervals between bifurcations decrease as a increases. Feigenbaum (1979) recognized a universal pattern in the bifurcation events of a range of nonlinear dynamical equations and characterized by the so-called Feigenbaum bifurcation constant δ as (Figure 6.5):

$$\delta = \frac{a_n - a_{n-1}}{a_{n+1} - a_n} \tag{6.4}$$

For the logistic map (Figure 6.3), the Feigenbaum constant δ has been estimated by Briggs (1991) to 84 decimal places and Briggs (1997) to 576 decimal places.

6.1.3 ATTRACTORS

The evolution of a dynamic system through time can be observed by tracing the instantaneous values of n-state variables in n-dimensional space, the phase-space. A system in a steady state will appear as a point in phase-space (that is, a stable equilibrium), while a stable oscillator traces a closed loop through phase-space (that is, a stable limit cycle). The point and the closed loop are both attractors for their respective systems; the systems develop toward those states regardless of a range of boundary conditions and perturbations. The stable equilibrium and the stable limit cycles are classical examples of attractors. A forced and damped oscillator (for example, a magnetically driven pendulum with friction) may be represented in a 2D phase-space by its instantaneous angular deflection and speed (the two-state variables). An overdamped oscillator will spiral toward a point attractor as it grinds to a halt, while under a range of forcing–damping ratios, the oscillating system will trace out a closed loop in phase-space. However, this nonlinear system also exhibits chaotic behavior under the right conditions, and it traces as a fractal in phase-space. This fractal is an attractor for the system in phase-space, termed a *strange attractor*. After introducing the Packard-Takens method used to visualize attractors (Section 6.1.3.1), these concepts are illustrated on the basis of the phase-space signatures of purely deterministic signals (Section 6.1.3.2), random signals (Section 6.1.3.3), and chaotic signals (Section 6.1.3.4).

6.1.3.1 Visualizing Attractors: Packard-Takens Method

Dissipative dynamical systems that exhibit chaotic behavior often have a strange attractor in phase-space (Grassberger and Procaccia 1983). It is, for instance, the case for the movements of atmospheric flows, which produce a specific phase-space trajectory now widely known as the Lorentz attractor (Lorenz 1963). More precisely, a strange attractor has orbits that lie within a defined region of phase-space but the orbits never intersect and never follow the same trajectory twice.

The phase-space attractor of a system is then a map of the changing conditions in the system; each point on the attractor is a summary of all the variables affecting the system at a moment in time. As the system evolves, changes in the variables result in a different location of the point in phase-space. The points in phase-space trace a trajectory that summarizes the changes of the system. Three-dimensional phase-space diagrams of the attractor describing the time series were produced using the "time delay" method (Packard et al. 1980; Takens 1981). In practice, a one-dimensional time series, and thus all the factors affecting it, can be represented by the trajectory of points in three-dimensional phase-space. The attractor is created by plotting each value as a function of its preceding value, or in other words, from the plot of $x(t + 1)$ vs. $x(t)$, where x is the actual value and t is the index of the point. Note that an attractor with a regular shape will also emerge in plots using $x(t + 2)$ or $x(t + 3)$, for example, or $x(t + n)$, with many n. This procedure is repeated for each successive point in the time series and the resultant points are connected producing the phase-space trajectory.

6.1.3.1.1 Periodic Attractors

Consider a periodic function Y_t with a single frequency, $Y_t = 0.4 \sin(100t)$, where t is time. The periodicity of Y_t is obvious in the raw data (Figure 6.6A) and results in a stable orbit in phase-space (Figure 6.6B,C). In contrast, in the case of a function without periodicity (that is, $Y_t = 0.4t + 0.4$ and $Y_t = 0.4t^2$), the phase-space signature will consistently be a straight line (not shown). Now consider a periodic function with two characteristic frequencies (Figure 6.6D), $Y_t = 0.4 \sin(100t) + 0.4 \sin(400t)$. An attractor with a regular shape will appear in phase-space using Y_{t+n}, with any n (Figure 6.6E,F). When the periodic function becomes more elaborate (Figure 6.6G), for example, $Y_t = 0.2 \sin(20t) + 0.3 \sin(100t) + 0.4 \sin(400t)$, a clear attractor may still emerge depending on the nature of the sine waves (Figure 6.6H,I), and may eventually need to be analyzed in more than three dimensions to show any regularity.

6.1.3.1.2 Random Attractors

When a random noise is added to an otherwise periodic relationship (Figure 6.6J), the characteristic shape of the attractors becomes very difficult to discern (Figure 6.6K,I). The phase-space portrait of a random distribution (Figure 6.6M), in turn, does not exhibit any specific signature with simulated points filling more or less the whole space evenly (Figure 6.6N,O).

6.1.3.1.3 Chaotic Attractors

Three of the most well-known and studied dynamical systems are briefly presented hereafter, the Hénon attractor and the Rössler and Lorenz attractors.

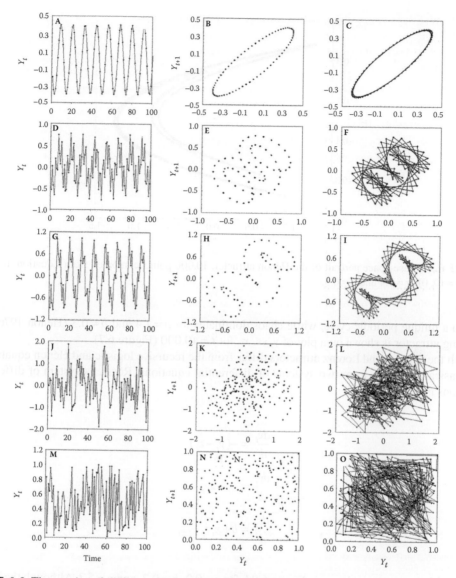

FIGURE 6.6 Time series of different simulated signals Y_t and the related phase-space portrait, Y_t vs. Y_{t+2}. From top to bottom, a sine wave, $Y_t = 0.4\sin(100t)$, a combination of two sine waves, $Y(t) = 0.4 \sin(100t) + 0.4 \sin(400t)$, a combination of three sine signals $Y(t) = 0.4 \sin(100t) + 0.3 \sin(100t) + 0.4 \sin(400t)$, a combination of the three sine signals and additive noise bounded between −1 and 1, and a random noise bounded between −1 and 1.

The Hénon attractor is defined by a recursive equation, $H(x, y) = (y + 1 - ax^2, bx)$, where a and b are adjustable parameters. An orbit, or trajectory, of the system consists of a starting point (x_0, y_0) and its iterated images:

$$(x_{t+1}, y_{t+1}) = (y_t + 1 - ax_t^2, bx_t) \qquad (6.5)$$

for $k = 0, 1, \dots n$. As previously discussed for the logistic equation, the dynamics of Equation (6.5) strongly rely on the choice of the constants a and b. For a range of values for a and b, most orbits

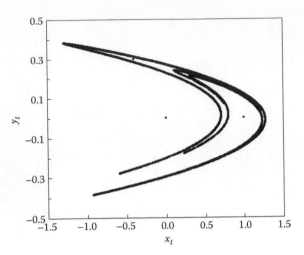

FIGURE 6.7 Phase-space portrait of the Hénon attractor. 10,000 points computed from Equation (6.5) with $(x_0, x_0) = (0, 0)$, $a = 1.4$ and $b = 0.3$.

tend to a unique periodic cycle, while chaos dominates for $a = 1.4$ and $b = 0.3$ (Hénon 1976). The resulting attractor is shown on a plot of y_t vs. x_t, for $t = 10,000$ (Figure 6.7).

Both the Rössler and Lorenz attractors differ from the recursive logistic and Hénon equations as they basically emerge from two systems of differential equations. Rössler's system of differential equations is (Rössler 1976):

$$\frac{dx_t}{dt} = -(y_t - z_t)$$

$$\frac{dy_t}{dt} = x_t + ay_t \qquad (6.6)$$

$$\frac{dz_t}{dt} = b + x_t z_t - cz_t$$

The resulting attractor is shown in Figure 6.8A for $a = 0.2$, $b = 0.2$, and $c = 5.7$. Although the Rössler system was artificially designed to create a model for a strange attractor, the Lorenz model was initially developed to simulate a global weather system (Lorenz 1963):

$$\frac{dx_t}{dt} = -\sigma x_t + \sigma y_t$$

$$\frac{dy_t}{dt} = -y R x_t - y_t - x_t y_t \qquad (6.7)$$

$$\frac{dz_t}{dt} = -B z_t + x_t y_t$$

where σ, B, and R are the weather system parameters, fixed at $\sigma = 10$, $B = 8/3$, and $R = 28$. The corresponding attractor is shown in Figure 6.8B.

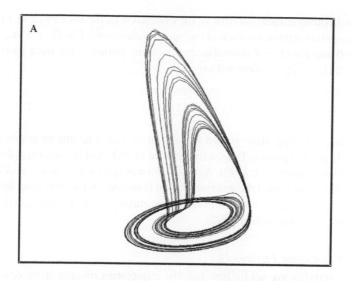

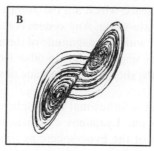 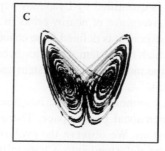

FIGURE 6.8 Three-dimensional phase-space portrait (x_t, y_t, z_t) of the Rössler attractor (A) and two-dimensional phase-space portraits (x_t, y_t) (B) and (x_t, z_t) (C) of the Lorenz attractor, shown for 5000 iterations.

6.1.3.2 Quantifying Attractors: Diagnostic Methods for Deterministic Chaos

In the following, our data sets are regarded as finite sets of time observations, $x(t)$, taken at regular intervals:

$$X(t) = \{x(1), x(2), \ldots, x(n)\} \tag{6.8}$$

where n is the total number of observations in each set. The time length of any observed period, T, is related to n as:

$$T = n\Delta t \tag{6.9}$$

More specifically, the three methods presented here—the Packard-Takens (PT) method (see Section 6.1.3.1), Lyapunov exponents estimates (Wolf et al. 1985), and the correlation integral method (Grassberger and Procaccia 1983)—are based on the assumption that the dynamics of any underlying dynamical systems can be described in some multidimensional phase-space from the knowledge of the time series of a single observation $x(t)$ by constructing E-dimensional vectors defined by:

$$\ddot{X}(t) = (x(t), x(t-\tau), \ldots, x(t-(D_E-1)\tau)) \tag{6.10}$$

where D_E is the embedding dimension (that is, the dimension of the vectors), and τ is the lag (that is, the number of data points separating each of the vector's elements). For $D_E = 3$ and $\tau = 1$, the vector $\vec{X}(t)$ consists of $x(t)$ and the $D_E - 1$ immediately preceding points of the time series; that is, the set of vectors $\{\vec{X}(3), \vec{X}(4), \ldots, \vec{X}(n)\}$ is denoted as:

$$\{(x(3), x(2), x(1), ((x(4), x(3), x(2)), \ldots, ((x(n), x(n-1), (x(n-2))\}.$$

In the above case, the delay time τ must be chosen so that it results in points that are not correlated to previously plotted points. Thus, a first choice of τ should be in terms of the decorrelation time of the time series (Tsonis et al. 1993). A straightforward procedure is to consider the decorrelation time equal to the lag at which the autocorrelation function for the first time attains the value of zero. Also note that no averaging or filtering should be employed since such data manipulations can obscure the presence of chaos (Ellner 1992).

6.1.3.2.1 Largest Lyapunov Exponents

The limits of predictability are set by how fast the trajectories diverge from nearby initial conditions. This feature is quantified by Lyapunov exponents, which are the average exponential rates of divergence or convergence of nearby orbits in phase-space. Any system containing at least one positive Lyapunov exponent is defined to be chaotic, with the magnitude of the exponent reflecting the time scale at which the system dynamics become unpredictable. In other words, the larger the positive exponent, the more chaotic the system and the shorter the time scale of system predictability (Wolf et al. 1985).

To define the Lyapunov exponents, imagine an infinitesimal hypersphere of initial conditions in the n-dimensional phase-space. There is one Lyapunov exponent for each degree of freedom of the system. We observe the evolution of the hypersphere as time progresses. The hypersphere will be deformed into a hyperellipsoid because of the evolution of the system. Then the ith Lyapunov exponent can be defined in terms of the length of the ith principal axis, p_i, of the ellipsoid as:

$$\lambda_L = \lim_{\tau \to \infty} \frac{1}{\tau} \ln \frac{p_i(\tau)}{p_i(0)} \tag{6.11}$$

where the λ_L are ordered from largest to smallest in an algebraic sense (Wolf et al. 1985; Mundt et al. 1991). A minimum condition for chaos is that the largest Lyapunov exponent (LLE), λ_L, is positive.

In practice, the algorithm developed by Wolf et al. (1985) is recommended to estimate the largest Lyapunov exponent, λ_L, from a time series, since it uses a relatively simple procedure and has been demonstrated to be robust over a large range of input parameters and relatively accurate for small, noisy data sets (Mundt et al. 1991). The delay time τ should be chosen as the decorrelation time of the time series, as previously mentioned.

6.1.3.2.2 Correlation Integral

Although the LLE is used to estimate the limits of predictability of a given time series, the complementary correlation integral (CI) algorithm is devoted to the quantitative characterization of the attractor of the series. In particular, this method can be regarded as a generalization of the correlation dimension described in Section 3.2.6. As demonstrated by Takens (1981), an attractor topologically equivalent to the attractor of the system producing the data is obtained for every value of τ and for D_E sufficiently greater than the fractal dimension, that is, $D_E \geq (2D + 1)$.

From the new multidimensional time series defined by Equation (6.10), the correlation integral (Grassberger and Procaccia 1983) is defined as:

$$C(r) = \lim_{N \to \infty} \frac{1}{N^2} \sum_{j=1}^{N} \sum_{i=j+1}^{N} \theta\left(r - \left|\vec{X}_i - \vec{X}_j\right|\right) \tag{6.12}$$

where N is the number of distinct pairs in the embedding space, $|\vec{X}_i - \vec{X}_j|$ is the Euclidean distance operator between the ith and jth sample, r is an arbitrary time called "lag time" (distance between vectors), and $\theta(\xi)$ is the Heaviside function, defined as follows:

$$\theta(\xi) = \begin{cases} 0 & \text{for } \xi < 0 \\ 1 & \text{for } \xi \geq 0 \end{cases} \tag{6.13}$$

The correlation integral $C(r)$ represents the probability that the distance between a pair of randomly chosen points on the D_E-dimensional reconstruction will be less than a distance r apart (Grassberger and Procaccia 1983). In the case of random processes, the phase-space trajectory is directly linked to the volume of the considered D_E-dimensional space as:

$$C(r) \underset{r \to 0}{\propto} r^{D_E} \tag{6.14}$$

while for an attractor, the phase-space trajectory is more compact and the correlation integral is then characterized by the following scaling properties:

$$C(r) \underset{r \to 0}{\propto} r^{\nu} \tag{6.15}$$

where the exponent ν is the correlation exponent (or correlation dimension); it can be estimated as the slope of the log-log plot of $C(r)$ vs. r, using a simple least-squares method.

For chaotic data, ν will approach a constant value as the embedding dimension E is increased. That constant value is an estimate of the correlation dimension that measures the local structure of the strange attractor. The dimension ν of the strange attractor indicates at least how many variables are necessary to describe evolution in time. For instance, $\nu = 2.5$ indicates that a given time series can be described by a system equation containing three independent variables.

6.1.3.2.3 Nonlinear Forecasting: Nearest-Neighbor Algorithm

Based on various prediction theories (Lorenz 1963; Tong and Lim 1980; Priestley 1980; Farmer and Sidorowich 1989), this method was initially developed for (1) making short-term predictions about the trajectories of chaotic dynamical systems, and (2) distinguishing between the complexity of natural dynamical systems where deterministic dynamics can lead to chaotic trajectories and the fluctuations intrinsically related to the sampling and measurement processes (Sugihara and May 1990b; Sugihara et al. 1990; Sugihara 1994). A direct consequence of the sensitive dependence on initial conditions of a chaotic system is that prediction will become exponentially less accurate as one attempts to predict further ahead. The nearest-neighbor (NN) algorithm quantifies this prediction accuracy as a function of the distance into the future that they are made. Specifically, the D_E-dimensional vectors introduced in Equation (6.10), that is, $\vec{X}(t) = (x(t), x(t-\tau), \ldots, x(t-(D_E-1)\tau))$, are used to calculate the distance between $\vec{X}(t)$ and the k

nearest-neighbors vectors $\vec{X}_i(t), \vec{X}_i(t) = (x_i(t), x_i(t-\tau), \ldots, x_i(t-(D_E-1)\tau))$, and the predicted value of each $x(t)$ is given as:

$$x(t+p) = \frac{1}{k}\sum_{i=1}^{k} x_i(t+p) \tag{6.16}$$

where p is the prediction distance, that is, the number of steps ahead that one is attempting to predict. The predicted values $x(t + p)$ are then plotted against the actual values $x(t)$, and the strength of the predictions evaluated through their coefficient of correlation r. Practically, the first half of a given data set $X(t)$, see Equation (6.8), is used to estimate $x(t + p)$, which is then plotted against the second half of $X(t)$.

Two sets of information are subsequently derived from the coefficient of correlation r, and used to classify the dynamics of the system under study. First, a plot of the coefficient of correlation r as a function of the prediction time provides information on the deterministic vs. chaotic nature of the system. For instance, purely deterministic (that is, nonchaotic) processes will return consistently high values of r whatever the prediction distance (Figure 6.9). In contrast, chaotic systems such as the logistic equation in the regime where $a > 3.5749$ lead to an exponential decrease of r as prediction time increases (Figure 6.9A). Note that the contamination of the logistic equation by external (white) noise* decreases uniformly the predictability of nonchaotic systems, depending on the amplitude of

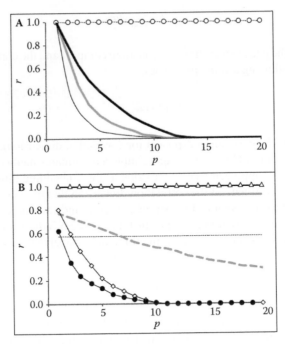

FIGURE 6.9 Nearest-neighbor technique applied to the logistic equation (see Equations 6.2 and 6.3). Top panel: $a = 2.5$ (A: dashed line), $a = 3.5$ (A: open dots) and $a = 4.0$ (thick black curve), $a = 4.0$ with external noise ε_t, $\varepsilon(t) = \pm0.005$ (thick gray curve), and $a = 4.0$ with external noise ε_t, $\varepsilon(t) = \pm0.05$, (thin black curve). Bottom panel: $a = 3.5$ with external noise ε_t, $\varepsilon(t) = \pm0.005$, (black line) and $\varepsilon(t) = \pm0.05$, (gray line) and $a = 3.5$ with process noise ε_t, $\varepsilon(t) = \pm0.005$ (open triangles) and $\varepsilon(t) = \pm0.05$ (open diamonds), and $a = 2.5$ with external noise ε_t, $\varepsilon(t) = \pm0.005$ (dashed line) and $\varepsilon(t) = \pm0.05$ (not shown, indistinguishable from $r = 0$) and $a = 25$ with process noise ε_t, $\varepsilon(t) = \pm0.005$ (dashed thick gray curve) and $\varepsilon(t) = \pm0.05$ (black dots).

---

* In which case, Equations (6.2) and (6.3), respectively, rewrite as $x_{t+1} = ax_t(1 - x_t) + \varepsilon(t)$ and $x_{t+1} = ax_t - ax_t^2 + \varepsilon(t)$.

the noise (Figure 6.9B). In contrast, noise-contaminated chaotic systems will steepen the exponential decay of predictability (Figure 6.9A). Internal noise,* in turn, leads to a nonexponential decay in the predictability of a nonchaotic system (Figure 6.9B), unless the magnitude of the noise is sufficient to drive the system into the unstable chaotic regime (Sugihara 1994) (Figure 6.9B). Second, the shape of a plot of r as a function of the embedding dimension D_E informs on the dimensionality of the system; chaotic systems show optimal predictability for low embedding dimensions, while random processes exhibit increasing predictability at higher embedding dimensions.

6.1.3.3 Case Study: Plankton Distribution in Turbulent Coastal Waters

6.1.3.3.1 *Ecological Framework*

A range of empirical and theoretical studies have demonstrated that fully developed turbulence is rather characterized by its multifractal properties (that is, high-order stochasticity and high dimensionality); see, for example, Frisch (1996) and references therein. Ruelle and Takens (1971) also showed that near the transition to turbulence, the many degrees of freedom of turbulence are coupled coherently and lead to an enormous reduction in dimension (that is, low-order deterministic chaos emerges). It is then likely that in aquatic environments characterized by fluctuating turbulent intensities (for example, shallow coastal and estuarine regions), the structure of both physical (temperature and salinity) and biological (phytoplankton biomass) parameters vary considerably. In other words, the dimensionality and predictability of a system might be related to the turbulent conditions.

Specifically, such transitions between low-order deterministic chaos and high-order stochasticity may be observed in the tidally mixed waters of the eastern English Channel (Figure 4.8), where turbulence intensities may vary by more than two orders of magnitude over one tidal cycle (Seuront 2005b; Seuront et al. 2002) and are generally thought to drive phytoplankton biomass variability (Seuront et al. 1996a, 1996b, 1999). Herein, the goal of this case study is, first, to find out whether time series of physical (temperature and salinity) and biological (phytoplankton biomass) parameters recorded in tidally mixed waters are chaotic or not, and second, to investigate the potential effects of differential tidal forcing on the chaotic or stochastic nature of the variables in question.

6.1.3.3.2 *Experimental Procedures and Data Analysis*

The sampling experiment was conducted during 60 hours (that is, five tidal cycles) in a period of spring tide, from March 28 to 30, 1998, at an anchor station located in the coastal waters of the eastern English Channel (50°47′300 N, 1°33′500 E) (Figure 4.8). The tidal range in this system is one of the largest in the world, ranging from 3 to 9 m, and the water column is believed to be fully homogenized by tide-generated turbulent mixing. Temperature, salinity, and *in vivo* fluorescence were simultaneously recorded at 2 Hz from a single depth (5 m) with a SBE 25 Sealogger CTD (conductivity-temperature-depth) probe, and a Sea Tech fluorometer, respectively. Every hour, samples of water were taken at 5 meters depth to estimate chlorophyll *a* concentrations, which appear significantly correlated with *in vivo* fluorescence (Kendall's $\tau = 0.778$, $p < 0.01$). In the following, the latter parameter will then be regarded as a direct estimate of phytoplankton biomass. To investigate the potential effect of varying turbulent forcings on the local structure of physical and biological parameters, the data analyzed here consist of 24 time series (labeled from S1 to S24) of 1 hour duration (7200 data points), resampled from the original data set in order to be representative of the different conditions of tidal current speed and direction, taken every 10 minutes, from the sampling depth (Table 6.1).

Time-series analysis requires the assumption of at least reduced stationarity; that is, the mean and the variance of a time series depend only on its length and not on the absolute time (Legendre

---

* Now Equations (6.2) and (6.3), respectively, rewrite as $x_{t+1} = a(x_t + \varepsilon(t))(1 - (x_t + \varepsilon(t))$ and $x_{t+1} = a(x_t + \varepsilon(t)) - a(x_t + \varepsilon(t))^2$.

TABLE 6.1
Tidal Velocity (m s⁻¹) and Direction (Dir, °), Water Column Depth (m) and Mean Values of Temperature (°C), Salinity (PSU), and *In Vivo* Fluorescence (Fluorescence Relative Units) for the 24 Studied Data Sets

| | Tidal Current | | | | |
	Speed	Dir (°)	Depth	T	S	F
S1	0.55	240	21.56	6.55	34.60	18.32
S2	0.45	220	22.49	6.53	34.60	15.20
S3	0.10	60	27.38	6.51	34.62	10.90
S4	0.95	15	28.28	6.50	34.66	9.39
S5	0.90	10	26.21	6.49	34.70	8.23
S6	0.15	10	23.25	6.51	34.65	10.25
S7	0.32	260	21.52	6.53	34.61	15.02
S8	0.62	230	22.21	6.52	34.62	17.24
S9	0.10	85	27.19	6.50	34.65	11.45
S10	0.98	10	28.47	6.49	34.72	6.80
S11	1.00	10	26.53	6.49	34.67	7.29
S12	0.30	10	23.66	6.50	34.64	11.00
S13	0.35	290	21.38	6.53	34.62	17.40
S14	0.30	200	21.72	6.55	34.62	15.82
S15	0.11	140	26.19	6.52	34.66	13.46
S16	0.80	10	28.65	6.50	34.69	10.75
S17	1.10	10	27.15	6.49	34.70	6.64
S18	0.40	10	24.15	6.51	34.68	7.35
S19	0.35	260	21.75	6.53	34.63	12.64
S20	0.87	250	21.68	6.55	34.62	17.69
S21	0.73	230	25.23	6.55	34.61	15.16
S22	0.18	10	28.65	6.53	34.71	8.30
S23	1.04	10	27.50	6.50	34.66	5.37
S24	0.60	10	25.95	6.50	34.62	3.87

and Legendre 2003). The existence and the significance of any potential linear trends were tested calculating Kendall's τ correlation, which does not require any hypothesis about the characteristics of the original data-set distribution. (Kendall's coefficient of correlation was used in preference to Spearman's coefficient of correlation ρ because Spearman's ρ gives greater weight to pairs of ranks that are further apart, while Kendall's τ weights each disagreement in rank equally) (see Sokal and Rohlf 1995 for further developments). We then eventually detrended the time series, fitting linear regressions to the original data by least squares, and used the regression residuals in further analysis. The purpose of this is to eliminate aliasing in further analysis due to large-scale structures present in the data sets, such as in monotonically increasing or decreasing trends.

In order to provide direct comparisons between the different parameters investigated here, the time observations, x_i, were converted into normalized, dimensionless descriptors, y_i, following:

$$y_i = \frac{x_i - x_{min}}{x_{max} - x_{min}} \tag{6.17}$$

where x_{max} and x_{min} are the maximum and minimum values of the series, respectively. Samples of the resulting time series are given in Figure 6.10, and the Packard-Takens, largest Lyapunov exponents, and correlation integral methods were used to investigate their properties.

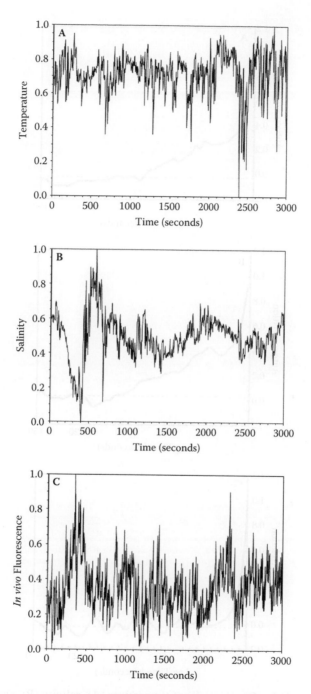

FIGURE 6.10 Samples of normalized temperature (A), salinity (B), and *in vivo* fluorescence (C) time series recorded in the inshore waters of the eastern English Channel; shown for data set S1. (Modified from Suelont, 2004.) (Kraichnan, R.H., 1967), Inertial ranges in two-dimensional turbulence, *Physics of Fluids*, 10, 1417–1423.

6.1.3.3.3 Results

The delay time τ has been chosen as the decorrelation time of the time series (Tsonis et al. 1993) as 75, 95, and 30 seconds for temperature, salinity, and *in vivo* fluorescence time series, respectively (Figure 6.11). This delay time was also used for the following calculations of Lyapunov exponents and correlation dimensions.

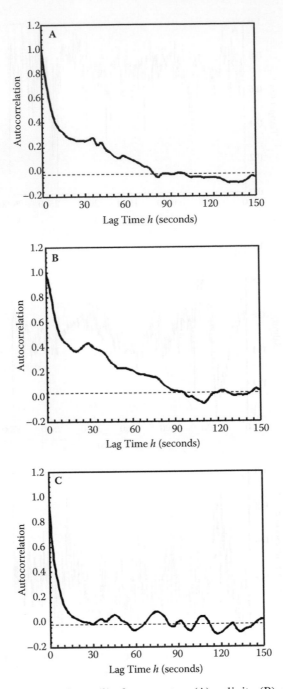

FIGURE 6.11 Autocorrelation functions $r(h)$ of temperature (A), salinity (B), and *in vivo* fluorescence (C) time series recorded in the inshore waters of the eastern English Channel; shown for data set S1. The dashed line is the special case $r(h) = 0$ leading to the decorrelation time of the time series.

The three-dimensional phase-space portraits of the attractors produced by the Packard-Takens method did not clearly exhibit any attractor (Figure 6.12). Note, however, the differences between the phase-space portraits of fluorescence on the one hand and temperature and salinity on the other hand. The phase-space portraits for temperature and salinity appear as somewhat elongated and

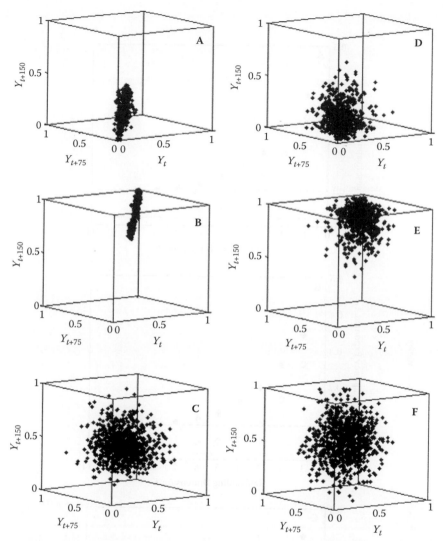

FIGURE 6.12 Three-dimensional phase-space portraits of normalized temperature, salinity, and *in vivo* fluorescence time series. (A, B, C) represent conditions of low hydrodynamic conditions for data set S3, and (D, E, F) represent conditions of high hydrodynamic conditions for data set S17.

relatively narrow spatial distributions (Figure 6.12A,B). Phase-space trajectories of *in vivo* fluorescence, however, did not exhibit any characteristic shape, suggesting a more space-filling—or "random"—behavior (Figure 6.12C). Moreover, comparison of phase-space portraits obtained from time series recorded in high and low hydrodynamic conditions leads to further results. Phase-space trajectories of temperature and salinity then appear clearly more structured in lower hydrodynamic conditions (Figure 6.12D,E), while the apparent randomness of *in vivo* fluorescence phase-space trajectories remains whatever the hydrodynamic conditions (Figure 6.12F).

The largest Lyapunov exponents, LLE, λ_L, calculated over a range of embedding dimensions E exhibit clearly different behaviors (Figure 6.13). By embedding dimension 8, the temperature and salinity LLE converge to positive values that are larger when the hydrodynamic conditions are high (Figure 6.13A,B). In other words, the higher the hydrodynamic conditions, the larger the positive exponent, the more chaotic the system, and the shorter the time scale of system predictability (Wolf et al. 1985). This is confirmed by the significant negative correlation between largest Lyapunov

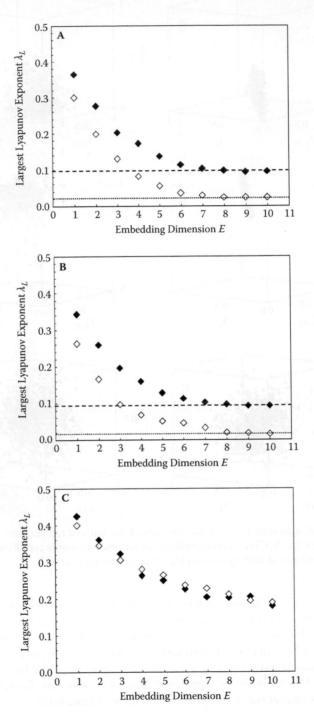

FIGURE 6.13 The largest Lyapunov exponent λ_L estimated for temperature (A), salinity (B), and *in vivo* fluorescence (C) time series under high (open diamonds: S15) and low (black diamonds: S23) hydrodynamic conditions. The dashed and dotted lines in (A, B) indicate the convergent values of λ_L under low and high hydrodynamic conditions, respectively. (Kraichnan, R.H., 1967, Inertial ranges in two-dimensional turbulence, *Physics of Fluids*, 10, 1417–1423.)

exponents of both temperature and salinity, and tidal current speed direction. The largest Lyapunov exponents and the associated time scale of predictability are shown in Table 6.2. In contrast, fluorescence LLE remain significantly higher than temperature and salinity LLE irrespective of the hydrodynamic conditions, but never converge to any constant value, even when the embedding dimension E is increased up to 10 (Figure 6.13C). This indicates more chaotic behavior and less predictability in phytoplankton biomass than in temperature and salinity fluctuations.

Figure 6.14 shows the correlation integral $C(r)$ on logarithmic scales as a function of distance r by varying embedding dimension E from 1 to 10. Estimates of the correlation dimension v (see Equation 6.15) for temperature and salinity did not converge to any constant value whatever the hydrodynamic conditions (Figure 6.15A,B) and indicate the lack of empirical evidence for deterministic chaos. Moreover, no significant differences were observed between temperature and salinity correlation dimensions, or between the different time series for either parameter, suggesting very similar behaviors of temperature and salinity time series in phase-space. The results for *in vivo* fluorescence time series are very similar with those of temperature and salinity. Clearly no saturation,

TABLE 6.2
Largest Lyapunov Exponents λ_L Estimates for Temperature, Salinity, and *in vivo* Fluorescence from the 24 Available Data Sets, and the Related Time Scale of System Predictability

	λ_L			Predictability (seconds)		
	T	S	F*	T	S	F
S1	0.048	0.045	0.212	20.83	22.22	4.72
S2	0.044	0.043	0.223	22.73	23.26	4.48
S3	0.012	0.009	0.225	83.33	111.11	4.44
S4	0.098	0.105	0.243	10.20	9.52	4.12
S5	0.092	0.094	0.172	10.87	10.64	5.81
S6	0.021	0.023	0.221	47.62	43.48	4.52
S7	0.031	0.035	0.236	32.26	28.57	4.24
S8	0.055	0.057	0.198	18.18	17.54	5.05
S9	0.011	0.009	0.217	90.91	111.11	4.61
S10	0.091	0.088	0.171	10.99	11.36	5.85
S11	0.095	0.084	0.223	10.53	11.90	4.48
S12	0.038	0.039	0.181	26.32	25.64	5.52
S13	0.041	0.039	0.234	24.39	25.64	4.27
S14	0.042	0.039	0.182	23.81	25.64	5.49
S15	0.012	0.016	0.234	83.33	62.50	4.27
S16	0.076	0.079	0.172	13.16	12.66	5.81
S17	0.121	0.133	0.228	8.26	7.52	4.39
S18	0.038	0.041	0.196	26.32	24.39	5.10
S19	0.032	0.034	0.253	31.25	29.41	3.95
S20	0.085	0.088	0.228	11.76	11.36	4.39
S21	0.076	0.074	0.234	13.16	13.51	4.27
S22	0.025	0.017	0.254	40.00	58.82	3.94
S23	0.097	0.096	0.174	10.31	10.42	5.75
S24	0.071	0.075	0.187	14.08	13.33	5.35
Mean	0.056	0.057	0.212	28.53	30.06	4.79
SD	0.032	0.033	0.027	24.36	28.84	0.64
Min	0.011	0.009	0.171	8.26	7.52	3.94
Max	0.121	0.133	0.254	90.91	111.11	5.85

* Following the absence of convergent behavior for the fluorescence Lyapunov exponents, we report here the λ_L estimated for $E = 10$.

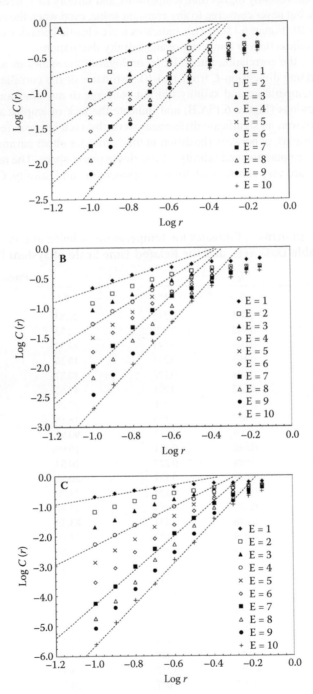

FIGURE 6.14 Log-log plots of correlation integral $C(r)$ versus distance r for various embedding dimensions E for temperature (A), salinity (B), and *in vivo* fluorescence (C) time series; shown for database S8. (Kraichnan, R.H., 1967, Inertial ranges in two-dimensional turbulence, *Physics of Fluids*, 10, 1417–1423.)

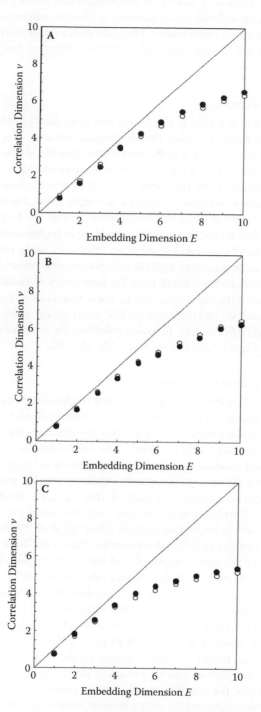

FIGURE 6.15 Correlation dimensions v vs. embedding dimensions E for temperature (A), salinity (B), and *in vivo* fluorescence (C) under high (open diamonds; dataset S15) and low (black diamonds; data set S23) hydrodynamic conditions. (Kraichnan, R.H., 1967, Inertial ranges in two-dimensional turbulence, *Physics of Fluids*, 10, 1417–1423.)

and therefore no evidence of low-order deterministic chaos, exists whatever the hydrodynamic conditions (Figure 6.15C). As previously shown for temperature and salinity time series, no significant differences exist between the correlation dimensions v. These results confirm the previous lack of convergence of fluorescence LLE (see Figure 6.13C), and indicate that there is no evidence for deterministic chaos in the temporal fluctuations of phytoplankton biomass time series.

6.1.3.3.4 Discussion

6.1.3.3.4.1 Phase-Space Portraits

The Packard-Takens method is probably the fastest and most direct method to infer the potential existence of deterministic chaos. Creating the phase-space attractor of a system with a computer is a very simple task. All that is needed is the copy of the data file, paste it shifted by one, two, or more places, and plot the data. Thus, a subjective assessment of the "degree of randomness" can be reached almost instantaneously from this kind of plot. It is nevertheless stressed that the characteristic shape of the attractor is not easy to describe in simple terms. Figure 6.12 shows projections of phase-space trajectories onto three-dimensional space, so that the fact that no attractors can be seen does not imply that they do not exist when embedding in higher-dimensional space. However, a strange attractor of higher-dimensional space often reflects its shape onto the lower-dimensional space as well. For instance, the trajectory onto the two-dimensional phase-space (embedding dimension E = 2 in Equation 6.10), reconstructed from the time series of variable x of the Lorenz equations, shows a clear strange attractor (Figure 6.8B). These results can then instead be regarded as a qualitative prerequisite analysis and demonstrate that inferring the existence of any deterministic structure beyond the highly fluctuating behavior exhibited by temperature, salinity, and *in vivo* fluorescence time series (Figure 6.10) is a far more difficult task.

6.1.3.3.4.2 Largest Lyapunov Exponents

The LLE estimates quantitatively confirm the subjective results of the Packard-Takens method, that is, a lower-dimensional behavior in low hydrodynamic conditions for temperature and salinity time series, and a higher-dimensional behavior for phytoplankton biomass time series that did not exhibit any convergent behavior of their LLE for values of the embedding dimension E up to 10 irrespective of the hydrodynamical conditions. What may be regarded as being very important for ecologists is that, unlike fractal dimensions, Lyapunov exponents remain well defined in the presence of dynamical noise and can be estimated by methods that explicitly incorporate noise (Ellner et al. 1991; Nychka et al. 1992). This leads us to consider that estimating Lyapunov exponents is the best approach for detecting chaos in ecological systems (Hastings et al. 1993). A number of limitations in Lyapunov exponent estimates to detect deterministic chaos can, however, be raised and regards both estimate accuracy and the minimum number of data points required in the analysis.

First, although the algorithm used in this chapter (Wolf et al. 1985) provides a good estimation of the largest Lyapunov exponents for noise-free, synthetically generated time series from chaotic dynamics, the estimation for experimental time series is still relatively imprecise (Rodriguez-Iturbe et al. 1989). Second, it has been stressed that to detect a chaotic attractor of dimension 3, at least 1,000 to 30,000 data points are needed (Wolf et al. 1985), while others (Ramsey and Yuan 1989) found that 5,000 data points is a lower bound for the detection of chaos on some simple dynamical systems known to display chaotic behaviors in certain regimes. Moreover, Vassilicos et al. (1993) demonstrated how the tests for chaos can give positive answers—for example, positive Lyapunov exponents—when subsamples with a smaller number of data points are used, and how these Lyapunov exponents converge to zero when the number of data points is increased.

The latter limitation has been specified through estimates of the largest Lyapunov exponents of the larger original time series (that is, 172,800 data points) of temperature, salinity, and *in vivo* fluorescence that were divided into 24 subsections of 7,200 points in the present work. Subsequent results (Figure 6.16A) then indicated that LLEs of temperature, salinity, and phytoplankton biomass time

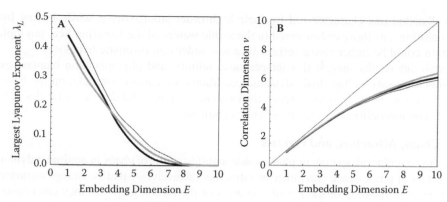

FIGURE 6.16 The largest Lyapunov exponent λ_L (A) and correlation dimensions v (B) plotted against embedding dimensions E for temperature (thick black line), salinity (thick gray line) and *in vivo* fluorescence (thin black line) from the original 172,800 data points time series. The dashed line is the correlation dimensions v expected for an uncorrelated noise (B).

series remain positive but converge to zero. As previously mentioned, a positive largest Lyapunov exponent indicates chaotic dynamics, but values quite close to zero should therefore only be interpreted as an order of magnitude. As a consequence, the different convergent positive values of the different LLEs estimated for temperature and salinity time series in high and low hydrodynamic conditions (Figure 6.13A,B) suggest a phenomenological shift between low-dimensional chaos and high-dimensional stochasticity as the one observed by Ruelle and Takens (1971) near the transition to turbulence. Alternatively, a positive largest Lyapunov exponent close to zero can be interpreted as having been derived from a stochastic time series with many degrees of freedom (Jeong and Rao 1996). More generally, systems with a Lyapunov exponent of zero are associated with a state called the edge of chaos, where complex behavior is the rule. The exact meaning of the edge of chaos depends on the context within which it is used but, roughly speaking, it describes the vicinity of some instability point separating a region of more ordered (or less random) behavior from a region of less ordered (or more random) behavior. The edge of chaos has attracted considerable interest among biologists and ecologists because processes such as evolution or adaptive behavior have been precisely shown to be just at the edge of chaos (Kauffman and Johnson 1991; Langdon 1992; Kauffman 1993). Such a critical state would increase the adaptive efficiency of a given system—for instance, in response to fluctuating environmental conditions—and could then be of prime interest in the future understanding of ecosystem functioning.

6.1.3.3.4.3 Correlation Integrals

Although it has been shown that if the data set is small, the correlation dimension v (see Equation 6.15) appears to converge toward a finite value even in the absence of chaos (Smith 1988), this is obviously not the case here (Figure 6.15). Moreover, correlation dimension v estimates for the 172,800 data-point time series (Figure 6.16B) did not converge to any constant value and confirm the lack of empirical evidence for deterministic chaos previously shown with smaller time series (Figure 6.15). Our results then cannot be associated with sampling limitation. A correlation dimension of 2 has thus been identified on the basis of a 1,200 values of chlorophyll transect recorded in the central waters of the Ligurian Sea (NW Mediterranean Sea; Ibanez 1986). This result then confirms the efficiency of the correlation algorithm to detect low deterministic chaos when applied to small data sets. This also confirms that different hydrodynamic conditions might be at the origin of differential space-time structures, in terms of low-order deterministic chaos or high-order stochasticity. Then, high hydrodynamic conditions, such as those occurring in the eastern English Channel, could be at the origin of temperature, salinity, and phytoplankton

biomass distributions characterized by their high-order stochasticity, while in low hydrodynamic conditions, as those encountered in the stable waters of the Ligurian Sea, phytoplankton
distribution could be rather characterized by a low-order deterministic behavior.

Although our results suggest that temperature, salinity, and phytoplankton biomass exhibit a
higher dimensionality in high hydrodynamic conditions, we cannot conclude, on the basis of the
three previously used analysis techniques, the existence of low-order deterministic chaos, but only
to a lower dimensionality in low hydrodynamic conditions.

6.1.3.4 Chaos, Attractors, and Fractals

Fractals can be temporal, spatial, or phase-space manifestations of chaos in nonlinear dynamic systems. Fractals in phase-space can either be attractors themselves—that is, strange attractors, such
as the bifurcation diagram of the logistic equation or the Hénon map—or they can constitute the
dividing line between separate attractor basins in phase-space (see, for example, Peitgen and Saupe
1988; Peitgen et al. 1992). The study of attractors is important because the geometry of an attractor
frequently captures much of the underlying dynamics and allows one-dimensional (fractal) description. We will nevertheless see hereafter that the geometry of strange attractors can be so complex
that it becomes impossible to describe them in terms of fractal dimensions and (low-order) deterministic chaos. In particular, this statement precludes the introduction of the multifractal, high-
order stochastic framework.

6.1.4 Chaos in Ecological Sciences

Since the seminal studies of chaos in discrete time models in population ecology (May 1974, 1975,
1976), the issue of chaotic dynamics in ecological systems has been widely controversial (Hassell
et al. 1976; Berryman and Millstein 1989; Pool 1989). Chaos in ecology has nevertheless been the
subject of an increasing amount of literature. In theoretical ecology, there are many examples of
temporal population models that exhibit chaos. The interaction of three variables in a predator–prey
nutrient system (Kot et al. 1992) is now a well-studied chaotic system, as chaotic dynamics expected
through a trophic coupling of three species (Hastings and Powell 1991). Recently, an ocean ecosystem model also exhibited chaotic properties related to external seasonal forcing (Popova et al. 1997).
In particular, the issues raised by chaos theory in ecology have been the subject of several reviews
(May 1980, 1987; Godfrey and Blythe 1991; Ellner 1992; Logan and Allen 1992; Hastings et al.
1993; Little et al. 1996).

As briefly suggested in the above section, the compelling reasons for the emerging chaos theory
to ecology are based on the hope that complex systems could be explained by relatively low-order
processes. This leads to the development of a suite of algorithms aimed at the detection of chaotic
behavior and the classification of system dynamics; see, for example, Hastings et al. (1993) and Ellner
and Turchin (1995) for reviews. While such approaches have been applied to a wide variety of time
series (Farmer and Sidorowich 1987; Ellner 1992; Theiler et al. 1992) to detect dynamic spatial chaos
(Rubin 1992; Rand 1994; Solé and Bascompte 1995), the development of nonlinear thinking to aquatic
ecology has a more recent history. Only a few studies have been devoted to detecting chaotic signature
in both marine time series and transects, and led to controversial results. Sugihara and May (1990b)
found evidence for chaotic dynamics in time series of weekly diatom counts, and Scheffer (1991)
argued that chaotic deterministic dynamics should be commonplace in plankton communities. Ascioti
et al. (1993), Strutton et al. (1996, 1997) and Seuront (2004), however, did not find any evidence of
chaotic dynamics in both zooplankton and phytoplankton time series, phytoplankton transects, and
temperature, salinity, and *in vivo* fluorescence time series, respectively. Ascioti et al. (1993) found a
significant level of predictability of zooplankton abundance from that of phytoplankton, indicative of
a deterministic trophic link. A recent application of the nearest-neighbor algorithm to time series of a
range of physical and biological variables for the north Pacific Ocean (Hsieh et al. 2005) showed that
physical variables were characterized by a high dimensionality (E bounded between 13 and 20) and

were best modeled as linear autoregressive processes of high order. In contrast, time series of biological variables such as Scripps Pier (California) diatoms and California Cooperative Oceanic Fisheries Investigations (CalCOFI) larval fish and zooplankton consistently exhibit a low-dimensional nonlinear signature ($3 \leq E \leq 8$). Recently, chaos has been identified in models of planktonic biodiversity (Huisman and Weissing 1999), the temporal dynamics of the deep chlorophyll maximum from Station ALOHA in the subtropical Pacific Ocean (Huisman et al. 2006), and the long-term dynamics of an experimental plankton community (Benincà et al. 2008)

6.1.5 A FEW MISCONCEPTIONS ABOUT CHAOS

Several misconceptions about chaos precisely pertain to its relationship to stochastic behavior (Hastings et al. 1993). Chaos and stochasticity are nevertheless not equivalent; not only do the underlying mechanisms differ, but the consequences for observers are also very different. In purely deterministic systems, predictions made from the governing equations will be perfect. Chaotic systems are predictable over short time scales because they are deterministic; the lack of predictive power over long time scales stems from the lack of complete information about the exact location of initial conditions. In contrast, purely stochastic systems are unpredictable over any time scale because of their probabilistic nature. In such approaches, the variability of a given descriptor is driven by "new" events, which represent exogenous variables—exogenous in the sense that they are not a part of an internal mechanism that drives the descriptor fluctuations. The branches of a tree move because of the wind, which is "exogenous" to the tree, and therefore "new" to it, whereas a chaotic model of the motion of trees would assume the existence of a simple deterministic "nonlinear" engine within the tree (that is, endogenous) that generates chaotic motion by a simple mechanism of feedback of the motion of the tree upon itself. Finally, the distinction between stochastic and deterministic dynamics has important practical implications. For instance, if fluctuations in population sizes are driven primarily by deterministic factors, and if those factors are understood, then the dynamics are predictable over short time scales. Management of such populations is feasible. On the other hand, if fluctuations are driven primarily by exogenous stochastic forces, then prediction and management become much more difficult.

Now, given that deterministic equations in a small number of variables can generate complicated behavior—see, for example, Equations (6.2), (6.5), (6.6), and (6.7), and Figure 6.3—the question arises: How much of the complicated behavior observed in nature can be described by a small number of variables? This question has been widely addressed in the framework of turbulence. Ruelle and Takens (1971) showed that near the transition to turbulence, the many degrees of freedom of turbulence are coupled coherently and lead to an enormous reduction in dimension (that is, low-order deterministic chaos). However, both empirical and theoretical studies have demonstrated that fully developed turbulence (for example, Frisch 1996 and references therein) was characterized by its multifractal properties (that is, high-order stochasticity). Moreover, the effects of both hydrodynamic and advective processes on the multifractal structure of both physical (temperature and salinity) and biological (phytoplankton biomass) parameters have been identified in a study of phytoplankton patchiness in turbulent environments (Seuront 2005b). This issue, together with the fundamental differences between fractals and multifractals, will be emphasized in Chapter 8.

6.1.6 THEN, WHAT IS CHAOS?

When we look at the changing world that we are living in, we can categorize observed changes into a few fundamental categories: growth and recession, stagnation, cyclic behavior, and unpredictable, erratic fluctuations. All of these phenomena can be described with very well-developed linear mathematical tools. Here *linear* refers to the result of an action being always proportional to its cause: if we double our effort, the outcome will also double. Patterns and processes descriptive in terms of linearity (for example, clocks, motion of planets) are referred to as being *ordered*. Their predictability is strong, and their characteristic attractors will be single points (that is, stable equilibrium), closed loops or tora through phase-space (that is, a stable limit cycles), thus revealing their finite dimensionality.

However, most of nature is nonlinear in the same sense that most of ecology is nonaquatic ecology. The situation where most of traditional science is focusing on linear systems can be compared to the story of the person who looks for the lost car keys under a street lamp because it is too dark to see anything at the place where the keys were lost. One whole class of phenomena that does not exist within the framework of linear theory has become known under the vague term of *chaos*. The modern notion of chaos describes irregular and highly complex structures in time and in space that follow deterministic laws and equations. This concept is specifically referred to as *low-order deterministic chaos*. Deterministic chaotic systems are characterized by a finite, short-term predictability, strange (that is, fractal) attractors, and then a low dimensionality. The dimension of the attractor estimated using specific techniques (see Section 6.1.3.2) indicates at least how many variables are necessary to describe evolution in time. For instance, a dimension of 2.5 indicates that the pattern or process of interest can be described by a system of equations containing three independent variables.

In contrast, the structure of a given system can be so complex, and its variability so violent (see Figure 2.2), that the methods devoted to the identification of (low-order) deterministic chaos become inefficient. No more evidence for organization in the phase-space (Figure 6.12), no more evidence for short-term predictability (Figure 6.13), no more evidence for low-order dimensionality (Figure 6.14): this is the signature of *high-order stochasticity*. In this framework, the systems do not show strange attractors in the phase-space, their predictability is extremely weak, and their description requires a large (even infinite) number of parameters. In the example shown in Figure 6.16B, the nonconvergence of the correlation dimension for embedding dimensions $D_E = 10$ means that the description of the studied temperature, salinity, and fluorescence time series would require a system of equations involving at least 10 independent variables for their description. These different features are illustrated in Figure 6.17.

6.2 FRACTALS AND SELF-ORGANIZATION

The scale-invariance of fractals is frequently related to self-organization in nonlinear dynamic systems consisting of large aggregations of interacting elements. As such a system moves on a strange attractor in phase-space, any particular length scale from external forcing is lost ("forgotten"), and instead, the smallest length scale of the individual elements propagates its effect across all scales. This generates pattern formation that may or may not exhibit fractal properties. An archetypical example is Rayleigh-Bénard convection, where individual fluid motions are chaotic while a pattern of convective cells is formed that scales with the viscosity of the fluid (instead of with the amount of heat dissipation or the dimension of the container). The emergence of structure, or order, in a system through its internal dynamics and feedback mechanisms is the essence of self-organization, as opposed to the generation of regularity as a result of external forcing. In a thermodynamic perspective, self-organization arises in nonlinear systems that are far from equilibrium and dissipative (irreversible). Coherent motion and patterns created in such systems are therefore called dissipative structures. Further thermodynamics interpretations of complex systems lead to principles of minimum entropy production in open systems and maximum entropy states in closed systems. The stochastic interpretation of these entropy principles in complex systems can in turn be related to information theory (Shannon and Weaver 1949; Jaynes 1957; Brillouin 1962; Kapur and Kesavan 1992).

6.3 FRACTALS AND SELF-ORGANIZED CRITICALITY

6.3.1 DEFINING SELF-ORGANIZED CRITICALITY

A variation of the self-organization concept is the model of self-organized criticality (Bak and Chen 1991; Bak et al. 1987, 1988). The archetypical example is the accumulating sand pile, in which the

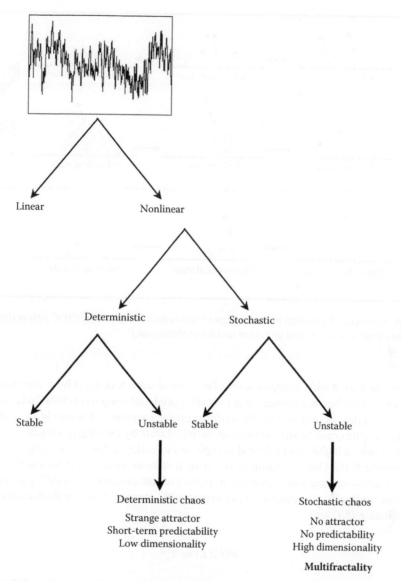

FIGURE 6.17 Schematic diagram showing the dichotomy leading to the characterization of a data set in terms of low-order deterministic chaos, or high-order stochasticity.

nonlinear dynamics between disturbed and avalanching sand grains retain the system in a critical state with the slopes of the pile at the angle of repose. The properties (such as angularity and size of the sediment) of the smallest element, the grain, determine the large-scale properties of the system as a whole (the critical angle of repose). A small disturbance (for example, the addition of another grain of sand at the top) can trigger avalanches that can attain any size, constrained only by the size of the pile itself. More specifically, when the pile is flat there is little interaction among the different regions of the pile and adding a single grain will only affect a few other grains nearby. The system is in a *subcritical state* (Figure 6.18). As the pile grows by adding grains of sand, avalanches of grains spill down the sides such that adding a single grain can initiate a cascade, affecting many other grains. Eventually, the slope of the pile grows until the angle of repose is reached. The pile reaches a *critical state* and essentially does not get any steeper (Figure 6.18). Now if grains are added,

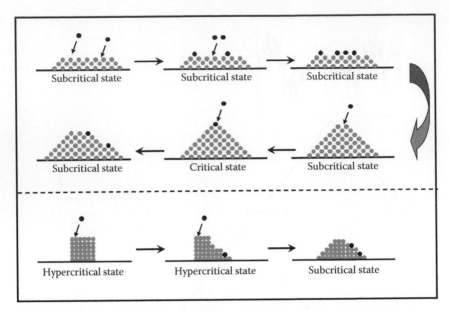

FIGURE 6.18 Schematic illustration of the concept of self-organized criticality (SOC) using the archetypical example of the dynamics of the sand pile. (See text for explanations.)

avalanches occur with a wide range of sizes. The critical state is defined by a stationary statistical distribution of avalanches that propagate across all spatial and temporal scales (only limited by the finite size of the pile), as opposed to the uniform power spectrum of a purely stochastic process. Alternatively, the pile could be started in a *supercritical state* by forming a vertical cylinder of sand. A supercritical pile is highly unstable and is expected to collapse down to a critical state as grains are added (Figure 6.18). One can think of the critical state as an attractor for the dynamics of the pile. Note that the dynamical and structural properties of self-organized criticality are characterized by a power law stating that the probability of events with intensity I greater than a critical threshold I_c follows Equation (5.1) as:

$$P(I \geq I_c) \propto I_c^{-\phi} \tag{6.18}$$

In the sand pile case, the events are avalanches of sand grains and the intensity of the events is the number of grains in an avalanche. Similarly, the number of grains $N(d)$ falling a distance d follows the power-law form $N(d) \propto d^{-D}$ where D is the fractal dimension of the avalanches. More generally, for a critical system, the distribution of fluctuation sizes s is described as:

$$F(s) \propto s^{-D} \tag{6.19}$$

where $F(s)$ is the frequency of s and D is the related fractal dimension. In the sand pile case, the events are avalanches of sand grains, and the size of an event is the number of grains in a particular avalanche. Frequency is estimated as the number of events of size s divided by the total number of events. Note that Equation (6.19) is conceptually similar to Equation (5.4), hence, $D = D_{pi}$ (see Section 5.2).

Because of the above-mentioned intrinsic scaling properties, self-organized structures can be described in terms of fractality. The negative exponent in Equation (6.19) leads to many small events or fluctuations punctuated by progressively larger events, hence the notion of intermittency that will be introduced in Chapter 8. From the previous statements, it also appears that self-organized criticality occurs in systems that build up stress and then release the stress in intermittent pulses, that is, large fluctuations interspersed among period of relative stasis. In that way, self-organized criticality can also be related to the concept of multifractals, basically designed to characterize the spectrum of the different intensity levels observed in the intermittent fluctuations of a given descriptor; see Chapter 8.

6.3.2 SELF-ORGANIZED CRITICALITY IN ECOLOGY AND AQUATIC SCIENCES

Self-criticality has been identified in a range of ecological areas, ranging from tree-fall gap formation in tropical rainforests (Manrubia and Solé 1996), bird population/extinction dynamics (Keitt and Marquet 1996), and species extinction observed in fossil records (Raup 1982) to models of ecosystems (Bak et al. 1989) and evolution (Kauffman and Johnsen 1991; Bak and Sneppen 1993; Flyvbjerg et al. 1993; Paczuski et al. 1995; de Boer et al. 1994). In aquatic ecology, the only known reference to self-organized criticality is a very recent work that found evidence for a critical state—more specifically, a critical biomass—in the microscale spatial organization of microphytobenthos in the sediment* (Seuront and Spilmont 2002) (Figure 3.21). It may be difficult to make the conceptual connection between the sand-pile model and the spatial patchiness of microphytobenthic organisms. Such an understanding is nevertheless a salient issue to bridge the gap still remaining between the physics of nonlinear, nonequilibrium systems and aquatic ecology. We try to make this point clearer hereafter.

The decrease in the number of patches above a critical biomass (see Figure 5.4) suggests that the dynamics of patch formation are structured by conflicting constraints. In the case of the sand pile model, the constraints are gravity, which acts to lower the height of the pile, and the addition of sand grains, which raises the height of the pile. The structure of the pile emerges from the interaction of these forces. It is a salient issue to realize that, although gravity acts uniformly on all grains in the pile, the probability of an avalanche is not spatially uniform across the pile. Some areas of the pile will have steeper slopes and thus a higher probability of sliding. Each avalanche changes the spatial pattern of slopes and thereby affects the size of subsequent avalanches, which in turn determine the structure of the pile yet again. It is this pattern of long-range correlations among avalanches that is the key to understanding self-organized criticality. The constraints, and their potential effects, that act on the structure and dynamics of a microphytobenthic assemblage are outlined hereafter. In the case of microphytobenthos biomass, the microscale distribution of patches is the result of both endogenous (for example, microphytobenthos growth, migration, and death) and exogenous processes (for example, tides, hydrodynamism, sediment quality, interspecific and intraspecific competition for nutrient, grazing) that can act to decrease or increase the microphytobenthos biomass. As illustrated in the sand pile model, these constraints do not act uniformly over the whole spatial domain. For instance, biomass losses related to grazing are dependent on both the spatial distribution and foraging abilities of predators. Growth and death are dependent on nutrient and light availability that is also a function of the burying depth of microphytobenthos cells, the density, and the spatial distribution of the sediment and the duration of the emersion. The microphytobenthic community at the sediment surface may be disturbed by turbulence and shear stress generated by tidal currents or wind waves and lead to microphytobenthos cell load in the water column. The degree of disturbance depends on the interplay of a number of factors, including sediment type, stability of the sediment surface, mean water depth, tidal height, magnitude of tidal currents, wave height, and macrofaunal abundance and activity. In particular, resuspension processes occur during immersion and lead to biomass losses for the microphytobenthic system. On the other hand, resettling of cells occurring at the beginning of emersion can be regarded as playing a major role in the observed patch pattern.

These constraints, acting quite obviously to increase or decrease microphytobenthos biomass, result in a dynamic balance as in the sand-pile model. However, the cause of patchiness, and

---

* See Section 3.2.4.2 for a description of the data and their analysis using the mass dimension method.

in particular the self-organized criticality observed in patch patterns, is less clear. A potential mechanism for patch formation is discussed hereafter, with specific reference to the critical biomass observed in the microphytobenthos patch pattern. A candidate mechanism for patchiness is competition among species. If competition is a driving force in structuring a microphytobenthos community, then the important dynamics would be observed in the niche space occupied by different species (MacArthur 1960; Hutchinson 1961; Odum 1971). Competitive pressure would be expected to be high in regions of niche space where species are densely packed, as would happen, for instance, when a number of species share the same food resource. It is possible that, like the steep region of the sand-pile, species occupying dense regions of niche space (that is, where chlorophyll concentration is higher than the observed critical biomass) are subject to higher extinction probabilities, and then reduce the probability of high-density patches. The loss of species would change the distribution of species in a niche space and, in turn, change the probability of extinction and patches, much like the dynamics of the sand pile model. The system is in a critical state. In contrast, species occupying sparse regions of the niche space (that is, where chlorophyll concentration is smaller than the observed critical biomass) are subject to weaker competition pressure and extinction probabilities. The system is then in a more stable, or subcritical state, and does not exhibit any fingerprints of self-organized criticality.

7 Estimating Dimensions with Confidence

As stated by Hastings and Sugihara (1993), the first key steps in fractal analysis are (1) the choice of an appropriate power law, (2) the application of log transforms, and (3) the use of linear regression to fit a log-transformed linear model. In Chapters 3 and 4, we thus summarize some of the more commonly used methods, together with more original ones, for estimating the fractal dimension D_F of both self-similar and self-affine natural objects. Formal mathematical derivations and proofs have been omitted; readers interested in fractal theory should consult Mandelbrot (1983), Voss (1985, 1988), Falconer (1985, 1993), Frontier (1987), Feder (1988), Sugihara and May (1990a), Schroeder (1991), Peitgen et al. (1992), Hastings and Sugihara (1993), Tricot (1995), and Gouyet (1996). Note, however, that some of the methods used to estimate the fractal dimension are empirically, not mathematically, derived. Other reviews that have summarized fractal dimension estimation methods include Loehle (1983), Frontier (1987), Milne (1988, 1991), Kaye (1989, 1994), Williamson and Lawton (1991), Kenkel and Walker (1993), Klinkenberg (1993), Nonnenmacher et al. (1994), Johnson et al. (1995), and Seuront (1998). Most of these reviews have been somewhat selective or have focused on a specific subdiscipline within the biological sciences. The diversity of available approaches for determining the fractal dimension reflects differences in objectives and in the type of data analyzed.

However, we stress here that these key steps are not as straightforward as might appear at first glance and that several intrinsic characteristics of fractal patterns and processes have to be clearly identified and carefully checked for identifying potential deviation from fractal behavior, and thus for the results of fractal analysis to be meaningful. It is clear that if the basic methodology is unreliable, a great deal of research effort is being compromised; briefly put, a dimension estimate is always produced, with no indication of its reliability of likely error. Reliable procedures giving some guidance as to the degree of confidence can be placed on its estimates are essential to preventing embarrassing errors. Such errors can have salient consequences, for instance, when dimension estimation is used as a possible diagnostic tool (see, for example, Zbilut 1988; Nunes Amaral et al. 1998; Ivanov et al. 1998, 1999). In this chapter, we thus address in detail several potential deviations from fractal behavior and propose some simple procedures to ensure the relevance of fractal dimension estimates.

7.1 SCALING OR NOT SCALING? THAT IS THE QUESTION

Fractal analysis is implicitly based on the identification of power laws in log-log plots and on the subsequent use of linear regression to fit a log-transformed linear model. However, as stated above, the apparent scaling can be simply the result of the generic property of the quantity to increase or decrease monotonically as the scale goes to zero irrespective of the geometry of the object. Consequently, one must question the validity of fitting a straight line over the whole range of available scales. We thus introduce here objective, statistically sound procedures for testing the existence of scaling properties, and then we briefly discuss the implications of finding multiple scaling properties.

7.1.1 IDENTIFYING SCALING PROPERTIES

When we are dealing with exact fractals (see, for example, Figures 2.5, 2.8, 3.1, and 3.3), there are no difficulties in calculating a fractal dimension. The log-log plots are very linear and we always recover an expected and *a priori* known result whatever the methods employed; see Seuront et al. (2004a, their Figures 6 and 7). Conversely, when we are dealing with any patterns and processes whose properties are not known *a priori* (for example, coastlines, time series of plankton abundance), complications begin to arise. In such cases, many analysts have implicitly made an assumption of linearity in the log-log plot (see, for example, Crist et al. 1992; With 1994; Erlandson and Kostylev 1995; Snover and Commito 1998; Dowling et al. 2000). As a result, the scaling region was estimated subjectively and its relevance simply related to the statistical significance of the coefficient of determination (r^2). We will nevertheless demonstrate here—on the basis of several very simple examples of log-log plots exhibiting extremely strong, and then *a priori* statistically significant, linearity (Figure 7.1)—that ensuring the significance of the coefficient of determination (r^2) is far from sufficient to conclude the presence of fractality in any patterns and processes.

Consider, for instance, the four test-case log-log plots shown in Figure 7.1. They correspond to the power law $M(\delta) \approx \delta^{-1.2}$ (Figure 7.1A; see Equation 2.1), to a smooth second-order polynomial, $\log M(\delta) = -0.32(\log \delta)^2 - 0.86 \log \delta - 0.06$ (Figure 7.1B), to a combination of a power-law behavior for the seven medium points and a second-order smooth polynomial (Figure 7.1C) for the extreme points, and to the addition of 10% random noise to the original power law (Figure 7.1D). They all exhibit an extremely strong and *a priori* significant ($p < 0.001$) linearity that could have led to the spurious conclusion of the presence of fractality in each of these four case studies. We will see hereafter why it is not the case and how this misinterpretation can be avoided.

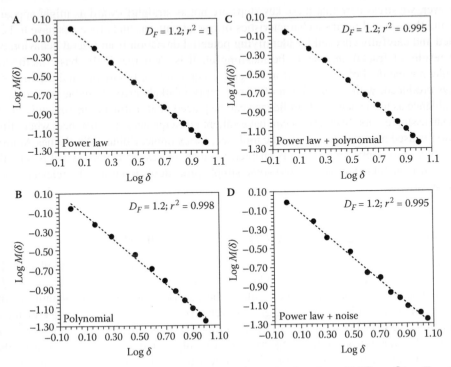

FIGURE 7.1 Scaling behavior of four distinct signals, shown as log-log plots of $M(\delta)$ vs. δ (see Equation 2.1): a power law (A; $M(\delta) = \delta^{1.2}$), a smooth second-order polynomial (B; $\log M(\delta) = -0.32(\log \delta)^2 - 0.86 \log \delta - 0.06$), a combination of the two previous signals (C), and the power law contaminated with additive noise (D). A first examination, based on the values of the coefficient of determination r^2, $r^2 \in [0.995 - 1]$, could lead to conclude strong linear behavior in all cases.

7.1.1.1 Procedure 1: $R^2 - SSR$ Procedure

Consider a regression window of a varying width that ranges from a minimum of five data points (the least number of data points to ensure the statistical relevance of a regression analysis) to the entire data set. The smallest windows are slid along the entire data set, with the whole procedure iterated $(n - 4)$ times, where n is the total number of available data points (Figure 7.2). Within each window and for each width, we estimate the coefficient of determination (r^2) and the sum of the squared residuals (SSR) for the regression. Finally, we use the values of δ (Equation 2.1), or more generally the scale values (see Chapters 3 and 4), which maximized the coefficient of determination and minimized the total sum of the squared residuals (Seuront and Lagadeuc 1997) to define the scaling range and to estimate the related fractal dimensions (Figure 7.3; Table 7.1). This optimization procedure will be referred hereafter to as the $R^2 - SSR$ criterion. Applying this procedure to cases B, C, and D shown in Figure 7.1 thus leads to identify scaling behaviors over a limited range of scales and over the whole range of available scales for case studies C and D, respectively (Figure 7.3 and Figure 7.4). On the other hand, the log-log plot of case study B never satisfies the $R^2 - SSR$ criterion, revealing the nonfractal nature of the underlying process.

However, the definition of "independent" and "dependent" variables required in least-squares regression analysis is not straightforward in power-law applications (see Zeide and Gresham 1991). This is a serious but largely unrecognized problem, and using least-squares regression in this way may result in biased slope estimates (Kenkel and Walker 1993; Loehle and Bai-Lian 1996). Two methods—(1) principal axis regression (equivalent to principal components analysis) and (2) reduced

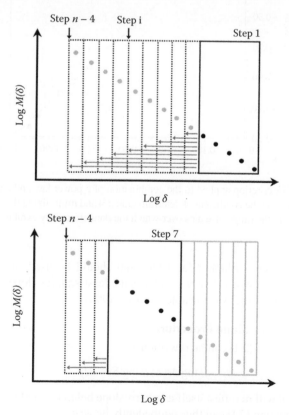

FIGURE 7.2 Schematic illustration of the $R^2 - SSR$ criterion: a regression window of a varying width that ranges from a minimum of five data points (dark lined rectangle) to the entire data set (dotted lines). The smallest windows are slid along the entire data set, with the whole procedure iterated $(n - 4)$ times, where n is the total number of available data points. Within each window and for each width, the coefficient of determination (r^2) and the sum of the squared residuals (SSR) for the regression are estimated.

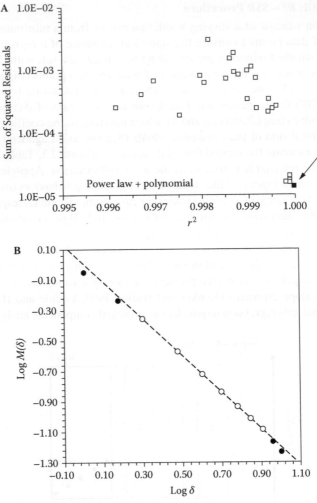

FIGURE 7.3 The $R^2 - SSR$ criterion applied to the combination of a power law and a polynomial (Figure 7.1C). The values of log δ maximizing the coefficient of determination (r^2) and minimizing the sum of the squared residuals (A; arrow), correspond to the range of scales over which the data effectively exhibit a scaling behavior, shown as open symbols (B).

major axis regression—are available (Zar 1996) and should be used instead. In particular, the reduced major axis slope is obtained very simply, as the least-squares slope divided by the product-moment correlation between the two variables (Niklas 1994).

7.1.1.2 Procedure 2: Zero-Slope Procedure

One may note that Equation (2.1) can be rewritten as:

$$d \log M(\delta)/d \log \delta = -D_F \qquad (7.1)$$

Then if scaling exists, it will manifest itself as a zero-slope behavior in plots of $d \log M(\delta)/d \log \delta$ vs. logδ (Figure 7.5A). Equation (7.1) can thus equivalently be written as:

$$d [d \log M(\delta)/d \log \delta]/\log \delta = 0 \qquad (7.2)$$

From Equations (7.1) and (7.2), it can be easily seen that the intersection of the range of scales exhibiting a zero-slope behavior with the y axis (that is, $d \log M(\delta)/d \log \delta$) provides an estimate of

TABLE 7.1
$R^2 - SSR$ **Criterion Applied to a Combination of Power Law and a Polynomial**

$i\text{-}j$	r^2	SSR	D_F
1-1	0.998	6.1E-04	1.120
1-2	0.998	7.2E-04	1.133
1-3	0.999	8.2E-04	1.142
1-4	0.999	9.6E-04	1.151
1-5	0.999	1.1E-03	1.157
1-6	0.999	1.9E-03	1.169
1-7	0.998	3.1E-03	1.183
2-1	0.999	2.0E-04	1.170
2-2	0.999	2.1E-04	1.174
2-3	1.000	2.4E-04	1.179
2-4	1.000	2.6E-04	1.183
2-5	0.999	7.3E-04	1.196
2-6	0.999	1.5E-03	1.211
3-1	1.000	1.7E-05	1.199
3-2	1.000	1.9E-05	1.201
3-3	1.000	1.4E-05	1.202
3-4	0.999	3.3E-04	1.218
3-5	0.999	9.0E-04	1.235
4-1	1.000	1.7E-05	1.196
4-2	1.000	1.6E-05	1.198
4-3	0.999	3.2E-04	1.224
4-4	0.998	8.0E-04	1.249
5-1	1.000	1.6E-05	1.198
5-2	0.998	2.8E-04	1.238
5-3	0.997	6.6E-04	1.273
6-1	0.997	1.8E-04	1.274
6-2	0.996	4.0E-04	1.317
7-1	0.996	2.5E-04	1.368

Note: See Figure 7.1C. In the first column, the first and second numbers (i and j) identify the window number and the number of data points n used to estimate the regression (that is, the width of the regression window; $n = j + 5$, because the minimum window width includes 5 points). For instance, $i = 1$ and $j = 2$ correspond to the first regression window with a 6-point width, while $i = 3$ and $j = 5$ correspond to the third regression window with a 9-point width. For each regression window and for each width, the coefficient of determination (r^2), the sum of the squared residuals (SSR), and the corresponding fractal dimension (D_F) are estimated. The shaded areas correspond to the window-width combination that satisfied the R^2-SSR criterion.

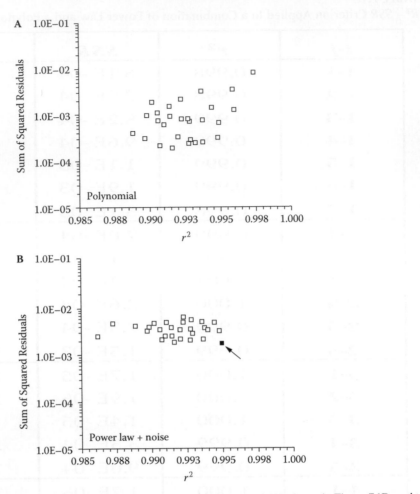

FIGURE 7.4 The $R^2 - SSR$ criterion applied to the second order polynomial (A), shown in Figure 7.1B; and to the power law contaminated with additive noise (B), shown in Figure 7.1D. In the former case, the $R^2 - SSR$ criterion is never satisfied, the values of log δ maximizing the coefficient of determination also maximizes the sum of squared residuals, undoubtedly showing the absence of a linear trend in Figure 7.1B. In the latter, the $R^2 - SSR$ criterion is fully satisfied (arrow), and confirms the linearity of the log-log plot of $M(\delta)$ vs. δ as observed in Figure 7.1D.

the fractal dimension D_F. Going back to Chapters 3 and 4, this framework can be generalized to any fractal length, surface, and volume. Equations (3.1) and (3.2), with Equations (3.6) and (3.7), can then be rewritten as:

$$d \log L(\delta)/d \log \delta = 1 - D_F$$
$$d \log S(\delta)/d \log \delta = 2 - D_F \qquad (7.3)$$
$$d \log V(\delta)/d \log \delta = 3 - D_F$$

or equivalently as:

$$d \, [d \log L(\delta)/d \log \delta]/\log\delta = 0$$
$$d \, [d \log S(\delta)/d \log \delta]/\log\delta = 0 \qquad (7.4)$$
$$d \, [d \log V(\delta)/d \log \delta]/\log\delta = 0$$

Scaling behaviors will thus be identified by a plateau in plots of $d \log L(\delta)/d \log\delta$ vs. $\log\delta$, $d \log S(\delta)/d \log\delta$ vs. $\log\delta$, and $d \log V(\delta)/d \log\delta$ vs. $\log\delta$, and the fractal dimensions D_F estimated as the intersection of the plateau with the $d \log M(\delta)/d \log\delta$ axis. To ensure the statistical relevance of this procedure, we use a sliding regression window similar to the one described in the $R^2 - SSR$ procedure. The significance of the differences between the slope of each regression and the expected zero-slope line is directly tested using standard statistical analysis for each window size; see, for example, Zar (1996). The scaling range will then be defined as the scales that statistically verify Equation (7.2). This procedure, hereafter referred to as the "zero-slope" criterion, has been successfully applied to the identification of scaling ranges in zooplankton swimming behavior (Seuront et al. 2004a, Figure 8).

However, the main disadvantage of this procedure is the intrinsic noise enhancement generated by taking the first derivative of any linear or pseudolinear trends (Figure 7.5B,C,D), that are challenging for standard statistical procedure (see, for example, Hirsch and Smale 1974). The application of this procedure to the test cases investigated here thus confirms the result obtained above from the $R^2 - SSR$ criterion for the second-order polynomial (no scaling regime—that is, plateau—detected; Figure 7.5B) and the combination of a power law and a second-order polynomial (scaling regime—that is, plateau—detected over a range of scales similar to the one obtained via the $R^2 - SSR$ criterion; Figure 7.5C). It might nevertheless lead to the spurious conclusion of the absence of a scaling behavior in the case of a power law with additive noise (Figure 7.5D). To overcome this limitation, we stress that the significance of the zero slope must be tested by simulation to explicitly include the potential effect of additive noise (see Section 7.3 for further details). To overcome the relative unreliability of the zero-slope procedure, an additional, more objective optimization criterion, referred to as the "compensated slope" procedure, is introduced hereafter.

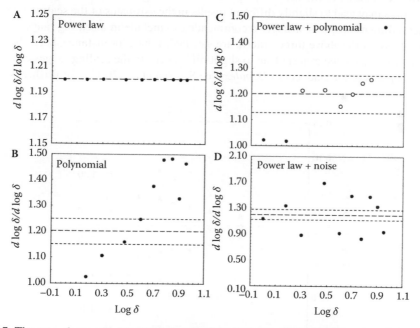

FIGURE 7.5 The zero-slope criterion applied to the four data sets shown in Figure 7.1. While the plateau behavior shown by the power-law behavior is obvious, no plateau (i.e., linear behavior) has been detected in the polynomial case (B). A statistically significant plateau has been shown for the combination of the power law and a polynomial over a restricted range of scale (C; open symbols). These results confirm the results of the $R^2 - SSR$ criterion. However, in the case of a power law with additive noise this criterion rather suggests the absence of a scaling behavior (D). The dashed and dotted lines represent the expected zero-slope and its 95% confidence interval, respectively.

7.1.1.3 Procedure 3: Compensated-Slope Procedure

To investigate both the existence and the nature of a scaling range, the scaling behavior generically illustrated by Equation (2.1) can be compensated by a scaling factor δ^c as:

$$M(\delta) \approx \delta^c \times \delta^{-D_F} \qquad (7.5)$$

where δ is the scale and c is the compensation exponent defined as $0 \le c \le 1$, $1 \le c \le 2$, and $2 \le c \le 3$ for a fractal curve, surface, and volume, respectively. More generally, varying c from $D_E - 1$ to D_E, where D_E is the Euclidean dimension of the embedding space, should converge to

$$M(\delta) \approx \delta^c \times \delta^{-D_F} \approx \delta^{c_e} \qquad (7.6)$$

where c_e is the compensated exponent that ultimately leads to $c_e = 0$ (that is, $M(\delta) = 1$) when $c = D_F$ (Figure 7.6 and Figure 7.7). A scaling behavior will then manifest itself as a straight line in a log-log plot of $[\delta^c \times \delta^{-D_F}]$ vs. δ. The range of scale to include in the analysis is subsequently chosen using the $R^2 - SSR$ procedure, and the significance of the potential differences between the slope of each regression and the expected zero-slope line is directly tested using standard statistical analysis for each window size. In that way, we ensure that the plateaus exhibited by the data points in Figure 7.5C, Figure 7.5D, Figure 7.7C, and Figure 7.7D are indeed a manifestation of scaling and not the result of random nonfractal structure. The value of c that fully satisfies the two previous criteria thus provides an estimate of the fractal dimension D_F. Finally, we stress that the extremely weak dispersion observed around the expected $\log[\delta^{c=1.2} \times \delta^{-D_F=1.2}] \approx \log \delta^{c_e=0}$ vs. $\log \delta$ (Figure 7.7C and Figure 7.7D) ensures the relevance and the robustness of this compensated-slope optimization criterion when compared to the zero-slope procedure (see Figure 7.5C and Figure 7.5D). Because these procedures may lead to slightly different results in the estimates of the scaling ranges and the related fractal dimensions, it is strongly recommended to include in a scaling range the scales for which at least two of the above three criteria are satisfied. It has, for instance, been shown that the $R^2 - SSR$ and the zero-slope criteria lead to slight differences in the scaling ranges estimated from 3D swimming paths of the cladoceran $D. pulex$; see Seuront et al. (2004a) for further details.

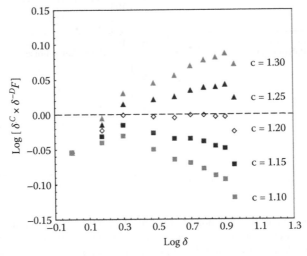

FIGURE 7.6 The compensated slope procedure, applied to the combination of a power law and a polynomial function for different values of the compensation exponent c. The value taken by the exponent c when the log-log plot of $[\delta^c \times \delta^{-D_F}]$ vs. δ exhibits a plateau behavior corresponds to the fractal dimension of the initial data set, that is, $c = D_F$.

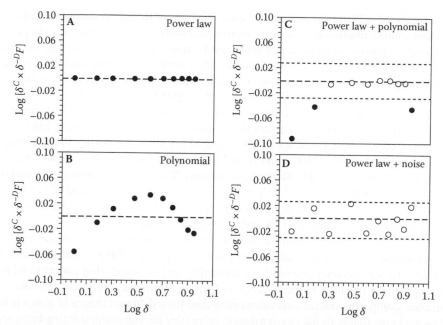

FIGURE 7.7 The compensated slope criterion applied to the four data sets shown in Figure 7.1. These results are fully consistent with the intrinsic nature of the four data sets: a pure plateau for the power law (A), an absence of plateau for the polynomial (B), a significant plateau for the combination of the power law and a polynomial over a restricted range of scale (C; open symbols), and a statistically significant plateau for the noise contaminated power law (D). The dashed and dotted lines represent the expected zero-slope and its 95% confidence interval, respectively.

7.1.2 SCALING, MULTIPLE SCALING, AND MULTISCALING: DEMIXING APPLES AND ORANGES

The intrinsic hypothesis of fractal theory is based on the existence of a single power law from smallest to largest available scales (see, for example, Mandelbrot 1983). We saw that this is indeed the case for theoretical fractals such as the Koch snowflake (Figure 2.5), the Sierpinski carpet and gasket (Figure 2.8), and the Cantor dust (Figure 3.3). However, there are powerful constraints on structure and function that do not allow organisms, as well as any observables in nature, to maintain the same geometric relationships among their components, as size (that is, scale) changes over several orders of magnitude. This has been further exemplified and discussed in Section 2.1. As a consequence, the potential scaling ranges of many patterns and processes are intrinsically characterized by lower and upper bounds.

This further questions the key assumption regarding the fractal dimension as a scale-independent parameter. Strictly speaking, this means that in a particular environment, if we calculate D_F for a fractal curve that is several meters long, we should obtain the same value of D_F for curves measured at a scale of hundreds of meters to kilometers. This converges on one of the central issues faced by landscape ecologists: understanding how to meaningfully extrapolate ecological information across spatial scales (Gardner et al. 1989; Turner and Gardner 1991). However, this is not as straightforward as it might appear at first glance. Indeed, if log-transformed data appear to be a piecewise linear model, linear regression can be used on separate intervals, overlapping only at endpoints, and endpoints can be varied so as to satisfy the above-mentioned optimization criteria. If linear regression then yields significantly different slopes on adjacent regions, one should reject the hypothesis of a single power law and replace it with the *multiple scaling* hypothesis of separate power laws over separate regions (see Figure 2.4).

The existence of self-similar hierarchies (that is, changes in fractal dimension when shifting between scales) also implies that in place of true self-similarity, we observe only *partial self-similarity* over a limited range of scales separated by transition zones, where the environmental properties or constraints acting upon organisms are probably changing rapidly (Frontier 1987; Seuront and Lagadeuc 1997; Seuront et al. 1999). Because different scales are necessarily related to different aspects of structure, fractal methods can be applied in order to detect self-similar hierarchies in ecology. Such hierarchical scaling has been observed, for instance, in coral reefs (Bradbury et al. 1984), from patch perimeter measures in deciduous forests (Krummel et al. 1987), vegetation patterns (Morse et al. 1985), landscapes (Wiens and Milne 1989; Scheuring 1991), phytoplankton patches (Seuront et al. 1996a, 1999), and from Eulerian and Lagrangian physical forcings in the coastal ocean (Seuront et al. 1996b; Seuront and Schmitt 2004). As a conclusion, the fractal dimension may have the desirable feature of only being constant over a finite, instead of an infinite, range of measurement scales; see, for example, Section 4.2.1.2 and Figure 4.6 for a discussion of the relevance of scaling regimes changing with scales in aquatic sciences.

In addition, upper and lower fractal limits are controlled by the size of the data set and should not be confused with scales where the fractal dimension of the pattern changes. This distinction is very useful for identifying characteristic scales of variability and for comparing patterns and processes that may respond, for instance, to the structure of their environment at different absolute scales. As a consequence, comparing fractal dimensions estimated from different ranges of scales is meaningless unless we know that both the environmental properties (or constraints) acting upon organisms and the organisms' physiology/biology/ecology are the same over these scales. Changes in the value of D_F with scale may indicate that a new set of processes is controlling the observed variability. Thus, the scale dependence of the fractal dimension over finite ranges of scales may carry more information, both in terms of driving processes and sampling limitation, than its scale independence over a hypothetical infinite range of scales. This issue is particularly relevant in aquatic sciences, where any divergence to a $-5/3$ power law in Fourier space is used to infer the nature of the processes controlling the observed variability (Section 4.2.1.2). Both whitening and reddening of the $-5/3$ power spectrum expected in cases of purely passive scalars advected by turbulent fluid motions are then a manifestation of various forms of biological activity; see, for example, Powell et al. (1975), Denman and Platt (1976), Denman et al. (1977), Lekan and Wilson (1978), Abbott et al. (1982), Weber et al. (1986), Powell and Okubo (1994), Seuront et al. (1996a, 1999), Seuront and Lagadeuc (2001), Lovejoy et al. (2001), and Currie and Roff (2006).

Alternatively, although the point of slope change may indicate the operational scale of different generative processes, it might simply reflect the limited spatial resolution of the data being analyzed (Hamilton et al. 1992; Kenkel and Walker 1993; Gautestad and Mysterud 1993). In order to distinguish these two situations, and thus to ensure the relevance of fractal analysis, one needs to be able to examine a given set at a variety of spatial (or temporal) scales. A data set has fractal limits and, as stated above, outside these limits methods to measure the fractal dimension will return a trivial value. Falconer (1993) recommends having at least three orders of magnitude between these limits to ensure the relevance of fractal analysis. However, this requirement can be reconsidered considering the extreme difficulty in gathering such a large number of discrete measurements in ecology, as well as the ecologically meaningful results obtained from data sets spanning between one and two orders of magnitude (Erlandson and Kostylev 1995; Seuront and Lagadeuc 1997, 1998; Commito and Rusignuolo 2000; Waters and Mitchell 2002). Unfortunately, there is no reliable way to test scale invariance and measure a fractal dimension of very small data sets.

Finally, multiple scaling should not be confused with multifractality, another possible deviation from fractal behavior, sometimes also referred to as multiscaling (Martinez et al. 1995; Seuront et al. 1999) or multifractal scaling (Rigaut 1991; Manrubia and Solé 1996), and extensively described hereafter (Chapter 8). This confusion is nevertheless quite common in the literature. For instance, Manrubia and Solé (1996) state that "the existence of several successive structures, reflected in the gentle change in D_F, constitutes evidence for multifractal scaling"; Millán and Orellana (2001) defined multifractals as a model whose "fractal dimension is a smooth function of scale"; and Stanley and Meakin (1988) considered that "such structures (termed multifractals) are characterized by fractional dimensions that vary in scale." However, the above examples all refer to multiple fractal scaling instead of multifractal structures. It will be shown in Chapter 8 that the term *fractal* strictly refers to the structure of a set S (for example, the spatial distribution of marine snails over a 1 m² domain) and does not provide any quantitative information related to the distribution of a measure μ (for example, the weight of those snails) on the set S. In that way, studying only the fractal structure of the set S would be equivalent to counting coins without referring to their relative values (Evertsz and Mandelbrot 1992). Hereafter, the term *multifractal* thus refers to the characterization of a measure μ on a set S.

7.2 ERRORS AFFECTING FRACTAL DIMENSION ESTIMATES

Although the previous section (Section 7.1) focused on the details of the most crucial limitation of fractal analysis that can lead to the consideration that many of the results presented in the literature are marred by a faulty application of linear regression and on the clarification of some terminological ambiguities that can arise from both self-similar and self-affine patterns, this section addresses some additional problems specifically related to self-similar patterns (Section 7.2.1) and to both self-similar and self-affine patterns (Sections 7.2.2 and 7.2.3).

7.2.1 GEOMETRICAL CONSTRAINT, SHAPE TOPOLOGY, AND DIGITIZATION BIASES

The constraints and limitations investigated in this section are all related to resolution and digitization issues usually disregarded as trivial details. It will nevertheless be briefly shown how the length, orientation, and placement of an image with respect to the initial box are all potential causes of error that can propagate into significant bias in the fractal dimension estimates.

Consider a box-counting procedure (such as those used in most self-similar methods; see Chapter 3) performed on three line segments of length L_1, L_2, and L_3 with respectively the following characteristics (Figure 7.8A):

- $L_1 = L_{i=0}$, where $L_{i=0}$ is the size of the larger box
- $L_2 = k\lambda$, where k is an integer and λ is the ratio that scales down each box side between two steps of the box-counting procedure ($\lambda = L_i/L_{i+1}$)
- $L_3 \neq k\lambda$

It is then straightforward from Equation (2.10) that the fractal dimension D_F returned by the algorithm will meet the theoretical expectation $D_F = 1$ for segments L_1 and L_2; see also Section 3.2.2 for more details on the box-counting algorithm. However, segment L_3 will be characterized by $D_F < 1$, because of the "overempty" boxes included in the computation. A box-counting algorithm will thus return the value $D_F = 1$ for a line segment only if (1) the box side and the line have an equal length (or the line length is a linear function of the scale ratio between two steps of the box-counting algorithm implementation), and (2) the line is either vertical or horizontal. Note that it is easy to remedy this cause of error by simply ensuring that the bounding box side coincides with the width of the segment line. The larger box must thus be framed so that the segment line is parallel and touches the edges. For more complex cases (for example, the distribution of a river network), the size of the largest box must be adjusted to the size of the largest component of a set S so that the shape touches as many edges as possible. In addition, the number of overempty boxes could also be significantly reduced using a scale ratio λ smaller than the value $\lambda = 2$ commonly used in the literature. It will nevertheless be shown hereafter that such a remedial procedure is not sufficient when considering more complex objects and their orientation.

From the above paragraph, one easily foresees that analogous effects are bound to happen when considering the effect of the orientation of the object relative to the initial box. Indeed, the number

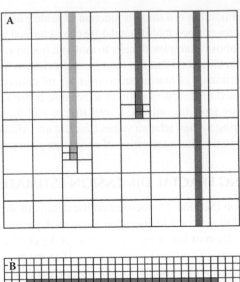

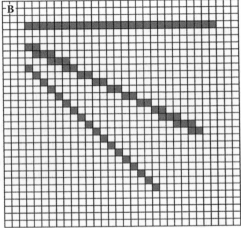

FIGURE 7.8 Source of bias in fractal dimension estimates. (A) Bounding box containing lines of correct width (dark gray) and a line segment that causes overempty boxes (light gray), and (B) digitized horizontal, straight, and tilted lines of same length, but characterized by different numbers of pixels.

of subboxes containing a part of a horizontal line will change if the line is tilted by any angle α (Appleby 1996) because (1) of basic geometrical considerations, and (2) a digitized tilted line is, in most cases, approximated by a staircase with (spurious) double pixels at most steps (Figure 7.8B). This effect is implicit to algorithms employed in digitization and is not removable per se. Contrary to the previous source of error that leads to a systematic underestimation of the fractal dimension D_F, different values of the angle α lead to different numbers of occupied boxes, that is, higher (overfull) boxes or lower (overempty) boxes than in the horizontal or vertical linear case (see Figure 7.8B), and then to both under- and overestimated fractal dimensions. Although this error can easily be corrected for simple shapes such as parallel lines (for example, by rotating the image before applying the box-counting procedure), it is unremovable for more complex shapes such as zooplankton trajectories (see, for example, Seuront et al., 2004a), (Figure 3.14). It is nevertheless possible to minimize this error using systematic replicates of grid orientation in the box-counting algorithm; see Section 3.2.2.3 for further details.

Finally, the resolution of the scanner used to digitize images can also be a source of bias itself. Contrary to computer-generated images, digitized images contain noise, do not necessarily fit to the resolution chosen in the box-counting algorithm, and are not all one-pixel-thick lines. This can have major consequences on the results of a box-counting procedure. Consider a line (one-dimensional) digitized with a poor resolution scanner and studied with a one-pixel-resolution box-counting algorithm. The digitized line might then be artificially more than one pixel thick and will thus return a trivial fractal dimension $D_F = 2$. Such a bias must then be avoided by (1) changing the resolution of the box-counting algorithm so that it fits the artificial one-dimensional resolution reached by the scanner, or (2) converting the digitized image into a vectorial image and then turning back the image into exactly one-pixel-thick lines and curves.

7.2.2 Isotropy

Strictly speaking, the fractal dimension of a section of a fractal surface is independent of the choice of the section. In other words, fractals are isotropic in space. Spatial isotropy can be checked by computing the fractal dimensions associated with sections through spatial pattern: For spatially anisotropic patterns, the fractal exponent is independent of the direction of the section. As an illustration, we considered the two-dimensional microscale distribution of microphytobenthos biomass in an estuarine tidal flat (Bay of Somme, France) over a one-meter-square sampling unit. The patch-intensity fractal dimensions (see Section 5.2) estimated for different kinds of transects (horizontal, vertical, and oblique) (Figure 7.9A,B) cannot be statistically regarded as being different ($p > 0.01$) (Figure 7.9C). The studied distribution is thus isotropic over the 1 m² domain.

While spatial anisotropy has been regarded as "a deviation from fractal behavior" (Hastings and Sugihara 1993), it could instead be regarded as an intrinsic property of a given pattern. This question has been specifically addressed in the three-dimensional framework of the analysis of zooplankton swimming behavior (Seuront et al. 2004a). The swimming behavior of the cladoceran *Daphnia pulex* (characterized in terms of divider and box dimensions; see Sections 3.2.1.3 and 3.2.2.4) has thus been shown to be significantly different in the vertical and the horizontal planes, revealing a spatial differential behavior related to biological and ecological processes such as food search or predator avoidance strategies.

7.2.3 Stationarity

This section will encompass two distinct, but complementary, aspects of the concept of stationarity that must be carefully studied to ensure the relevance of fractal analysis. The distinction between them is not necessarily straightforward and might unfortunately lead to spurious results. The first one, which strictly refers to the basic statistical concept of stationarity used in time-series analyses, will be referred to as *statistical stationarity* hereafter and is a fundamental prerequisite to any fractal analysis. The second one, referred to as *fractal stationarity*, has seldom been investigated and is a fundamental property of fractal patterns and processes. It can provide extremely valuable information to ensure the relevance and increase the ecological meaning of fractal analysis.

7.2.3.1 Statistical Stationarity

Fractal analysis requires the assumption of at least reduced stationarity; that is, the mean and the variance of a time series depend only on its length and not on the absolute time (Legendre and Legendre 2003). Indeed, many time series and transects show pronounced linear or cyclic trends. Such trends are highly detrimental to the estimates of fractal dimensions and must be removed prior to performing a fractal analysis.

For instance, the presence of a linear trend (in the framework of a transect studies of macroalgae distribution) can be tested calculating Kendall's coefficient of rank correlation τ, between the series and the x axis values in order to detect the presence of a linear trend. Kendall's coefficient of correlation was used in preference to Spearman's coefficient of correlation ρ, although advised in Kendall (1976), because Spearman's ρ gives greater weight to pairs of ranks that are further apart, while Kendall's τ weights each disagreement in rank equally (Seuront and Lagadeuc 1997).

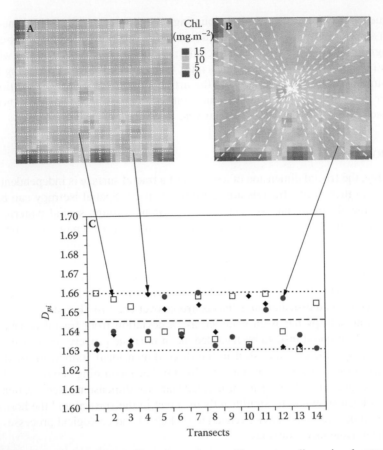

FIGURE 7.9 Spatial isotropy of fractal dimension estimates. From a two-dimensional pattern of microphy-tobenthos biomass (recorded in the Bay of Somme, April 25, 2002; Seuront and Spilmont, unpublished data) one-dimensional fractal dimensions have been estimated for horizontal and vertical sections (A) and different diagonal sections (B) of the initial pattern. The resulting patch-intensity dimensions (see Section 5.2) cannot be regarded as significantly different from the two-dimensional estimate (C; $p > 0.05$), showing the isotropic character of the initial distribution. The dashed and dotted lines represent the two-dimensional patch-intensity dimension and its 95% confidence interval, respectively. **(See color insert following page 80.)**

Many natural phenomena also exhibit periodic trends, such as the annual cycle of temperature and rainfall data, the daily cycle of sea-surface temperature, or the tidal cycle of chlorophyll and salinity data. These cycles must first be appropriately identified and then removed using techniques such as sinusoidal or polynomial regressions, moving averages, principal component filtering, or spline curve fitting procedures that can be found in most statistical textbooks (for example, Hamilton 1994; Chatfield 1996, 2003; Legendre and Legendre 2003), as well as in most actual data-analysis software.

Time series and transects will then eventually be detrended by fitting linear or nonlinear models to the original data and using the residuals from the trend in further analysis, which can be measured as the difference between actual values and the trend, or as the ratio of actual values to the trends.

7.2.3.2 Fractal Stationarity

One crucial characteristic of fractal patterns and processes is that their increments are stationary (that is, independent) in time or in space. Practically, this means that any subset of a fractal set has the same fractal dimensions as the original set. For time series, stationarity is easily tested by comput-ing the fractal dimensions for different subsets of the original time series: The fractal dimension of a temporally stationary pattern or process is independent of the time period selected. Stationarity can be tested similarly in spatial patterns by computing the fractal dimensions of spatially distinct subsets: The fractal dimension of a spatially stationary pattern or process is independent of the area selected.

As an illustration, we have considered time series of temperature, salinity, and *in vivo* fluorescence (a proxy of phytoplankton biomass) recorded using a Sea-Bird 25 Sealogger CTD probe and a Sea Tech fluorometer at a frequency of 2 Hz during a period of neap tide (September 23, 1997) in the offshore waters of the eastern English Channel (Figure 4.8). The resulting time series includes 28,777 data points. The fractal dimensions, obtained via spectral analysis (see Section 4.2.1), are $D_{FFF_T} = 1.67$, $D_{FFT_S} = 1.80$, and $D_{FFT_F} = 1.67$ for temperature, salinity, and fluorescence time series, respectively. Each of the three initial time series has subsequently been divided into 24 sections of 1199 data points. The results of this local analysis are shown on Figure 7.10 and demonstrate that the fractal dimensions of the temperature and salinity subsections do not exhibit any significant difference in their distribution ($p > 0.01$). The fluorescence fractal dimensions, however, strongly fluctuate from one subseries to the other, revealing the fractal nonstationarity of the original time series. This behavior, showing the nonpassive character of phytoplankton biomass (see also Seuront and Schmitt 2005b), can be related to the combination of the gradual and periodic changes in phytoplankton species composition and the periodic changes in turbulence intensities related to tidal advective and hydrodynamic processes, respectively (Seuront 1999, 2005b).

Thus, strictly speaking, the original temperature and salinity time series are fractal, while the fluorescence time series cannot be regarded as being fractal. Fractal nonstationarity could be regarded as a deviation from fractal behavior, especially in the field of dynamical systems. There are no *a priori* known reasons why an isolated nonlinearly oscillating pendulum would present fractal nonstationarity. In contrast, in ecology, such a property should instead be regarded as a main source of information regarding the underlying dynamics that rule the observed variability. Thus, a global analysis would have led to the spurious conclusions that the fractal dimensions of temperature and *in vivo* fluorescence cannot be significantly regarded as being different. However, the local analysis led us to consider that (1) phytoplankton biomass cannot be regarded as always being a

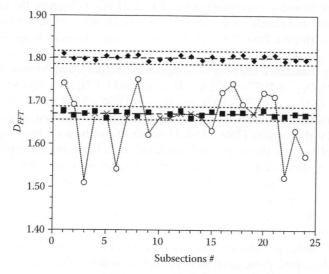

FIGURE 7.10 Fractal stationarity and nonstationarity shown from the Fourier fractal dimension of subsections of temperature (squares), salinity (diamonds), and *in vivo* fluorescence (open circles and crosses) time series. Temperature and salinity fractal dimensions cannot be regarded as different from one subsection to the other one, nor from the original time series. On the opposite, the strong fluctuations of the local fractal dimensions of fluorescence subsections indicate the nonstationary character of phytoplankton biomass variability. The open circles and the crosses indicate fluorescence fractal dimensions that are and are not significantly different from the temperature fractal dimension. The dashed and dotted lines represent the Fourier dimensions of the initial time series of temperature and salinity and their 95% confidence intervals, respectively.

purely passive scalar (see Figure 7.10), and (2) the local phytoplankton biomass fractal structure is driven by differential biological and physical processes. In ecology, fractal stationarity should then not be regarded as a condition to ensure the relevance of fractal analysis but as a diagnostic criterion to identify potential differential properties, in space or in time, of the system under interest.

7.3 DEFINING THE CONFIDENCE LIMITS OF FRACTAL DIMENSION ESTIMATES

The methods for estimating fractal dimensions have disadvantages that have previously been recognized and discussed. Now the question is to know how relevant a fractal dimension estimate is, in terms of statistical significance. Indeed, a rapid survey of the fractal studies conducted in ecology in general, and aquatic ecology in particular, would prove easily that a fractal estimate is always produced, but most of the time with no indication of its reliability or likely error. We briefly show here that once all the potential biases in the fractal dimension estimate have been identified and corrected, the confidence limits for fractal dimension (as well as their zero-slope behavior after an appropriate derivation; see Section 7.1.1) can be readily obtained with simulation methods.

We will not discuss in this section the different simulation methods available in the literature, as excellent and exhaustive reviews are available to interested readers (see, for example, Peitgen and Saupe 1988; Voss 1988; Peitgen et al. 1992). Instead, we will briefly illustrate (on the basis of simulations based on random midpoint displacements, successive random additions and the fast Fourier transform filtering algorithms; see Peitgen et al. 1992) how any fractal dimension estimate can be given a confidence interval. Generally speaking, assuming a given hypothesis, if the smallest and greatest values for the fractal dimension D_F obtained in 950 of 1,000 (or, even better, 990 of 1,000) replicate simulations are respectively $D_{F_{min}}$ and $D_{F_{max}}$, then D_F is bounded between $D_{F_{min}}$ and $D_{F_{max}}$ with probability 0.95 (or 0.99) under that hypothesis. Note that if a fractal dimension D_F has been empirically obtained from a self-affine trace of n data points, it is recommended basing the simulation procedure on 1,000 replicates of same length and fractal dimension. This estimate is chosen instead of choosing to estimate out of 100 realizations because the tail of the distribution is sampled more adequately in the former way. As an illustration, we have considered the fractal dimensions estimated from the 28,777 data-point time series of temperature, salinity, and fluorescence investigated in Section 7.2.3.2, that is, $D_{F_T} = 1.67$, $D_{F_S} = 1.80$, and $D_{F_F} = 1.67$, respectively. The resulting confidence intervals for temperature, salinity, and fluorescence time series are $D_{FFT_T} \in [1.63 - 1.71]$, $D_{FFT_S} \in [1.77 - 1.83]$, and $D_{FFT_F} \in [1.63 - 1.71]$ at the 95% significance level and $D_{FFT_T} \in [1.64 - 1.69]$, $D_{FFT_S} \in [1.78 - 1.82]$, and $D_{FFT_F} \in [1.64 - 1.69]$ at the 99% significance level. The reliability of fractal simulations has sometimes been put into question, especially for very high and very low fractal dimensions (Tate 1998). However, it can be seen from Figure 7.11 than the mean difference between the theoretical and simulated fractal dimensions (expressed as a percentage of the theoretical dimension) never exceeds 5% whatever the fractal dimension for simulation lengths of 10^3, 10^4, and 10^5. Such minute differences cannot be reasonably regarded as being significant or a serious source of bias.

The simulation route to ascribe confidence limits to fractal dimension estimates is specifically illustrated here in the self-affine framework. However, similar simulations techniques can be used to simulate fractal landscapes and surfaces (for example, Peitgen and Saupe 1988). An approach similar to the one described above can readily be applied to define the confidence limits of the self-similar fractal dimension estimates.

7.4 PERFORMING A CORRECT ANALYSIS

With the problems and limitations raised in the above sections (Sections 7.1 to 7.3) in mind, practical and step-by-step ways to conduct both self-similar and self-affine fractal analyses are described hereafter. The crucial step of properly identifying the range of scales over which the data actually

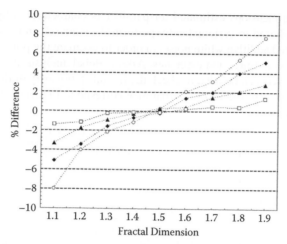

FIGURE 7.11 Mean difference between theoretical and simulated fractal dimensions (expressed as percentage of the theoretical dimension) using the successive random addition algorithm for 10^2 (open circles), 10^3 (black diamonds), 10^4 (black triangles), and 10^5 (open squares) simulated data points.

exhibit a scaling behavior (see Section 7.1) will not be discussed here to avoid restatement of previous arguments, but nevertheless remains one of the most important conditions to fulfill for a fractal analysis to be successful and meaningful.

7.4.1 Self-Similar Case

First, one needs a high-quality map. Second, we shall not proceed manually but use computer software. Although some of the methods described in Chapter 3 are quite easy to code in basic computer languages, others are not straightforward and may require a higher level of programming skills. Some of the most basic techniques (for example, divider and box-counting methods) can easily be found as freeware packages on the World Wide Web. Third, the map has to be digitized, if possible with a high-resolution scanner (that is, 1200 DPI or more), and then a proper vectorization algorithm has to be applied to the map to ensure one-pixel thickness of the curve. For surfaces, it is then necessary to check isotropy and stationarity over the studied domain. In the case of disconnected objects (for example, a set of islands), the bounding box must be chosen so as to limit the number of empty pixels around the set, and subsequently rotated at an angle α ($\alpha \in [0 - 45°]$), first to minimize the effects of overempty and overfull boxes related to angle bias, and second to define confidence limits to the resulting fractal dimension estimates. Ultimately, all unconnected parts of the set could also be analyzed separately; the fractal dimension of the whole set will then be considered as the mean of the fractal dimensions characterizing the unconnected parts.

7.4.2 Self-Affine Case

It is relatively easy to conduct a successful self-affine fractal analysis. Several diagnostic calculations nevertheless need to be systematically performed in an effort to avoid spurious features in the results caused by the analyses. First, one needs to look for the presence of any linear, periodic, or aperiodic trends in the data, and eventually remove them. In the absence of any trends, before performing the calculations, the data can advantageously be normalized (nondimensionalized) by dividing all values by the average of the total series; see also Section 6.1.3.3.2, Equation (6.17). This procedure can avoid some "density-dependent" artifacts that can virtually bias the results of comparing fractal dimensions estimated for parameters characterized by different units. As a refinement, it is also advised to systematically perform the analyses on the original data sets with and without uncorrelated additive white noise, in order to estimate the potential effects of environmental noise on the fractal dimension estimates. After a global analysis, a local analysis of subsections of the initial data sets can be tried to check the fractal stationarity of the series and to identify potential differential levels of organization that may correspond to different biological or physical forcing/driving processes.

8 From Fractals to Multifractals

Since the seminal work of Mandelbrot (1977, 1983), many patterns and processes have proven to be efficiently described by fractals in many fields of the natural sciences. They include coastline and mountain topography (Phillips 1985; Burrough 1981; Mandelbrot 1977, 1983), percolation theory (Abdusalam 2000), soil structure (Anderson and McBratney 1995; Anderson et al. 1996), water transport in soil (Pachepsky and Timlin 1998), river networks (Claps and Olivetto 1994), cosmology (Luo and Schramm 1992; Argyris et al. 2001), hydrothermal emission (Barat et al. 1999), seismicity (Khattri 1995), human vision (Billock et al. 2001), rainfall dynamics (Lovejoy and Mandelbrot 1985; Breslin and Belward 1999), heart rate variability (Mäkikallio et al. 2001), the sequence structure of DNA (Provata and Almirantis 2000), diffusion-limited aggregation processes (Meakin 1983), exchange rates fluctuations (Richards 2000), electrochemical deposition (Fleury 1997), smoke properties (Snegirev et al. 2001), cloud shapes (Lovejoy 1982), sea-surface geometry (Glazman 1991), and breaking waves (Longuet-Higgins 1994).

In aquatic ecology, the concept of fractals has been successfully applied to the structure of coral reefs (Bradbury et al. 1984), the structure of marine snow (Li et al. 1998), structural complexity in mussel beds (Snover and Commito 1998; Commito and Rusignuolo 2000; Kostylev and Erlandson 2001), the spatial distribution of intertidal benthic communities (Azovsky et al. 2000), the behavior of marine (Erlandson and Kostylev 1995; Seuront et al. 2004b) and freshwater (Seuront et al. 2004a; Uttieri et al. 2007) invertebrates, the behavior of marine vertebrates (Dowling et al. 2000; Mouillot and Viale 2001), species diversity (Frontier 1987), and zooplankton (Tsuda 1995) and phytoplankton (Seuront and Lagadeuc 1997, 1998; Waters and Mitchell 2002) patchiness.

However, many patterns and processes are now widely acknowledged as being highly intermittent (that is, characterized by a few hotspots dispersed over a wide range of low-density areas) as the distribution of microscale fluctuations of turbulent kinetic energy dissipation rate (Figure 8.1A) or the distance traveled by the calanoid copepod *Temora longicornis* (Figure 8.1B). In particular, there are clear differences arising from the comparison of these intermittent patterns (Figure 8.2A,B) with standard fractal processes such as Brownian motions (Figure 8.2C), which raises the question of the reality of describing such processes in terms of fractals and scaling behavior.

The variety of multifractal formalisms, described in Sections 8.3 and 8.4, is directly derived from the theory of complex systems and fully developed turbulence. As a direct (and unfortunate) consequence, they are far from comprehensible (and then usable), at least for ecologists without a reasonable mathematical and statistical background. In this context, the next section (Section 8.1) provides a qualitative introduction to the concepts of multifractality and intermittency as well as a nonexhaustive list of applications of multifractality in a range of scientific fields, and introduces a very simple and intuitive algorithm that might become usable as a standard procedure by ecologists.

8.1 A RANDOM WALK TOWARD MULTIFRACTALITY

8.1.1 A QUALITATIVE APPROACH TO MULTIFRACTALITY

What are multifractals? It is not easy to give a succinct definition of multifractals. In addition, it is the author's belief (as discussed in Section 7.1.2) that different authors use different names for this phenomenon, which can lead to terminology ambiguities. Following Feder (1988), one may distinguish a measure (of probability, or of some physical quantity such as mass, energy, or a number of

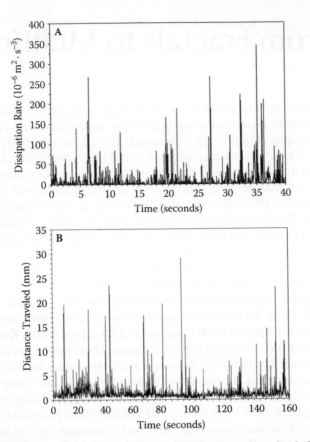

FIGURE 8.1 Intermittency, shown from a high resolution (100 Hz) time series of turbulent kinetic energy dissipation rates (A) and distance traveled by the calanoid copepod *Temora longicornis* every 0.08 second (B).

individuals) from its geometric support, which might or might not have a fractal geometry. Then, if a measure has different fractal dimensions on different parts of the support, the measure is a multifractal.

To make this point clearer, consider a modern city viewed directly from above from a plane. From this point of view, one may consider this city as black and white objects: in black are buildings and in white are streets and parks (Figure 8.3A). The only information one can get is the distribution of the built and the unbuilt areas. This is the so-called *geometric support* of the city. Now, if one changes the angle of vision by taking a position still in the air but not directly above the city, the view is from the side (Figure 8.3B). The black and white city is now a set of buildings of different heights. This is the so-called *measure* we are interested in. It could also have been the color, the width, or the age of the buildings. It is now possible to estimate the distribution of a wide range of building heights. Each height will (eventually) be characterized by a fractal dimension, thus the concept of multifractal.

8.1.2 MULTIFRACTALITY SO FAR

The concept of multifractality can implicitly be found in the formulation of self-organized criticality (Section 6.3) and the related cumulative frequency distributions (Chapter 5). It is nevertheless specifically the study of intermittency in the framework of dynamical systems (Grassberger 1983; Grassberger and Procaccia 1983; Henstchel and Procaccia 1983; Halsey et al. 1986) and

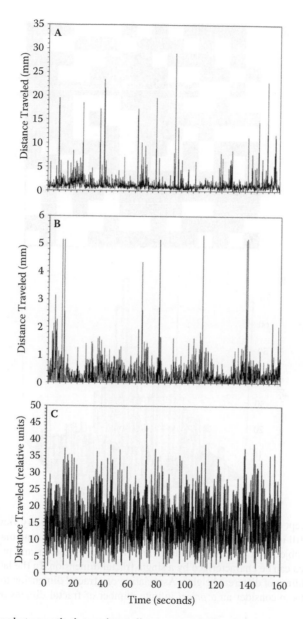

FIGURE 8.2 Differences between the intermittent distance traveled by the marine copepod *T. longicornis* (A) and the freshwater cladoceran *Daphnia pulex* (B), and the nonintermittent character of the distance traveled by a random walker (C).

fully developed turbulence (Schertzer and Lovejoy 1983; Parisi and Frisch 1985) that led to the introduction of multifractality. Recent multifractal studies include characterization of river flows and networks (De Bartolo et al. 2006; Koscielny-Bunde et al. 2006; Livina et al. 2007), asteroid belts (Campo Bagatin et al. 2002), seismicity and volcanic activity (Telesca et al. 2002; Dellino and Liotino 2002), ocean circulation (Chu 2004; Isern-Fontanet et al. 2007), diffusion limited reactions (Chaudhari et al. 2002), pore and particle distributions in soil (Martín and Montero 2002; Bird et al. 2006; Chun et al. 2008), Internet traffic (Masugi and Takuma 2007), rainfall (Labat et al. 2002; Lovejoy and Schertzer 2006), exchange currency markets (Alvarez-Ramirez

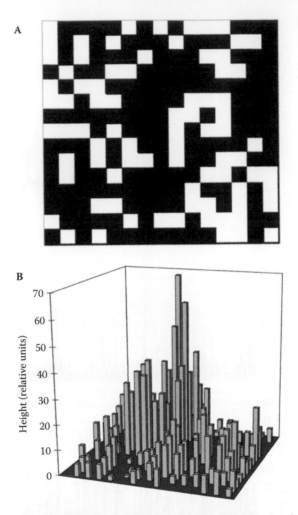

FIGURE 8.3 The concept of multifractality, illustrated considering a modern city viewed directly from above from a plane (A) and still in the air, but from the side (B). In the former case, one may consider this city as a set of black and white objects: in black are buildings and in white are streets and parks. The distribution of the built (or inbuilt) area can be characterized by a single fractal dimension. In the latter, the black and white city is now a set of buildings of different heights. Associating one fractal dimension to the distribution of each building height leads us to consider an *a priori* infinite number of fractal dimensions, referred to as being multifractal.

2002), stock markets (Turiel and Pérez-Vicente 2003), Dow Jones fluctuations (Andreadis and Sertelis 2002), stock exchange prices (Ausloos and Ivanova 2002), surface properties (Stach et al. 2001; Moktadir et al. 2008), DNA sequences (Tiňo 2002) and vascular branching (Zamir 2001; Grasman et al. 2003).

Applications of multifractals to ecology still remain extremely anecdotic, limited to forest ecology (Scheuring and Riedi 1994; Solé and Manrubia 1995a, 1995b, 1996; Manrubia and Solé 1996; Drake and Weishampel 2000, 2001), population dynamics (Ozik et al. 2005), the characterization of species-area relationship, species diversity, and species abundance distribution (Ricotta 2000; Ricotta et al. 2002; Borda-de-Água et al. 2002; Iudin and Gelashvily 2003; Laurie and Perrier 2007), and the characterization of nutrient, phyto- and zooplankton patchiness (Pascual et al. 1995; Seuront et al. 1996a, 1996b, 1999, 2002; Lovejoy et al. 2001).

8.1.3 FROM FRACTALITY TO MULTIFRACTALITY: INTERMITTENCY

8.1.3.1 A Bit of History

The concept of intermittency finds its origin in the early measurements of turbulent velocity fluctuations of Batchelor and Townsend (1949), who recognized that "as the wavenumber is increased the fluctuations seem to tend to an approximate on–off, or intermittent, variation." Two decades later, Stewart (1969) was more specific and identified that "the non-Gaussian, intermittent character of the small scale structure becomes more marked as the Reynolds number increases." He also acknowledged that while intermittency "seems to be fundamental to the nature of the turbulence cascade... we do not have a fully satisfactory theoretical explanation" (Stewart 1969). This limitation still stands today. Until very recently (Baumert et al. 2005), intermittency has seldom been referred to, or defined precisely, even in monographs devoted to turbulent processes (Tennekes and Lumley 1972; Pond and Pickard 1983; Summerhayes and Thorpe 1996; Bohr 1998; Kantha and Clayson 2000; Pope 2000). Surprisingly, in a 74-page chapter devoted to intermittency, Frisch (1996) only states that a process is intermittent when it "displays activity during only a fraction of the time, which decreases with the scale under consideration." Intermittency has similarly been described as "the active turbulent regions do not fill the whole volume, but only a subvolume in a very irregular way" (Jiménez 1997) and "active regions occupy tiny fractions of the space available" (Seuront et al. 1999).

8.1.3.2 Intermittency in Ecology and Aquatic Sciences

Intermittency, as a word and as a concept, does not seem to have ever been used, or even introduced, as such in terrestrial and landscape ecology in spite of the plethora of published papers on space-time heterogeneity and related scaling properties. Similarly, intermittency has seldom been described in aquatic sciences. In physical oceanography, intermittency has mainly been discussed in terms of its consequences on sampling, data processing, and statistics (Baker and Gibson 1987; Gibson 1991; Yamazaki 1990; Bohle-Carbonel 1992). The situation is similar in biological oceanography where turbulent intermittency and its potential effects are often not discussed. Defining patchiness, variability, and heterogeneity (see Section 8.1.4 for terminological details) has now been a issue for more than a century (Haeckel 1891; Hardy 1936; Cassie 1959, 1963; Cushing 1962). Uneven plankton distributions have subsequently been widely described; see, for example, Mitchell and Furhman (1989), Bjørnsen and Nielsen (1991), Seuront and Lagadeuc (1997, 1998), Waters and Mitchell (2002), and Waters et al. (2003). However, intermittency has seldom been quantified, despite increasing evidence of the intermittent nature of plankton distributions (Pascual et al. 1995; Seuront et al. 1996a, 1999; Seuront 2005b; Seuront and Lagadeuc 2001; Lovejoy et al. 2001). Instead, turbulent intermittency has recurrently been considered as irrelevant to marine life. For instance, intermittent events have been described as "very intense from the point of view of plankton, but calculations show that their probability is small" (Estrada and Berdalet 1997). Similarly, intermittent bursts "must certainly be spectacular events from the point of view of plankton, comparable to the passing of a tornado at our scale, and probably with similar consequences on the individual involved" but "they are sufficiently rare that they can be neglected in most calculations" (Jiménez 1997). The issue of the relevance of rare events to biological and ecological fluxes will be thoroughly investigated in Section 8.6. Intermittency has been widely observed; however, it has still escaped the confines of a narrow, precise definition.

8.1.3.3 Defining Intermittency

The definition of intermittency greatly varies from author to author and from field to field, leading to a largely scattered and nonunified framework. For instance, in rainfall and river flow studies, intermittency refers to the episodic nature of the underlying process, often considered an "on–off" basis, especially in arid environments (Chesson et al. 2004). A similar use of the term *intermittency* can also be found in energy resources (Asmus 2003; Anderson and Leach 2004). In the field of dynamical systems, intermittency has been related to several types of transitions to chaos and classified as types I, II, and III intermittency when the system under consideration is in proximity to the saddle node, Hopf, and reverse period doubling bifurcation (Pomeau and Manneville 1980). By analogy to the bifurcation diagrams detailed in Section 6.1.1 (Figure 6.3 and Figure 6.4), in these three types of intermittency, the temporal evolution of a system can be divided into ranges of the time in which the behavior of the system is almost periodic (that is, laminar phases) and exhibits chaotic bursts. Chaos-chaos intermittency is due to crisis phenomena occurring in the system (Ott 1993, 2002) and the on-off intermittency is due to a symmetry breaking bifurcation (Pikovsky 1984; Platt et al. 1993). Practically, the identification of the type of the intermittency observed may yield important information about a system by defining the bifurcations possible for its dynamics (see, for example, Żebrowski and Baranowski 2004; Alvarez-Llamoza et al. 2008).

The phenomenon of intermittency has widely been mixed up with its statistical consequences, and thus generally poorly defined even in specialized monographs. The literature, hence, recurrently refers to intermittency through statements such as "the kurtosis is a useful measure of intermittency for signals having a bursty aspect" (Frisch 1996), "the signals tended to become bursty when the order of differentiation is increased" (Frisch 1996), "most of the time the gradients would still be of the order of magnitude of their standard deviation, but occasionally we should expect stronger bursts, more often than in the Gaussian case" (Jiménez 1997), "the discrepancies between the Kolmogorov predictions and the experimental values of the high-order moments" (Pope 2000), and "we occasionally should expect stronger bursts than expected in a non-intermittent, homogeneous turbulence, which accentuate the skewness of a given probability distribution, causing it to deviate from Gaussianity" (Seuront et al. 2001).

The production of turbulence is not a continuous process but usually has an intermittent character and the turbulence appears as bursts (Svendsen 1997). This intermittency has been acknowledged as "a common phenomenon in many complex systems, and a natural consequence of cascades" (Jiménez 2000). Intermittency has also been related to the coherent nature of turbulence and the presence of strong vortices, with diameters on the order of 10 times the Kolmogorov length scale l_k, $l_k = (v^3/\varepsilon)^{1/4}$ where v is the kinematic viscosity (m² s⁻¹) and ε the turbulent kinetic energy dissipation rate (m² s⁻³) (Siggia 1981; Jiménez et al. 1993; Jiménez and Wray 1994). The term *intermittency* has alternatively been coined to describe "the phenomena connected with the local variability of the dissipation" (Jiménez 1998) as well as "instantaneous gradients of scalars such as temperature, salinity or nutrients, greatest at scales similar to the Kolmogorov microscale" (Gargett 1997).

Pope (2000), and more recently Jiménez (2006) in the *Encyclopedia of Mathematical Physics*, distinguished external from internal intermittencies. External intermittency refers to the coexistence of turbulent and laminar regions in inhomogeneous turbulent flows, such as in boundary layers or in free-shear layers. The interface between laminar irrotational flow and turbulent vortical fluid is typically sharp and corrugated (Jiménez 2006). As a consequence, an observer sitting near the edge of the layer is immersed in turbulent fluid only part of the time and hence experiences an intermittently turbulent flow. In this context, an intermittent flow is characterized by a fluid motion that is "sometimes laminar and sometimes turbulent" (Pope 2000). For the engineering community in fluid mechanics, intermittency is also viewed as a transition between laminar and turbulent flows. Specifically, Wilcox (1998) considers that "approaching the freestream from within the boundary layer, the flow is not always turbulent. Rather, it is sometimes laminar and sometimes turbulent, that is, it is intermittent." Internal intermittency (Pope 2000; Jiménez 2006) is specifically related to the increasingly non-Gaussian properties of velocity fluctuations as spatial separation increases. This property is responsible for the long tails of the probability distributions of the velocity derivatives.

A more intuitive definition that can directly be applied in ecology stated that "this form of variability reflects heterogeneous distributions with a few dense patches and a wide range of low density patches" (Seuront et al. 2001). Most of the previously published work referred to intermittency in the framework of turbulent flows, including wave turbulence (Biven et al. 2001; Newell et al. 2001; Bouruet-Aubertot et al. 2004), plasma turbulence (Sorriso-Valvo et al. 2001; Hidalgo et al. 2006), and solar wind turbulence (Bruno et al. 2001; Chapman et al. 2005). However, a general consensus can be reached considering that a given pattern or process is intermittent in space or in time if (1) it is characterized by sharp local fluctuations, (2) it is responsible for a skewed probability distribution, and (3) it has a long-term memory signature, perceptible from the power-law form of its autocorrelation function.

8.1.4 Variability, Inhomogeneity, and Heterogeneity: Terminological Considerations

In ecology, the term *variability* refers to changes in the values of a given quantitative or qualitative descriptor; it is distinct from heterogeneity, which refers to a composition of different entities or kinds of elements (Kolasa and Rollo 1991; *Merriam-Webster's Collegiate Dictionary* 2008). This distinction is, however, not as clear as may appear at first glance, with meanings essentially dependent on the choice of approach (Naeem and Colwell 1991; Shashack and Brand 1991). Even papers devoted to the synthesis of these concepts (Kolasa and Rollo 1991; Naeem and Colwell 1991; Shashack and Brand 1991) may be misleading in that spatial and temporal heterogeneity are used to describe spatial or temporal variability, respectively, irrespective of the basic previous definitions. Definitions themselves appear to be highly variable even within a collective synthetic work on the subject (Levin et al. 1993). Spatial heterogeneity was then defined as an equivalent of spatial autocorrelation (van Hes 1993), despite the stress for a clear distinction between these two concepts (Davis 1993). Fractal geometry and the resulting scaling properties have also been suggested as a way to characterize space-time heterogeneity in ecology (Milne 1991). In aquatic sciences, both physical and biological patterns and processes have been referred to in terms of "temporal intermittency" and "spatial heterogeneity'" (Platt et al. 1989).

Terminological ambiguities are potentially detrimental to scientific progress (Popper 2002) and are not limited to ecological sciences (Box 8.1). The term *inhomogeneity*, seldom used in the literature, but as a synonym of *variability*, *nonhomogeneity*, or *nonuniformity* (Coplen and Krouse 1998; Blundell and Rawlings 1999; Jiménez 2006), is suggested hereafter as a way to describe the variability of a descriptor structured in space or in time in terms of scaling, and discussed as a way to reach a terminological consensus. A descriptor exhibiting nonscaling properties cannot be distinguished from observational "white" noise (Figure 8.4A) and as such is characterized by its variability. In contrast, a scaling descriptor exhibiting scaling properties will be in homogeneous in space or in time (Figure 8.4B). Now, considering that an ecological entity is a pattern bounded in space or in time (Cousins 1988), an inhomogeneous descriptor can then be regarded as a structural ecological entity. As a consequence, *heterogeneity* will not be applied to the variability of a given descriptor in space or in time as widely done (see, for example, Kolasa and Pickett 1991) but instead to patterns or processes exhibiting different levels of structures (that is, inhomogeneity) over space or time and hence corresponding to different driving processes.

BOX 8.1   TERMINOLOGICAL AMBIGUITIES IN SCIENCE

Terminological ambiguities are not specific to ecological sciences but seem to be the rule in science in general. To assess this issue, all papers reporting the term *variability, homogeneity, heterogeneity, inhomogeneity,* and *intermittency* that appeared in the journals *Nature* and *Science*, and under the ScienceDirect banner of Elsevier from 1998 to 2008, were scrutinized.

TABLE 8.B1.1
Bibliographic Survey of All Papers Reporting the Terms *Variability,*
Heterogeneity, Inhomogeneity, **and** *Intermittency* **in Their Title or**
Abstract from 1998 to 2008

	Nature		Science		ScienceDirect	
	Title	Abstract	Title	Abstract	Title	Abstract
Variability	402	—	57	232	6,799	29,070
Homogeneity	25	—	2	11	445	7,194
Heterogeneity	218	—	17	70	2,620	13,188
Inhomogeneity	3	—	0	10	321	1,967
Intermittency	1	—	0	3	135	443

From the 63,233 papers resulting from this survey (Table 8.B1.1), it appears that *variability* and *homogeneity* are consistently used to described the fluctuations and the absence of fluctuations in the distribution of any parameter; see, for example, Porter and Semenov (1999) and Schindell et al. (1999). In contrast, the terms *heterogeneity* and *inhomogeneity* are vaguely defined, while the seldom-used term *intermittency* has a more constant meaning (see Section 8.1.3.3). The term *heterogeneity* is, however, mostly used to describe the fluctuations, that is, the *variability* (Kolasa and Rollo 1991; *Merriam-Webster's Collegiate Dictionary* 2008), of a given process, as "velocity heterogeneity" or "temperature heterogeneity" in the Earth's core (Sumita and Olson 1999; Vidale and Earle 2000), as widely done by many biologists and ecologists (Mitchell and Furhman 1989; Rainey and Travisano 1998). *Heterogeneity* seldom fits the basic definition of different entities or kinds of elements (Kolasa and Rollo 1991; *Merriam-Webster's Collegiate Dictionary* 2008), and when it is the case, the corresponding papers are most of the time related to ecological sciences (Guegan et al. 1998). In particular, this demonstrates that such terminological ambiguities are far from being an ecological specificity, and that ecological sciences are finally not so badly off. On the other hand, the very violent and *a priori* unpredictable fluctuations perceptible in turbulent velocity and scalar fields, financial markets fluctuations, or medical sciences are systematically described in terms of intermittency, thus describing a specific kind of variability, and opposed to *homogeneity* (Shraiman and Siggia 2000; Helmlinger et al. 2000). Finally, the concept related to inhomogeneity refers without distinction to the variability of a given descriptor, for example, "the isotopic inhomogeneity of this material: the variability in its sulfur-34/sulphur-32 isotope ratio" (Coplen and Krouse 1998), its heterogeneity (Bonn et al. 1998) *sensu* Kolasa and Rollo (1991), its nonhomogeneity (Wu et al. 1999), or its intermittency (Jiménez 2006).

This is exemplified using some of the results obtained in the previous sections. For instance, the distributions of temperature and salinity investigated in the inshore waters of the eastern English Channel (Section 4.2.1.3.2, Figure 4.9A,B,D,E) are homogeneous as they exhibit a scaling behavior (or a lack of scaling behavior) over the whole range of investigated scales (Figure 4.9A,B and Figure 8.4A,B). In contrast, the distribution of *in vivo* fluorescence is heterogeneous as it exhibits two distinct scaling ranges (Figure 4.9C and Figure 8.4C,D). The isotropy of the fractal dimensions derived from the two-dimensional pattern of microphytobenthos biomass shown in Figure 7.9A,B (Section 7.2.2) leads us to consider this pattern as homogeneous (Figure 8.4E). A pattern leading to fractal anisotropy would be, in turn, qualified as being heterogeneous (Figure 8.4F). Finally, the fractal stationarity and nonstationarity exhibited respectively by temperature and salinity time series and by fluorescence time series (Figure 7.10; Section 7.2.3.2) lead us to consider the temporal pattern of temperature and salinity as temporally homogeneous (Figure 8.4E), while fluorescence is temporally heterogeneous (Figure 8.4F). These previous examples clearly indicate that a descriptor inhomogeneous (or not) in space or in time can be either homogeneous or heterogeneous in space or in time.

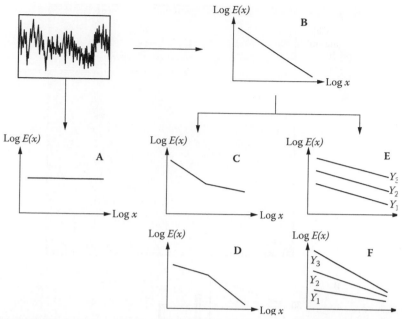

FIGURE 8.4 Schematic illustration of the concepts related to homogeneity, heterogeneity, and inhomogeneity, based on the results of a spectral analysis, where $E(x)$ is the spectral density related to a frequency or a wave number x following that the descriptor is considered in time of in space. A homogeneous descriptor can either exhibit (A) nonscaling properties (noninhomogeneous descriptor) or (B) scaling properties (inhomogeneous descriptor) in time or in space. More specifically, an inhomogeneous descriptor can be heterogeneous in time or in space when it exhibits different scaling regimes, shown in (C, D). A descriptor exhibiting the same scaling properties (that is, inhomogeneous) over the range of available scales can be homogeneous in time or in space if its inhomogeneous properties depend on the sampling time or location (E). Similarly, an inhomogeneous descriptor can be heterogeneous in space or in time if its inhomogeneous properties evolve with space or time (F). Y_i represents data sets sampled at different times or locations.

The previous propositions suggest some terminological specifications in comparison with basic systemic approaches. For instance, hierarchical approaches, initially developed in the framework of landscape analysis, have been devoted to describe "how heterogeneity changes with scale" (Allen and Starr 1982). On the other hand, following our approach, a system considered as being hierarchical must be viewed as a heterogeneous system presenting different scales of inhomogeneity. In that way, the main point of hierarchical theory should be instead regarded as the way to describe *how inhomogeneity changes with scales*. Moreover, the concepts developed in the present chapter could also be regarded as a way to complement hierarchical approaches in the sense that they allow us to describe how the structure of a given descriptor, hierarchical (Figure 8.4C,D) or not (Figure 8.4E,F), evolves in time or in space. These concepts could subsequently provide an efficient framework to reconcile space- and time-oriented approaches. Indeed, a descriptor exhibiting different inhomogeneous structure will be regarded as being heterogeneous, the inhomogeneity fluctuating either in space or in time, which is still actually not widely done (Kolasa and Rollo 1991; Naeem and Colwell 1991; Shachak and Brand 1991; Levin et al. 1993).

8.1.5 INTUITIVE MULTIFRACTALS FOR ECOLOGISTS

Before reviewing and illustrating the traditional multifractal methods in Section 8.2, a method that is believed to be much more intuitive to nonmathematically oriented readers will be introduced. Consider the spatial distribution of microphytobenthos biomass at spatial scales ranging from

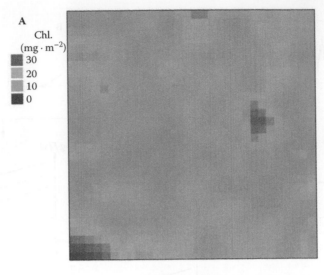

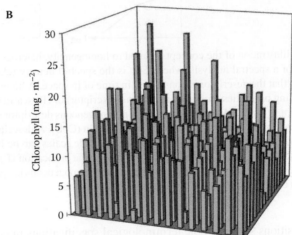

FIGURE 8.5 Concept of multifractality, illustrated considering a two-dimensional pattern of microphytoben-thos biomass (A) and the related three-dimensional view of the different biomass values (B). Note the similar-ity with Figure 8.3. The understanding of such a pattern could be limited to "traditional" fractal analysis such as the patch-intensity dimension (Section 5.2) or surface dimensions (Section 3.2.9), or focus on the distribu-tion of each different biomass (see Figure 8.6).

6.67 cm to 1 m (Figure 8.5). Here, a top view of the distribution does not appear as a black and white "cityscape" but instead as a two-dimensional mosaic (Figure 8.5A). This is because the chlorophyll concentration is a spatial continuous function rather than a binary one. A three-dimensional view of the chlorophyll distribution (Figure 8.5B) nevertheless shows similar features with the city side view (cf. Figure 8.4B). Now the question is: How do we describe simultaneously the different chlorophyll concentrations and their related scaling properties? The algorithm proposed hereafter can be regarded as a generalization of the box-counting methods (see Section 3.2.2).

First, one needs to rethink the initial distribution S in terms of n distinct subsets S_c such as:

$$S = \sum_{c=C_{min}}^{C_{max}} S_c \qquad (8.1)$$

where C_{min} and C_{max} are the minimum and maximum chlorophyll concentrations of the distribution, and S_c is given as:

$$S_c \subset (C \geq c) \tag{8.2}$$

where C is the actual chlorophyll concentration and c a given chlorophyll threshold. This procedure results in the creation of a set of "black and white" patterns where the white areas correspond to areas where $C \geq c$ (Figure 8.6A).

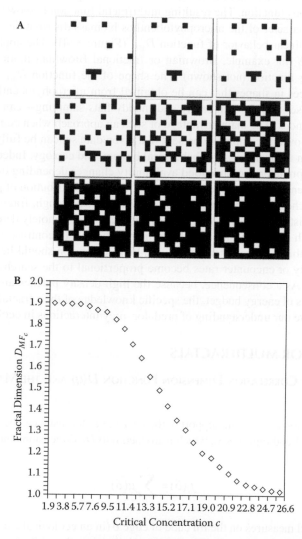

FIGURE 8.6 Multifractals for ecologists. Using different thresholds, the two-dimensional distribution shown in Figure 8.5 can be divided in a series of black and white patterns: in black are the areas where the biomass exceeds a given threshold, and in white are the "empty" areas (A). Each concentration threshold can then be related to one fractal dimension, the ultimate result being a multifractal function (B) that can be thought as a characteristic structural property of the pattern. (Modified from Seuront and Spilmont, 2002.)

Second, standard algorithms aimed at counting the number of boxes of length δ occupied by at least a part of a given set S (see Section 3.2.2) are modified as:

$$N_\delta(C \geq c) \propto \delta^{-D_{MF_c}} \tag{8.3}$$

where $N_\delta(C \geq c)$ corresponds to the number of boxes of length δ containing a chlorophyll concentration greater than a threshold concentration c, and D_{MF_c} the multifractal function associated to the threshold concentration c. For each threshold concentration c, the slope of the log-log plot of $N_\delta(C \geq c)$ vs. δ is an estimate of D_{MF_c}; each threshold concentration c is thus characterized by its own fractal dimension. Although the increment for threshold c can be arbitrarily selected, it is strongly recommended that we choose it close to the smallest observed chlorophyll concentration difference in order (1) to capture the maximum, if not all, of the structure present in set S, and (2) for D_{MF_c} to converge to a continuous function. The resulting multifractal function is shown on Figure 8.6B.

The multifractal character of the microphytobenthos biomass distribution shown in Figure 8.5 is ensured by the nonlinear behavior of function D_{MF_c} (Figure 8.6B). The application of Equation (8.3) to a fractal set S (for example, Brownian or fractional Brownian motions) would have led to a linear decreasing function (not shown). The shape of the function D_{MF_c}—and in particular the potential differences in shapes that can be obtained from microphytobenthos distribution, for example, at different seasons or from sandy and muddy flats sampling—can then be used as an ecological structural index. The main advantage of such an approach when compared to the fractal framework is that the overall complexity present in a given data set can be fully quantified. This is particularly relevant in the framework of behavioral biology and ecology. Indeed, prey distribution is very important for predators because food availability changes depending on the fractal dimension. Low fractal dimensions indicate a smooth and predictable distribution of particles gathered in small numbers of patches, while high fractal dimensions indicate rough, fragmented, space-filling and less predictable distributions. Therefore, when a predator can remotely detect its surroundings, prey distributions with low dimension should be more efficient. In contrast, when a predator has no remote detection ability, prey distributions with high dimension should be preferable, because available food quantity or encounter rates become proportional to the searched volume as fractal dimension increases. As a consequence, because the high-density patches are expected to be the most valuable in terms of energy budget, the specific knowledge of their fractal dimension can be a crucial way to improve our understanding of predator–prey interactions in ecology.

8.2 METHODS FOR MULTIFRACTALS

8.2.1 Generalized Correlation Dimension Function $D(q)$ and the Mass Exponents $\tau(q)$

8.2.1.1 Theory

Consider a D_E-dimensional space (the support of the measure μ, for instance, the height of buildings in a city; see Section 8.1.1 and Figure 8.3), which is divided into D_E-dimensional boxes of size δ, then:

$$f_i(\delta) = \sum_i^N \mu(\delta) \tag{8.4}$$

denotes the integrated measures on the ith cube of edge δ (in an ecological example, the probability of finding an organism in this volume). One can then define the qth-order moment of the probability distribution (or "partition function") $M_q(\delta)$ (Halsey et al. 1986):

$$M_q(\delta) = \sum_{i=1}^{N(\delta)} f_i(\delta)^q \tag{8.5}$$

where $N(\delta)$ in the number of boxes of size δ. The information dimension (Equation 3.34) and the correlation dimension (Equation 3.41) can then be written as:

$$I_q(\delta) = \frac{1}{1-q} \log M_q(\delta) \tag{8.6}$$

and

$$D(q) = \lim_{\delta \to 0} \frac{I_q(\delta)}{\log(1/\delta)} \tag{8.7}$$

where $I_q(\delta)$ and $D(q)$ are the generalized information dimension (or Reyni information of qth order; Rényi 1970) and the generalized correlation dimension (Grassberger 1983; Hentschel and Procaccia 1983), respectively. Note that the generalized correlation dimension $D(q)$ is conceptually similar to the multifractal function D_{MF_c} introduced in Section 8.1.5.

From Equation (8.5) through Equation (8.7), one readily finds the previously defined box dimension (see Section 3.2.2, Equation 3.15), the information dimension (see Section 3.2.5, Equation 3.34), and the correlation dimension (see Section 3.2.6, Equation 3.41) for integer q as special cases. The box dimension, the information dimension, and the correlation dimension are then respectively given as:

$$D_b = \lim_{q \to 0} D(q) = D(0) \tag{8.8}$$

$$D_i = \lim_{q \to 1} D(q) = D(1) \tag{8.9}$$

and

$$D_{cor} = \lim_{q \to 2} D(q) = D(2) \tag{8.10}$$

The generalized dimension function $D(q)$ is defined for all real values of q and is estimated as the slope of the log-log plot of $I_q(\delta)$ vs. δ. For monofractal sets, the function $D(q)$ is a linear function of q; in other words, no additional information is gained by examining higher moments, that is, more extreme values of the measure μ. Alternatively, for multifractal sets, $D(q)$ is a nonlinear function of q. Note that there are lower and upper limiting dimensions, $D_{-\infty}$ and $D_{+\infty}$, which are related to the regions of the set where the measure μ is sparser and denser, respectively. For positive values of q, $D(q)$ reflects the scaling of the large fluctuations and strong singularities. In contrast, for negative values of q, $D(q)$ reflects the scaling of the small fluctuations and weak singularities (Vicsek 1993; Takayasu 1997).

Remember that for a given value of δ, the mass $m(\delta)$ (see Equation 3.26) is expressed as the first moment (that is, the mean) of the probability distribution, Equation (8.5) can be equivalently written as:

$$M_q(\delta) \propto \delta^{-\tau(q)} \tag{8.11}$$

where $\tau(q)$ is a mass exponent function. For monofractal sets, $\tau(q)$ is linear, $\tau(q) = qH - 1$, where H is the Hurst exponent. In contrast, multifractal sets display a nonlinear function $\tau(q)$. Note that $\tau(q)$ is related to the spectral exponent β as:

$$\beta = 2 + \tau(2) \tag{8.12}$$

Moreover, for multifractal sets, Equations (8.6), (8.7), and (8.11) lead to:

$$\tau(q) = (q-1)D(q) \tag{8.13}$$

Note that for $q = 0$, $D(q) = \tau(q) = D(0)$ and $\tau(1) = 0$. Combining Equations (8.12) and (8.13) finally leads to:

$$\beta = 2 + D(2) \tag{8.14}$$

where $D(q)$ is the generalized correlation dimension (see Equation 8.7), and β is the spectral exponent.

8.2.1.2 Application: Salinity Stress in the Cladoceran *Daphniopsis Australis*

The cladoceran *Daphniopsis australis* (Sergeev and Williams 1985) is endemic to saline lakes and swamps of the southeast of Australia and is adapted to salinities ranging from 4 to 30 PSU (Sergeev and Williams 1985). Despite the potential keystone role of this species in the structure and function of South Australian inland water ecosystems, there have been few ecological studies on this genus and none on this particular species. In particular, nothing is known on the effect of extreme salinities (that is, up to 160 PSU) (Schapira et al. 2009) occurring in summer in relation to evaporation on their biology and ecology. Here the swimming behavior of *D. australis* is investigated at a known optimal salinity (22 PSU) and in the case where S = 50 Practical Salinity Unit (PSU).

Individuals of *D. australis* were continuously cultured at Flinders University (South Australia) in 20-liter containers and fed on the phytoplankton *Isochrysis Tahitian* under constant conditions of temperature (22°C) and salinity (22 PSU) on a 12-hour light–dark cycle. Behavioral experiments were conducted on individual males following Seuront (2006), and their successive displacements analyzed using the generalized dimension function $D(q)$ (Equation 8.7) and the mass exponent function $\tau(q)$ (Equation 8.11). The functions $D(q)$ and $\tau(q)$ clearly exhibit a multifractal signature under optimal salinity conditions (S = 22 PSU) (Figure 8.7A,B). In contrast, for S = 50 PSU, both functions indicate that the successive displacements of male *D. australis* have weaker multifractal properties (Figure 8.7C,D). This is consistent with and generalizes previous results showing a decrease in the complexity of behavioral sequences under stressful conditions for a range of organisms (Section 4.2.2.2) (Alados et al. 1996; María et al. 2004; Alados and Huffman 2000; Seuront and Leterme 2007). A shift between multifractal and fractal properties—or, more generally, a change in multifractal properties in animal behavior—is then suggested as a potential diagnostic tool to assess animal stress levels and health. This specific issue is explored further in Section 8.2.2.

8.2.2 MULTIFRACTAL SPECTRUM $f(\alpha)$

8.2.2.1 Theory

The number $\alpha_i = \frac{\log \mu_i(\delta)}{\log \delta}$, also referred as to the Hölder exponent, is the singularity strength of the ith box. This exponent may be interpreted as a crowding index of a measure of the concentration (accumulation) of μ: the greater α_i is, the smaller the concentration of the measure, and *vice versa*. For every box size δ, the numbers of cells $N_\alpha(\delta)$ in which the Hölder exponent α_i has a value within the range $[\alpha, \alpha + d\alpha]$ behave like:

$$N_\alpha(\delta) \propto \delta^{-f(\alpha)} \tag{8.15}$$

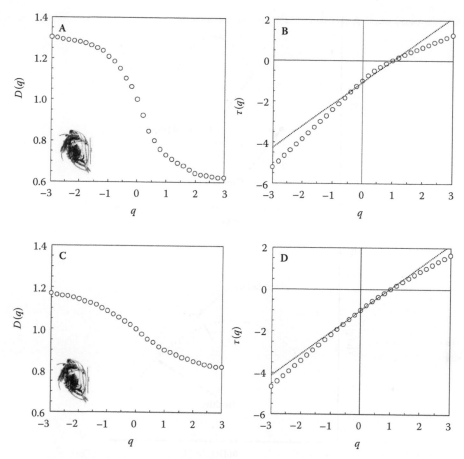

FIGURE 8.7 Generalized dimension function $D(q)$ and the mass exponent function $\tau(q)$ estimated from the successive displacements of male *Daphniopsis australis* under contrasted conditions of salinity; that is, (A, B) S = 22 PSU, and (C, D) S = 50 PSU. The multifractality of *D. australis* swimming behavior—that is, the nonlinearity of $D(q)$ and $\tau(q)$—is stronger under low salinity conditions. Note that for $q = 0$, $D(q) = \tau(q) = D(0)$, $D(0)$ being the fractal dimension of the support of the measure, and $q = 1$, $\tau(q) = 0$. The dashed lines (B, D) are the functions $\tau(q)$ expected in case of monofractality.

Thus, $f(\alpha)$ corresponds to the fractal dimension of the subset in which α_i equals α; that is, $f(\alpha)$ characterizes the abundance of cells with Hölder exponent α and is called the singularity spectrum of the distribution. The singularity spectrum $f(\alpha)$ and the mass exponent function $\tau(q)$ are connected via a Legendre transform as (Evertsz and Mandelbrot 1992):

$$\alpha(q) = \frac{d\tau(q)}{dq} \tag{8.16}$$

and

$$f(\alpha(q)) = q\alpha(q) - \tau(q) \tag{8.17}$$

Considering the relationship between the mass exponent function $\tau(q)$ and the generalized dimension function $D(q)$ (cf. Equation 8.13), the singularity spectrum $f(\alpha)$ contains exactly the same information as $\tau(q)$ and $D(q)$.

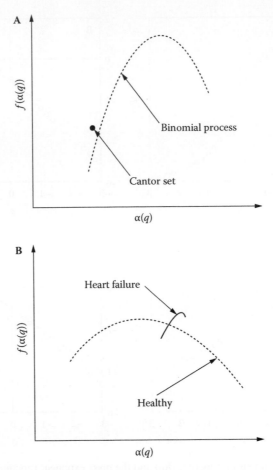

FIGURE 8.8 Schematic illustration of the multifractal spectrum $f(\alpha(q))$ vs. $\alpha(q)$ for a fractal set and a simple multifractal pattern (A), and for healthy and unhealthy heartbeat fluctuations (B). The narrow range of $\alpha(q)$ values for heart failure is indicative of monofractality. (Redrawn from Ivanov et al., 1999.)

Because the calculation of $f(\alpha(q))$ from Equations (8.16) and (8.17) can be highly problematic, Chhabra and Jensen (1989) developed a much simpler method for the calculation of $f(\alpha(q))$ and $\alpha(q)$ for multifractal structures as:

$$\alpha(q) = \frac{\displaystyle\sum_{i=1}^{N(\delta)} \mu_i(q,\delta)\log\mu_i(\delta)}{\log\delta} \tag{8.18}$$

and

$$f(\alpha(q)) = \frac{\displaystyle\sum_{i=1}^{N(\delta)} \mu_i(q,\delta)\log\mu_i(q,\delta)}{\log\delta} \tag{8.19}$$

where the quantity $\mu_i(q,\delta) = \mu_i(\delta)^q / \sum_{i=1}^{N(\delta)} \mu_i(\delta)^q$. The parameter q provides a scanning tool to scrutinize the denser and rarer regions of the measure μ. For $q > 1$, regions where μ has a high degree of concentration are amplified, while for $q < -1$ regions with a small degree of concentration are magnified. Finally, for $q = 1$, the measure itself is replicated; see Evertsz and Mandelbrot (1992) and Chhabra and Jensen (1989) for further details. The function $f(\alpha(q))$ thus gives the entropy dimension of the distorted measure $\mu(q, \delta)$ and characterizes the original measure μ by analyzing the variation under successive distortions driven by the parameter q. The singularity spectrum $f(\alpha(q))$ takes its maximum value for $q = 0$ and typically has a parabolic shape around this point (Figure 8.8). The number $f(\alpha(0)) = \alpha(0) = D(0)$ is the box dimension of the measure μ, and the number $f(\alpha(1)) = \alpha(1) = D(1)$ is the information dimension. As an illustration, Figure 8.8A shows a comparison of the shapes of the singularity spectrum $f(\alpha(q))$ obtained from monofractal and multifractal sets, the Cantor set and a multiplicative binomial process, respectively. Figure 8.8B illustrates how differences in the shape of the spectrum $f(\alpha(q))$ can be indicative of heart failure and then proposes an additional diagnostic tool.

8.2.2.2 Application: Temperature Stress in the Calanoid Copepod *Temora Longicornis*

The swimming behavior of the calanoid copepod *Temora longicornis* (Figure 4.18A, Section 4.2.3.2.1) has been investigated under different conditions of temperatures representative of the most extreme range of temperature that the species may encounter in its natural environment, that is, from 4° to 28°C. *T. longicornis* were collected from the inshore waters of the eastern English Channel using a WP2 net (200-μm mesh size) at a temperature of 16°C and a salinity of 32.5 PSU. Specimens were diluted in buckets with surface waters, transported to the laboratory, and acclimatized for 24 hours in 5-liter beakers filled with natural seawater. Prior to the experiments, adult females (1.1 ± 0.1 mm, $\bar{x} \pm$ SD) were sorted by pipette under a dissecting microscope and left in the behavioral container (a 2-liter, $20 \times 20 \times 5$ cm Plexiglass container) filled with 0.45 μm filtered natural seawater to acclimatize for 10 minutes at the experimental temperature (Seuront 2006). The temperature treatments were randomized, and the resulting sequence of temperature treatments was 16, 8, 20, 4, 28, and 24°C. Groups of 5 individual females were considered for each temperature treatment and their activity videotaped for 20 minutes.

The function $f(\alpha(q))$ obtained for the successive displacements of *T. longicornis* generally exhibits the single-humped shape, typical of multifractal patterns (Figure 8.9). For extreme (low and high) temperatures, $f(\alpha(q))$ are narrower, suggesting monofractality (Figure 8.9). As previously stressed (Section 8.2.1.2), this shift between multifractality and monofractality may reflect perturbation of the physiological control mechanisms of motion behavior.

8.2.3 CODIMENSION FUNCTION $c(\gamma)$ AND SCALING MOMENT FUNCTION $K(q)$

Consider an intermittent quantity—for example, the turbulent kinetic energy dissipation rate ε_λ (see Figure 2.2 and Figure 8.1A)—where the subscript λ refers to the scale ratio $\lambda = L/l$ where L and l are the largest external scale and the resolution of the measurements, respectively. When $\lambda \gg 1$, intermittency can be characterized by the statistical distribution of singularities (that is, intensities) γ:

$$\varepsilon_\lambda \propto \lambda^\gamma \tag{8.20}$$

and by the related probability density distribution (Schertzer and Lovejoy 1987):

$$\Pr(\varepsilon_\lambda \geq \lambda^\gamma) \propto \lambda^{-c(\gamma)} \tag{8.21}$$

where $c(\gamma)$ is a codimension function characterizing the singularities distribution. $c(\gamma)$ can be expressed as a generalization of Equation (2.12) following:

$$c(\gamma) = D_T - D(\gamma) \tag{8.22}$$

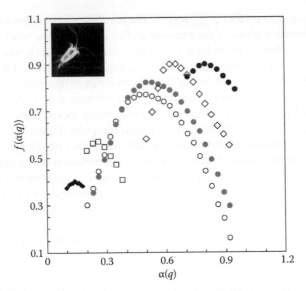

FIGURE 8.9 Multifractal spectrum $f(\alpha(q))$ vs. $\alpha(q)$ for the successive displacements female *Temora longicornis* for different temperatures: 4°C (black diamonds), 8°C (open squares), 16°C (open dots), 20°C (gray dots), 24°C (open diamonds), and 28°C (black dots). Note that the multifractal signature (the nonlinearity of the function $f(\alpha(q))$ over a range of $\alpha(q)$ values) is weakened for extreme temperatures.

where $D(\gamma)$ characterizes the hierarchy of fractal dimensions associated with the different intermittency levels. The codimension function $c(\gamma)$ diverges for high thresholds. This divergence is related to the divergence of moments. The absolute slope of the algebraic tail of the probability density distribution is directly related to the moment of divergence, q_D, as:

$$\Pr(\varepsilon_\lambda \geq \varepsilon_{th}) \propto \varepsilon_{th}^{-q_D} \tag{8.23}$$

where ε_{th} is a given threshold. Note the similarity between Equations (8.23) and (5.4). This implies that the typical signature of self-organized criticality (that is, Equation 5.4) may also be indicative of multifractality.

Under fairly general conditions, the properties of the probability distribution of a random variable are equivalently specified by its statistical moments. The latter corresponds to the introduction of the scaling moment function $K(q)$, which describes the multiscaling of the statistical moments of order q of the field ε_λ:

$$\langle (\varepsilon_\lambda)^q \rangle \propto \lambda^{K(q)} \tag{8.24}$$

where "$\langle \cdot \rangle$" indicates statistical averaging.

The relation existing between the two scaling functions $c(\gamma)$ and $K(q)$ reduces to the Legendre transform (Parisi and Frisch 1985) for large-scale ratios (that is, $\lambda \gg 1$):

$$K(q) = \max_\gamma (q\gamma - c(\gamma)) \Leftrightarrow c(\gamma) = \max_q (q\gamma - K(q)) \tag{8.25}$$

Equation (8.25) implies that there is a one-to-one correspondence between singularities and orders of moments: to any order q is associated the singularity γ_q that maximizes $q\gamma - c(\gamma)$ and is the solution of $c'(\gamma_q) = q_\gamma$. Similarly to any singularity γ is associated the order of moment q_γ that maximizes

$q\gamma - K(q)$ and is the solution of $K'(q_\gamma) = \gamma_q \cdot c(\gamma)$ and $K(q)$ exhibit several general properties of multifractals as convexity and nonlinearity. In particular, for conservative multifractal processes (that is, $\langle \varepsilon_\gamma \rangle = \langle \varepsilon_1 \rangle$, $\forall \lambda$), since $K(1) = 0$ corresponds via the Legendre transform to the fact that the corresponding mean singularity of the process, $C_1 = K'(1)$ is a fixed point of $c(\gamma)$, the latter is therefore tangential to the first bisectrix line ($c(\gamma) = \gamma$) in $\gamma_1 = c(\gamma_1) = C_1$, hence $c'(C_1) = 1$ (see Seuront et al. 1999, Figures 8, 9). The determination of the probability distribution would require the determination of moments at all scales. With the assumption of scaling, it reduces to the determination of a hierarchy of exponents that remain nevertheless *a priori* infinite, and therefore indeterminable, especially for the highest orders, which correspond to the most extreme variability. However, in the framework of universal multifractals (Schertzer and Lovejoy 1987), the calculation complexity induced by the hierarchy previously described is included in two fundamental parameters, C_1 and α, which describe the multiscaling behavior of the functions $K(q)$:

$$\begin{cases} K(q) = \dfrac{C_1}{\alpha - 1}(q^\alpha - q) & \alpha \neq 1 \\[2mm] K(q) = C_1 q \ln(q) & \alpha = 1 \end{cases} \tag{8.26}$$

and $c(\gamma)$:

$$\begin{cases} c(\gamma) = C_1 \left(\dfrac{\gamma}{C_1 \alpha'} + \dfrac{1}{\alpha} \right)^{\alpha'} & \alpha \neq 1 \\[3mm] c(\gamma) = C_1 \exp\left(\dfrac{\gamma}{C_1} - 1 \right) & \alpha = 1 \end{cases} \tag{8.27}$$

with $\frac{1}{\alpha} + \frac{1}{\alpha'} = 1$.

C_1 is the mean singularity of the process and also the codimension of the mean singularity; it therefore measures the mean fractality of the process. It satisfies $0 \leq C_1 \leq D_E$, where D_E is the Euclidean dimension of the observation space (for example, $D_E = 1$ for time series and transects); $C_1 = 0$ for a homogeneous process, and $C_1 = d$ for a process so heterogeneous that the fractal dimension of the set contributing to the mean is zero. It then characterizes a mean inhomogeneity and can be regarded as the measure of the sparseness of a given field: the higher the C_1, the fewer the field values corresponding to any given singularity. The index α, called the Lévy index, is the degree of multifractality bounded between $\alpha = 0$ and $\alpha = 2$, which correspond respectively to the monofractal model and to the lognormal model (see Section 8.3). It defines how fast the fractality is increasing with higher and higher singularities: As α decreases, the high values of the field do not dominate as much as for larger values of α; there, the functions are larger deviations from the mean. As an illustration, the functions $K(q)$ estimated for the original time series of temperature, salinity, and *in vivo* fluorescence time series studied in Section 6.1.3.3 are shown in Figure 8.10. Finally note that the functions $K(q)$ relate to the generalized correlation function $D(q)$ (Section 8.2.1) as:

$$D(q) = D_E - K(q) \tag{8.28}$$

where D_E is still the Euclidean dimension of the observation space.

Finally, even if the functions $c(\gamma)$ and $K(q)$ were originally introduced in the framework of fully developed turbulence, and as such illustrated above using the turbulent energy dissipation rate ε_λ as

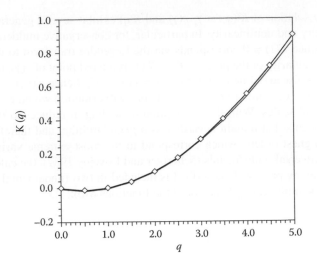

FIGURE 8.10 Functions $K(q)$ estimated for the 172,800 data-point time series of temperature (black curve), salinity (open diamonds), and *in vivo* fluorescence (gray curve) studied in Section 6.1.3.3.

an example, they can be applied to any intermittent field ϕ_λ. Also note that the field ϕ_λ relates to the fluctuations of a given quantity Q as:

$$\Delta Q_\lambda \propto \phi_\lambda^a l^H \tag{8.29}$$

where $\Delta Q_\lambda = |Q(x+l) - Q(x)|$ is the gradient of the quantity Q at scale l and H is the Hurst exponent defined through structure function analysis; see Section 8.2.4. A direct consequence of Equation (8.29) is that it is possible to obtain the field ϕ_λ directly from a scalar quantity Q through a fractional differentiation of order H; see, for example, Schertzer et al. (1998) for further details. This is easy to perform through a multiplication by k^H in Fourier space, and strictly equivalent to power-law filtering.

8.2.4 Structure Function Exponents $\zeta(q)$

8.2.4.1 Theory
This analysis technique is devoted to the direct study of the multifractal properties of the fluctuations of any scalar field S, and is based on the qth-order structure functions:

$$\langle \Delta S_\tau \rangle = \langle | S(t+\tau) - S(t) | \rangle \tag{8.30}$$

where for a given time lag τ the fluctuations of the scalar S are averaged over all the available values ("$\langle \cdot \rangle$" indicates statistical averaging). For scaling processes, one way to statistically characterize intermittency is based on the study of the scale-invariant structure exponent $\zeta(q)$ defined by the following:

$$\langle (\Delta S_\tau) \rangle^q = \langle (\Delta S_T)^q \rangle \left(\frac{\tau}{T} \right)^{\zeta(q)} \tag{8.31}$$

where T is the largest period (external scale) of the scaling regime. The scaling exponent $\zeta(q)$ is estimated by the slope of the linear trends of $\langle(\Delta S_\tau)^q\rangle$ vs. τ in a log-log plot. The first moment $\zeta(1)$, characterizing the scaling of the average absolute fluctuations, corresponds to the scaling Hurst exponent $H = \zeta(1)$, characterizing the degree of nonconservation of a given field. The second moment is linked to the power spectrum exponent β as:

$$\beta = 1 + \zeta(2) \tag{8.32}$$

For simple (monofractal) processes, the scaling exponent of the structure function $\zeta(q)$ is linear; that is, $\zeta(q) = qH$. In particular, $\zeta(q) = q/2$ for Brownian motion, and $\zeta(q) = q/3$ for nonintermittent turbulence. For multifractal processes, this exponent is nonlinear and concave, and relates to the function $\beta = 1 + \zeta(2)$ as:

$$\zeta(q) = qH - K(q) \tag{8.33}$$

$K(q)$ is then an intermittent correction, hence expresses the deviation of the function $\zeta(q)$ from linearity due to intermittency. The different forms taken by the functions $\zeta(q)$ and $K(q)$ are detailed in Section 8.3.

The potential effect of varying turbulent forcings on the local structure of physical and biological parameters investigated in Section 6.1.3.3 has been specified through the analysis of the 24 time series of temperature, salinity, and *in vivo* fluorescence with structure functions. The resulting functions $\zeta(q)$ estimated for temperature, salinity, and fluorescence are clearly nonlinear, showing the multifractal nature of their distributions (Figure 8.11 and Figure 8.12). More specifically, the function $\zeta(q)$ estimated for the temperature and salinity time series remained the same across variable turbulence and tidal conditions (Figures 8.11 and 8.12). In contrast, for a given turbulence level,

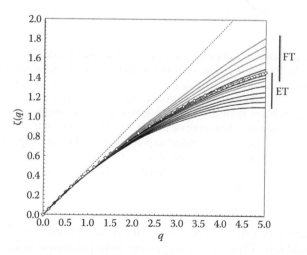

FIGURE 8.11 Scaling function $\zeta(q)$ estimated for *in vivo* fluorescence during the second half of flood tide (FT: gray) and ebb tide (ET: black), compared to the statistically undistinguishable empirical exponents estimated for temperature and salinity (open diamonds) and the theoretical homogeneous linear case $\zeta(q) = qH$ (dashed line). It is clear from the nonlinearity of $\zeta(q)$ that phytoplankton distributions are far from being homogeneously distributed, and in both tidal conditions patchiness increases with decreasing values of the turbulent energy dissipation rates. The energy dissipation rates considered here from bottom to top are: 5.07×10^{-7}, 1.11×10^{-6}, 5.17×10^{-6}, 1.43×10^{-5}, 5.13×10^{-5}, and 3.52×10^{-4} m² s⁻³ during ebb tide, and 2.60×10^{-7}, 1.07×10^{-6}, 5.16×10^{-6}, 1.24×10^{-5}, 6.32×10^{-5}, 1.05×10^{-4}, and 2.98×10^{-4} m² s⁻³ during flood tide. (Modified from Seuront, 2005b.)

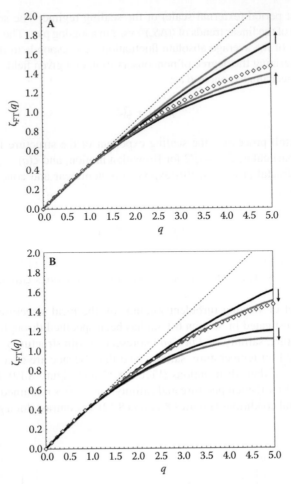

FIGURE 8.12 The scaling function $\zeta(q)$ estimated for *in vivo* fluorescence during the first and second half of flood tide (A) and ebb tide (B) over the same range of turbulence intensities. During the first and second half of flood tide, dissipation rates range respectively from 1.54×10^{-7} to 2.55×10^{-4} m^2 s^{-3} (from top to bottom: black) and from 2.60×10^{-7} to 2.98×10^{-4} m^2 s^{-3} (from top to bottom: gray). During the first and second half of ebb tide, dissipation rates range from 1.91×10^{-7} to 3.06×10^{-4} m^2 s^{-3} (from top to bottom: black) and the second half of ebb tide for dissipation rates ranging from 5.07×10^{-7} to 3.52×10^{-4} m^2 s^{-3} (from top to bottom: gray). The statistically undistinguishable empirical exponents estimated for temperature and salinity (open diamonds) and the theoretical homogeneous linear case $\zeta(q) = qH$ (dashed line) are shown for comparison. The vertical arrows indicate the shift in phytoplankton patchiness between the first and second half of flood and ebb tide. (Modified from Seuront, 2005b.)

phytoplankton biomass was always more patchy during the second half of ebb tide than during the second half of flood tide (Figure 8.11), suggesting more patchiness in inshore than in offshore waters. The difference between the function $\zeta(q)$ estimated during flood tide (Figure 8.12A) and ebb tide (Figure 8.12B) indicates that for the same range of turbulence intensities phytoplankton patchiness decreased during the flood tide (Figure 8.12A) and increased during the ebb tide (Figure 8.12B). In addition, phytoplankton patchiness was higher at the beginning of the ebb tide than at the beginning of the flood tide (Figure 8.12). This is indicative of mixing between inshore and offshore water masses during the transition between flood and ebb tides. Flood and ebb tide phytoplankton populations can nevertheless exhibit very similar levels of patchiness (for example, for the highest and lowest turbulence conditions considered for flood and ebb tides, respectively, in Figure 8.11). Finally, phytoplankton distributions were more and less patchy than would be a purely passive scalar in ebb and flood tide conditions, respectively. The gradients observed in the phytoplankton concentrations were thus higher and lower than temperature and salinity gradients in ebb and flood tide conditions, respectively, whatever the intensity of turbulence. As a consequence, inshore and offshore phytoplankton populations must be respectively considered as more and less homogeneously distributed than purely passive scalars. This is indicative of the predominant influence of the biological properties of phytoplankton cells on turbulence processes.

8.2.4.2 Eulerian and Lagrangian Multiscaling Relations for Turbulent Velocity and Passive Scalars

8.2.4.2.1 Eulerian Multiscaling Relations for Turbulent Velocity and Passive Scalars

In Section 4.2.1.3, Eulerian and Lagrangian fluctuations of turbulent velocity and passive scalars were described under the homogeneity assumption (Kolmogorov 1941; Obukhov 1941, 1949; Corrsin 1951). However, the flux of energy ε (Figure 2.2) and the flux of scalar variance χ exhibit sharp intermittent fluctuations at all scales (Figure 8.13). More specifically, Batchelor and Towsend (1949) showed that instantaneous dissipation rates intermittently reached very high values and that this intermittency was more important when the scale ratio, hence the Reynolds number, was large. The original assumption of homogeneity then becomes untenable, and turbulent fields had to be thought of as intermittent and scale-dependent processes. This led to the refined similarity hypothesis (Kolmogorov 1962, Obukhov 1962), stating that velocity fluctuations are influenced by the

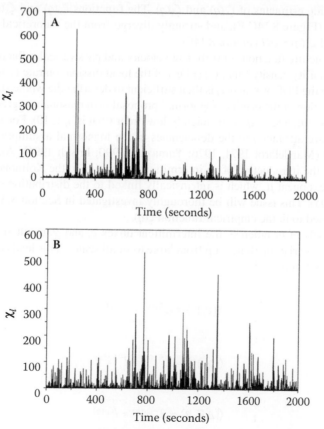

FIGURE 8.13 Time series of the variance fluxes χ_l estimated from time series of (A) temperature ($\times 10^{-6} \, °C^2 \, s^{-1}$) and (B) *in vivo* fluorescence (fluorescence^2 s^{-1}) recorded in the inshore waters of the eastern English Channel with a Sea-Bird Sealogger CTD and a Sea Tech fluorometer, respectively. (Modified from Seuront, 2008.)

local value of the dissipation averaged over a distance l, ε_l. The "refined similarity hypothesis" (Kolmogorov 1962, Obukhov 1962) then leads to the introduction of the subscript l in Equations (4.19) and (4.20) that now read:

$$\varepsilon_l \approx \frac{(\Delta V_l)^3}{l} \tag{8.34}$$

and

$$\chi_l \approx \frac{(\Delta S_l)^2 (\Delta V_l)}{l} \tag{8.35}$$

A lognormal distribution was originally introduced for ε_l (Kolmogorov 1962, Obukhov 1962), leading to describe all the statistics of a turbulent field with only two parameters, the mean and the variance. However, since the use and the relevance of second-order statistics can be restrictive and characterizes intermittent fluctuations very poorly, the velocity structure functions were introduced in order to study the statistical properties of turbulence (Monin and Yaglom 1975) (see Section 8.2.4.1). Figure 8.14 illustrates the structure function analysis of turbulent velocity fluctuations (Figure 8.14A) and oceanic *in vivo* fluorescence fluctuations (Figure 8.14D). Figure 8.4B,E show the scaling of the structure functions for various orders of moments: In a log-log plot, the straight lines provide estimates of $\zeta_V(q)$ and $\zeta_S(q)$. The functions $\zeta_V(q)$ and $\zeta_S(q)$ are clearly nonlinear and convex (Figure 8.14C,F), and strongly diverge from the theoretical shapes $\zeta_V(q) = q/3$ (Figure 8.14C) and $\zeta_S(q) = qH$ (Figure 8.14F).

Note that the structure functions of turbulent velocity and passive scalar fluctuations are directly related to the probability density function (PDF) of the local dissipation rate ε_l and variance flux χ_l. A proper model for the PDF of ε_l and χ_l is then sufficient to describe the whole statistics of turbulent velocity. The prediction of the original lognormal proposal is in reasonable agreement with empirical data for statistical moments q of sufficiently low order (that is, $q \leq 10$). For large order ($q > 10$), the discrepancies are attributed to the deficiencies in the lognormal assumption, which has been severely criticized (Mandelbrot 1974, 1976; Yamazaki 1990; Frisch 1996). Another consequence of intermittency is the introduction of an intermittent correction to the Kolmogorov spectral slope, the intermittency exponent μ, which is intrinsically linked to the distribution chosen for ε_l and χ_l (Seuront et al. 2005). This issue will be thoroughly investigated in Section 8.3.2 in relation to the different models used to fit the empirical function $\zeta(q)$.

In cascade models of turbulence, the intermittent fluxes ε_l and χ_l result from a multiplicative process in which the variability builds up from large to small scales. This leads to multifractal fields with (Seuront and Schmitt 2005a):

$$\langle (\varepsilon_l)^q \rangle \approx \lambda^{K_\varepsilon(q)} \approx l^{-K_\varepsilon(q)} \tag{8.36}$$

$$\langle (\chi_l)^q \rangle \approx \lambda^{K_\chi(q)} \approx l^{-K_\chi(q)} \tag{8.37}$$

$$\langle |\Delta V_l|^q \rangle \approx \lambda^{-\zeta_V(q)} \approx l^{\zeta_V(q)} \tag{8.38}$$

$$\langle |(\Delta S_l)^2 \Delta V_l|^q \rangle \approx \lambda^{-\zeta_{V,S}(3q)} \approx l^{\zeta_{V,S}(3q)} \tag{8.39}$$

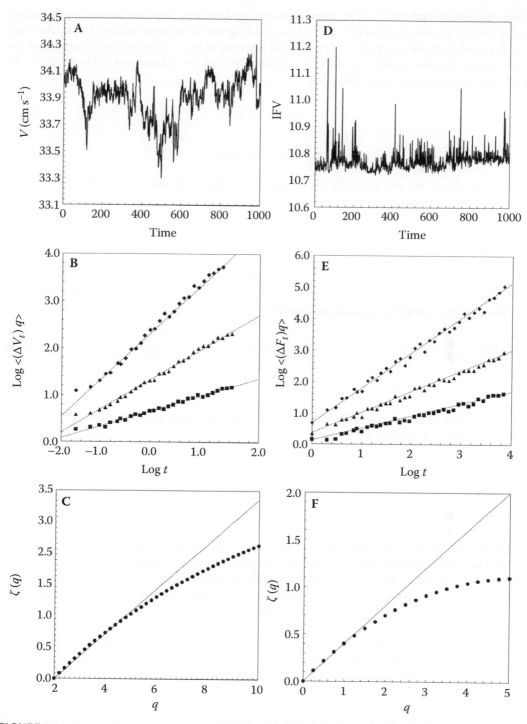

FIGURE 8.14 Time series of grid-generated turbulent velocity recorded by hot-wire velocimetry at 100 Hz in a circular flume (A) and *in vivo* fluorescence (D) recorded in the coastal waters of the eastern English Channel. The corresponding structure function analyses are shown in (B) and (E) for different values of q ($q = 1, 2,$ and 3 from bottom to top). The related structure functions exponents $\zeta(q)$ are clearly nonlinear and convex, illustrating the intermittent nature of velocity and fluorescence fluctuations, shown in (C, F). The dashed lines are the theoretical, nonintermittent exponents for turbulent velocity $\zeta(q) = q/3$ (C) and fluorescence $\zeta(q) = qH$ (F). (Modified from Seuront, 2008.)

where the angle brackets "$\langle \cdot \rangle$" indicate ensemble averaging, λ the scale ratio between the largest external scale L and the actual scale l (that is, $\lambda = L/l$), $K_\varepsilon(q)$ and $K_\chi(q)$ the scaling moment functions for the fluxes ε_l and χ_l, and $\zeta_V(q)$ and $\zeta_{V,S}(q)$ the scaling moment functions of the velocity structure function and the joint structure function scaling exponent of the product $(\Delta S_l)^2 \Delta V_l$. From Equations (8.34) and (8.35), the functions $K_\varepsilon(q)$ and $K_\chi(q)$ can be defined as:

$$K_\varepsilon(q) = q - \zeta_V(3q) \tag{8.40}$$

$$K_\chi(q) = q - \zeta_{V,S}(3q) \tag{8.41}$$

Because the fluxes are conserved by the equation of motion over the inertial subrange, they are assumed to be scale-independent:

$$\langle \varepsilon_l \rangle \approx \langle \varepsilon_1 \rangle \tag{8.42}$$

$$\langle \chi_l \rangle \approx \langle \chi_1 \rangle \tag{8.43}$$

Equations (8.36) and (8.37) subsequently lead to:

$$K_\varepsilon(1) = 0 \tag{8.44}$$

$$K_\chi(1) = 0 \tag{8.45}$$

Such multifractal fields are called "conservative multifractals," and the conservation of the fluxes ε_l and χ_l (Equations 8.44 and 8.45) lead to:

$$\zeta_V(3) = 1 \tag{8.46}$$

$$\zeta_{V,S}(3) = 1 \tag{8.47}$$

Equations (8.46) and (8.47) correspond to the exact formulations for the small-scale dissipation fields. The scaling moment functions $K_\varepsilon(q)$, $K_\chi(q)$, $\zeta_V(q)$, and $\zeta_{V,S}(q)$ characterize all the fluctuations of the fluxes of energy and scalar variance, and the fluctuations of the velocity shear and scalar gradient. In other words, as under fairly general conditions, the probability distribution of a random variable is equivalently specified by its statistical moments, the scaling moment functions $K(q)$ and $\zeta(q)$ describe the scale dependence of the statistical moments of order q.

Equation (8.36) through Equation (8.39) characterize all the fluctuations of the energy flux ε_l (Equation 8.34) and scalar variance flux χ_l (Equation 8.35) through the scaling moment functions $K(q)$ and $\zeta(q)$. The fluctuations of a passive scalar are defined by the scaling moment function $\zeta_S(q)$ as:

$$\langle |\Delta S_l|^q \rangle \approx \lambda^{-\zeta_S(q)} \tag{8.48}$$

However, because the passive scalar flux φ_l is a nonconservative mixed flux of energy ε_l and scalar variance χ_l as:

$$\varphi_l = \varepsilon_l^{-1/2} \chi_l^{3/2} \tag{8.49}$$

which are intrinsically correlated, the scaling moment function $\zeta_S(q)$ is related to the structure function of velocity fluctuations and scalar gradients as:

$$\zeta_S(q) = \zeta_{V,S}(3q/2) - \zeta_V(q/2) \tag{8.50}$$

where $\zeta_V(q) = q/3 + K_\varepsilon(q/6) - K_\chi(q/2)$; see Seuront and Schmitt (2005a). Equations (8.35) and (8.49) also lead to expressing the squared passive scalar fluctuations, $\langle \Delta S_l^2 \rangle = \langle (S_{x+l} = S_x)^2 \rangle$, as:

$$\langle \Delta S_l^2 \rangle \approx \varphi_l^{2/3} l^{2/3} \tag{8.51}$$

With the introduction of discrete and continuous cascade models, a wide variety of distributions (including improvement of the initial lognormal proposal) has been proposed for ε_l and φ_l. A brief review of the distributions found in the literature, and a test of their performance at fitting oceanic temperature fluctuations (that is, a proxy for phytoplankton biomass), are proposed in Section 8.3.3.

8.2.4.2.2 Lagrangian Multiscaling Relations for Turbulent Velocity and Passive Scalars

In an intermittent framework, the Lagrangian scaling relations given by Equations (4.25) and (4.26) are rewritten as:

$$\varepsilon_t \approx \frac{\Delta V_l^2}{t} \tag{8.52}$$

and

$$\chi_t \approx \frac{\Delta S_l^2}{t} \tag{8.53}$$

where $\Delta V_t = |V(\tau+t) - V(\tau)|$ and $\Delta S_t = |S(\tau+t) - S(\tau)|$ are the velocity shear and passive scalar gradients for an element of fluid at the scale t. The scalar variance flux in Equation (8.53) does not depend any more on a cross-product of velocity and passive scalar fluxes as in the Eulerian framework; see Equation (8.35). The related scaling relations then come as:

$$\langle (\varepsilon_t)^q \rangle \approx \Lambda^{K_\varepsilon(q)} \approx t^{-K_\varepsilon(q)} \tag{8.54}$$

$$\langle (\chi_t)^q \rangle \approx \Lambda^{K_\chi(q)} \approx t^{-K_\chi(q)} \tag{8.55}$$

$$\langle |\Delta V_t|^q \rangle \approx \Lambda^{-\zeta_V(q)} \approx t^{\zeta_V(q)} \tag{8.56}$$

$$\langle |\Delta S_t|^q \rangle \approx \Lambda^{-\zeta_S(q)} \approx t^{\zeta_S(q)} \tag{8.57}$$

where $\Lambda = T/t$ is the scale ration between the fixed outer time scale T and the actual time scale t, and the Lagrangian scaling moment functions for the velocity and passive scalar fluxes $K_\varepsilon(q)$ and $K_\chi(q)$ are given by:

$$K_\varepsilon(q) = q - \zeta_V(2q) \tag{8.58}$$

$$K_\chi(q) = q - \zeta_S(2q) \tag{8.59}$$

The fluxes are still assumed to be conservative, that is, $K_\varepsilon(1)=0$ and $K_\chi(1)=0$, which implies $\zeta_V(2)=1$ and $\zeta_S(2)=1$ (Seuront et al. 1996b). As Equation (8.32) is still valid in the Lagrangian framework, there is no intermittency correction for the second moment, corresponding exactly to a power spectrum $E(f) \approx f^{-2}$. The difference in spectral slope can then be used to identify Lagrangian and Eulerian regimes in oceanic data, as illustrated in Section 4.2.1.3.3 (Figure 4.10).

The Lagrangian velocity structure function scaling exponents $\zeta_V(q)$ and the Lagrangian passive scalar structure function scaling exponents $\zeta_S(q)$ are estimated from Equations (8.56) and (8.57), and are respectively given by:

$$\zeta_V(q) = \frac{q}{2} - K_\varepsilon\left(\frac{q}{2}\right) \tag{8.60}$$

$$\zeta_S(q) = \frac{q}{2} - K_\chi\left(\frac{q}{2}\right) \tag{8.61}$$

where $K_\varepsilon(q)$ and $K_\chi(q)$ are estimated from the intermittent field ε_t and χ_t. The structure function exponent $\zeta(q)$ estimated from the Lagrangian regime of the time series of temperature and salinity shown in Figure 4.10 exhibit a nonlinear signature indicative of multifractality (Figure 8.15).

8.3 CASCADE MODELS FOR INTERMITTENCY

8.3.1 HISTORICAL BACKGROUND

The first description of the turbulence cascade came with the intuitive scheme of Richardson (1922), who recognized that "big whirls have little whirls that feed on their velocity, and little whirls have lesser whirls and so on to viscosity" in the molecular sense. This was later formalized by the self-similarity hypothesis (Kolmogorov 1941), which states that velocity fluctuations between two points separated by a distance l depend only on the average dissipation rate ε. The squared velocity fluctuation, $\langle \Delta V_l^2 \rangle = \langle [V_{x+l} - V_x]^2 \rangle$, thus writes as:

$$\langle \Delta V_l^2 \rangle \propto \varepsilon^{2/3} l^{2/3} \tag{8.62}$$

where x and $x + l$ are two points separated by a distance l. In Fourier space, Equation (8.62) is strictly equivalent to Equation (4.23) for turbulent velocity fluctuations and Equation (8.51) to Equation (4.24) for passive scalar fluctuations. The generalized structure functions for moments of order $q > 0$ of the absolute velocity increments are defined as:

$$\langle |\Delta V_l|^q \rangle \propto \varepsilon^{q/3} l^{q/3} \tag{8.63}$$

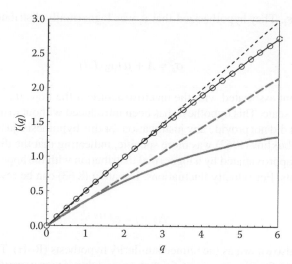

FIGURE 8.15 Empirical values of the Lagrangian structure function exponents $\zeta(q)$ for temperature (black continuous line) and salinity (open dots) time series recorded in the eastern English Channel from a drifting boat 24.9 m long. The dotted black line is the theoretical exponents $\zeta(q) = q/2$ expected in case of nonintermittent Lagrangian turbulence. The thick continuous line is the empirical Eulerian $\zeta(q)$ for temperature and salinity estimated over Eulerian scales, and the thick dashed gray line the theoretical case $\zeta(q) = qH$ expected in case of nonintermittent Eulerian turbulence, respectively.

Equation (8.63) leads to the K41 linear law (that is, nonintermittent, homogeneous turbulence):

$$\zeta_V(q) = q/3 \tag{8.64}$$

where $\zeta_V(q)$ is the scaling exponent of the velocity structure functions:

$$\langle |\Delta V_l|^q \rangle \propto l^{\zeta_V(q)} \tag{8.65}$$

For any passive scalar advected by turbulent flows, Equation (8.51) rewrites as:

$$\langle |\Delta S_l|^q \rangle \propto l^{\zeta_S(q)} \tag{8.66}$$

where $\zeta_S(q)$ is the scaling moment function of scalar gradients; see Equation (8.50). Furthermore, Equation (8.62) and Equation (8.63) lead to a general relationship between the spectral exponent β and the second-order moment structure function exponent $\zeta(2)$:

$$\beta_i = 1 + \zeta_i(2) \tag{8.67}$$

where $\beta_i = \beta_V$ for velocity fluctuations (see Equation 2.5) and $\beta_i = \beta_S$ for passive scalar fluctuations (see Equation 4.23), and $\zeta_i(2) = \zeta_V(2)$ for velocity fluctuations (see Equation 8.65) and $\zeta_i(2) = \zeta_S(2)$ for passive scalar fluctuations; see Equation (8.66).

Kolmogorov (1962) and Obukhov (1962) took intermittency into account considering that the structure function of velocity fluctuations is a function of a locally averaged dissipation rate for a sphere of radius l, ε_l. They hypothesized that ε_l was lognormally distributed with the variance σ_l^2 of $\log \varepsilon_l$ given by:

$$\sigma_l^2 = A + \mu \log(L/l) \tag{8.68}$$

where A is a constant associated with the macrostructure of the flow, μ a universal constant, and L the largest external scale. This hypothesis has been introduced without firm theoretical foundations; Kolmogorov (1962) did not provide any justification for this hypothesis, just stating that "it is natural to suppose that." Obukhov (1962) was more specific, indicating that the distribution of any positive quantity should be approximated by a lognormal distribution with the appropriate values for the first and second moments. For velocity fluctuations, Equation (8.63) can be rewritten as:

$$\langle |\Delta V_l|^q \rangle \propto \varepsilon_l^{q/3} l^{q/3} \tag{8.69}$$

Equation (8.69) is also known as the refined similarity hypothesis (RSH). This leads to the nonlinear form for the structure function exponent $\zeta_V(q)$ characterizing intermittency.

8.3.2 Cascade Models for Turbulence

Since the first attempt to provide a quantitative description of the Richardson cascade was made by Yaglom (1966) and Gurvich and Yaglom (1967), a range of discrete and continuous cascade models have been introduced to describe intermittent fluxes (see Seuront et al. [2005] for an exhaustive review).

A first family of models is composed of discrete models, for which the scale ratio between a structure and the daughter structure is a discrete integer. Due to their discrete nature, these models are not realistic but have been introduced for their simplicity and ability to reproduce experimental intermittency. These models include the lognormal model, the mono-fractal β-model, the α-model, the p-model, the random β-model, and the B-model. Detailed reviews of these models may be found in Paladin and Vulpiani (1987), Meneveau and Sreenivasan (1991), Frisch (1996), and Seuront et al. (2005). In addition, the limitations of these models and their limited ability to fit experimental data, especially for the higher orders of moment q, are detailed in Frisch (1996).

The continuous log-infinitely divisible (log-ID) stochastic models represent a more realistic family of cascade models. Specifically, infinite divisibility specifies that any random variable belonging to this law may be written as a sum of an arbitrarily large number of independent random variables, each having the same law (independent identically distributed) (see, for example, Feller, 1971). This property intrinsically limits the number of probability laws; the most known ID laws are the Gaussian, Lévy stable, Poisson, and Gamma. The corresponding log-ID continuous cascade models for velocity fluctuations and passive scalar fluctuations are briefly reviewed and their ability to fit the fluctuations of oceanic *in vivo* fluorescence (that is, a proxy of phytoplankton biomass) critically assessed.

8.3.2.1 Lognormal Model

The lognormal model (Kolmogorov 1962; Obukhov 1962) corresponds to a quadratic form for $\zeta_V(q)$ and $\zeta_S(q)$. For velocity fluctuations, Equation (8.46) and the condition $\zeta_V(0) = 0$ lead to expressing $\zeta_V(q)$ as:

$$\zeta_V(q) = \frac{q}{3} - \frac{\mu}{2}\left(\left(\frac{q}{3}\right)^2 - \frac{q}{3}\right) \tag{8.70}$$

where the intermittency parameter $\mu = K_\varepsilon(2)$, and $K_\varepsilon(q) = \frac{\mu}{2}[(q/3)^2 - (q/3)]$. Similarly, for passive scalar fluctuations, the function $\zeta_S(q)$ writes as:

$$\zeta_S(q) = qH - \frac{\mu}{2}(q^2 - q)$$ (8.71)

where the parameter H is the degree of nonconservation of the average field; that is, $H = \zeta_S(1)$, $\mu = K_\varphi(2)$, and $K_\varphi(q) = \frac{\mu}{2}(q^2 - q)$.

8.3.2.2 The Log-Lévy Model

Because the Gaussian law belongs to the Lévy-stable family of random distributions, that is, stable and attractive processes under addition (Feller 1971), the lognormal cascade model has been generalized to log-stable cascades (Schertzer and Lovejoy 1987; Kida 1991). In this context, the structure function scaling exponents $\zeta(q)$ have a precise theoretical shape defined as (Seuront et al. 1996a):

$$\zeta(q) = Aq + Bq^\alpha$$ (8.72)

where A and B are empirical constants, and α is the Lévy exponent for stable variables (Feller 1971). Equation (8.72) defines a family of distributions defined according to the value of α, $0 < \alpha \leq 2$. When $\alpha = 2$, Equation (8.72) recovers the lognormal model, and the β-model when $\alpha \to 0$ (Seuront et al. 2005). For $1 \leq \alpha < 2$, the scaling is superdiffusive (Shlesinger et al. 1996), while the value $\alpha = 1$ indicates that the scaling becomes quadratic in time and corresponds to the lower limit of superdiffusive processes, that is, Lévy flight (Shlesinger et al. 1996). In contrast, values $\alpha \leq 0$ do not correspond to probability distributions that can be normalized. When $\alpha \leq 2$, the variance of the process diverges, and when $\alpha \leq 1$ the mean is not defined. For turbulent velocity, the normalization condition $\zeta_V(3) = 0$ leads to the reformulation of Equation (8.70) as:

$$\zeta_V(q) = \frac{q}{3} - \frac{C_{1\varepsilon}}{\alpha_\varepsilon - 1}\left(\left(\frac{q}{3}\right)^{\alpha_\varepsilon} - \frac{q}{3}\right)$$ (8.73)

where $K_\varepsilon(q) = \frac{C_{1\varepsilon}}{\alpha_\varepsilon - 1}((\frac{q}{3})^{\alpha_\varepsilon} - \frac{q}{3})$, and the parameter $C_{1\varepsilon}(C_{1\varepsilon} = \mu/2)$ is an intermittency parameter characterizing the fractal dimension of the mean. It satisfies $0 \leq C_{1\varepsilon} \leq 1$ (the larger $C_{1\varepsilon}$, the more intermittent the process): $C_{1\varepsilon} = 0$ for a homogeneous process and $C_{1\varepsilon} = 0$ for a process so heterogeneous that the fractal dimension of the set contributing to the mean is zero (Seuront et al. 1999). For passive scalar fluctuations, the lack of a known condition of normalization leads to:

$$\zeta_S(q) = qH - \frac{C_{1\varphi}}{\alpha_\varphi - 1}(q^{\alpha_\varphi} - q)$$ (8.74)

where $K_\varphi(q) = \frac{C_{1\varphi}}{\alpha_\varphi - 1}(q^{\alpha_\varphi} - q)$. We can estimate $C_{1\varphi}$ and α_φ from Equation (8.74). If Equation (8.74) is differentiated and evaluated at $q = 0$, simple algebra shows that:

$$q\zeta'(0) - \zeta(q) = \frac{C_{1\varphi}q^\alpha}{\alpha_\varphi - 1}$$ (8.75)

α_φ is given by the slope of $[q\zeta'(0) - \zeta(q)]$ vs. q in a log-log plot, and $C_{1\varphi}$ can be estimated from the intercept.

8.3.2.3 Log-Poisson Model

The log-Poisson model (Dubrulle 1994, 2006; She and Waymire 1994; Castaing and Dubrulle 1995) describes the functions $\zeta_V(q)$ and $\zeta_S(q)$ as:

$$\zeta_V(q) = \frac{q}{3} - c\left((1-\gamma)\frac{q}{3} - 1 + \gamma^{q/3}\right) \tag{8.76}$$

and

$$\zeta_S(q) = qH - c((1-\gamma)q - 1 + \gamma^q) \tag{8.77}$$

where $K_\varepsilon(q) = c((1-\gamma)\frac{q}{3} - 1 + \gamma^{q/3})$ and $K_\phi(q) = c((1-\gamma)q - 1 + \gamma^q)$. γ is linked to a maximum intermittency (that is, the most extreme event reachable from a finite sampling), and c is an analogue of C_1 and characterizes the heterogeneity of this maximum intermittency. To estimate the parameters γ and c, one needs to note that:

$$\zeta_S'(q+1) = \left(1 - \frac{\gamma}{c}\right)\zeta_S'(q) + \frac{\gamma}{c}(H - \gamma) \tag{8.78}$$

The slope and the intercept of a $\zeta'(q+1)$ vs. $\zeta'(q)$ plot give respectively $(1 - \gamma/c)$ and $\frac{\gamma}{c}(H-\gamma)$; then considering the parameter H (that is, $H = \zeta(1)$), the estimation of γ and c is straightforward.

8.3.3 Assessment of Cascade Models for Passive Scalars in a Turbulent Flow

The strength of the lognormal (LN), log-Lévy (LL), and log-Poisson (LP) models to describe the intermittent properties of passive scalars passively advected by fully developed turbulence was assessed against the functions $\zeta_S(q)$ estimated from a time series of temperature* sampled simultaneously with the *in vivo* fluorescence time series shown in Figure 8.14D. Figure 8.16 shows the function $\zeta_S(q)$, for moment up to order 8 (with a 0.1 increment), together with the theoretical fits given by Equations (8.71), (8.74), and (8.77), for the lognormal, log-Lévy, and log-Poisson models, with the values proposed in Table 8.1. The lognormal model fits the empirical data for statistical moments $q < 2.5$. This shows that the lognormal model is only compatible with the data up to relatively low order of moments. In contrast, the empirical curves for both the log-Lévy and log-Poisson models are indistinguishable for moments up to order 3 to 5 (Figure 8.16). The scaling exponents $\zeta_S(q)$ were linear after a critical moment q_c because of sampling limitations; see Seuront et al. (1999, 2005) and Seuront (2008) for further details on multifractal phase transitions. Briefly, phase transitions relate to the occurrence of a maximum intermittency γ_{max}. For first-order phase transitions, γ_{max} is the maximum value taken by a given scalar associated with the occurrence of very rare and violent intermittencies. In contrast, for a second-order phase transition, γ_{max} corresponds to the maximum intermittency effectively detected from a finite sample size. In both cases, for $q \geq q_c$, the function $\zeta_S(q)$ has the following linear asymptotic behavior:

$$\zeta_S(q) = 1 - \gamma_{max}q \tag{8.79}$$

---

* The reader is referred to (Seuront et al. 2005) for an evaluation of continuous cascade models to characterize the intermittency of turbulent velocity fluctuations in the atmosphere and the ocean.

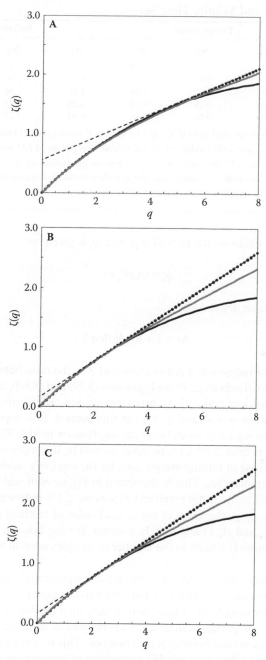

FIGURE 8.16 The structure function exponents $\zeta(q)$ estimated from 500 realizations (A), 100 realizations (B) and 1 realization (C) of segments of 128 points using a temperature time series of 65,536 data points. The critical order of moment q_s (above which $\zeta(q)$ becomes linear due to sampling limitations) decreases with decreasing independent realizations. The log-Lévy (black) and log-Poisson (gray) models fit the nonlinear part of the function $\zeta(q)$ very well (that is, for $q \leq q_s$), but the log-Lévy exhibits a universal shape whatever the number of samples taken into account. In contrast, the log-Poisson model is intrinsically dependent on the number of samples used in the analysis.

TABLE 8.1
Temperature and Salinity Time Series

	Temperature			Salinity		
	N_1	N_2	N_3	N_1	N_2	N_3
H	0.41	0.42	0.42	0.43	0.43	0.42
C_1	0.04	0.04	0.04	0.06	0.06	0.06
α	1.80	1.82	1.80	1.80	1.80	1.80
γ	0.25	0.20	0.18	0.28	0.23	0.20
c	0.79	0.56	0.46	0.82	0.63	0.52

Note: Parameters of the log-Lévy (H, C_1, and α) and log-Poisson (γ and c) models obtained for 500 (N_1), 100 (N_2), and 1 (N_3) independent realizations of segments of 128 points using a temperature time series of 65,536 data points. The log-Poisson parameters are strongly dependent to the number of independent realizations considered in the structure function analysis. In contrast, the log-Lévy model returns universal values for H, C_1, and α.

In the case of sampling limitations, the critical exponent q_s is given by:

$$q_s = (\Delta s/C_1)^{1/\alpha} \tag{8.80}$$

Δs is the sampling dimension defined as:

$$\Delta s = 1 + \log N_s/\log \lambda \tag{8.81}$$

where N_s is the number of independent realizations and λ is the ratio between the largest and the smallest scales of the inertial subrange. From Equations (8.79) and (8.80), it comes that q_s increases with the number of independent realizations. First- and second-order multifractal phase transitions then respectively occur when $q_c < q_s$ and $q_c \geq q_s$. In this context, it is important to mention a fundamental advantage of the log-Lévy model over the log-Poisson model. The log-Poisson model, as recently discussed (Seuront et al. 2005), is implicitly limited by the upper bound on the intermittencies it can detect, even in case of infinite sample size; for the log-Lévy model, there is no theoretical upper bound for the intermittencies. This is illustrated in Figure 8.16 and Table 8.1. The log-Lévy models capture the nonlinearity of the empirical exponents $\zeta_S(q)$ independently of the number of samples used to estimate $\zeta_S(q)$, the value of the critical order of moment q_s (Figure 8.16), and the related parameters H, $C_{1\varphi}$ and α_φ (Table 8.1). In contrast, the log-Poisson fit and the values of the parameters γ and c are strongly linked to the number of samples used in the analyses (Figure 8.16 and Table 8.1).

These results have strong implications in terms of both appropriately sampling the marine environment and understanding the nature of the related biophysical patterns. The previous results demonstrate that the log-Lévy models are much more robust than the log-Poisson models to reliably quantify the intermittency properties of marine scalars because of their ability to capture their statistical properties even with a limited number of observations. This is critical for the implementation of sampling strategies and the subsequent reliable assessment of intermittent patterns and processes.

8.4 MULTIFRACTALS: MISCONCEPTIONS AND AMBIGUITIES

8.4.1 Spikes, Intermittency, and Power Spectral Analysis

The limitations of power spectral analysis and the fact that the spectral slope is not necessarily a strong test of any model of physical-biological interactions without a preliminary careful examination of the raw data have recently been emphasized by Franks (2005). For instance, different phase relationships of the various sines and cosines making up the Fourier decomposition of the data may indeed lead qualitatively very distinct signals to return a $k^{-5/3}$ spectral slope (see Franks 2005, Figure 13). It is, however, acknowledged that a careful examination of the raw data to identify specific features (for example, first- and second-order stationarity) that might be driving the spectral slope is an absolute prerequisite to spectral analysis (Bloomfield 2000), and ultimately to any type of time series analysis; see, for example, Chatfield (2003), Kantz and Schreiber (2004), and Wei (2005).

Franks (2005) found that the power spectra of images of phytoplankton fluorescence gathered with an imaging fluorometer were "white" ($\beta = 0$) over nearly two decades. He subsequently claimed that "the few, large, intensely fluorescent cells" that occur over a background of low fluorescence—and might somehow be related to the fluorescence intermittent hot spots clearly visible in Figure 8.14d—"cause the spectrum to be flat, or white." To support this statement and his empirical findings, under the assumption that fluorescence hotspots behave as delta functions, he created 400 data series of 256 points with 20 randomly placed delta functions, calculated their spectra, averaged them, and plotted the average spectrum, which was flat. Similarly, he showed that randomly adding 30 delta functions of increasing amplitudes to a synthetic data set 256 points long created from 30 sine waves whose amplitude was determined by a $k^{-5/2}$ spectral slope lowers the slope of the spectrum from $k^{-5/2}$ with no spikes to $k^{-2/3}$ when the spikes were 5× the amplitude of the largest sine wave. As stated by Franks (2005), it is agreed that "spikes in plankton could arise from any number of causes that have nothing to do with mixing in 3D isotropic turbulence." However, to consider that fluorescence hotspots—whether they are created by large cells, aggregates, pieces of seaweed, or zooplankton guts—behave as randomly distributed delta functions is a very strong assumption. This implies that the distribution of these hotspots follows a Markovian process, thus returning a "white," memoryless power spectrum, which contradicts (1) many empirical works that have found spectral slopes significantly different from zero for nutrients, phyto- and zooplankton distributions (Tsuda et al. 1993; Seuront et al. 1996a, 1999, 2002; Mountain and Taylor 1996; Wiebe et al. 1996; Lovejoy et al. 2001; Pershing et al. 2001); and (2) more specific investigations specifically dealing with the scaling properties of intermittent behaviors in nutrient, phytoplankton, and zooplankton (Seuront et al. 1996a, 1996b, 1999, 2002; Seuront and Lagadeuc 2001; Lovejoy et al. 2001; Seuront 2005b). In addition, from a purely methodological point of view, the power spectrum resulting from randomly placed delta functions will intrinsically return a "white" behavior because the Fourier transform of a delta function is a constant, that is, a white spectrum. In contrast, there is no assumption related to the use of structure functions, which would pick up the stochastic properties of any intermittent field whatever they are, that is, flat or steep spectra.

The role of intermittent fluctuations on power spectral slopes is clarified hereafter on both theoretical and empirical grounds. From Equations (8.67) and (8.73), Equation (4.23) can be rewritten as:

$$E_V(k) \approx k^{-\left[5/3-(C_{1\varepsilon}/(\alpha_\varepsilon-1))((2/3)^{\alpha_\varepsilon}-2/3)\right]} \qquad (8.82)$$

This leads to a slope steeper than 5/3 because $\zeta_v(2) < 0$ (see Equation 8.73). Using $C_{1\varepsilon} = 0.15$ and $\alpha_\varepsilon = 1.50$ for atmospheric turbulence and $C_{1\varepsilon} = 0.16$ and $\alpha_\varepsilon = 1.55$ for oceanic turbulence (Seuront et al. 2005) in Equation (8.82) leads to $\beta_V(k) = 1.70$. Similarly, from Equation (8.67) and (8.74), Equation (4.24) is rewritten as:

$$E_s(k) \approx k^{-[1+2H-(C_{1\varphi}/(\alpha_\varphi-1))((2)^{\alpha_\varphi}-2)]} \qquad (8.83)$$

Using $C_{1\varphi} = 0.04$ and $\alpha_\varphi = 1.70$ for *in vivo* fluorescence advected by fully developed turbulence in the coastal waters of the eastern English Channel in Equation (8.83) leads to $\beta_s(k) = 1.77$. Both intermittent turbulent velocity fluctuations and intermittent *in vivo* fluorescence fluctuations lead to a spectral slope steeper than the theoretical $\beta_s(k) = 5/3$ expected under nonintermittent turbulent conditions. It is stressed here that Equation (8.83) does not involve any assumption on phytoplankton concentration being considered as a passive conservative tracer or a biologically active tracer. Equation (8.83) is instead general and would lead to a steepening of any power spectrum due to intermittency correction, whether it follows a $k^{-5/3}$ power law or not.

It is finally stressed that the divergences observed between Franks's conclusions (Franks 2005) and those presented here might also stem from the differences in the approaches used. First, the observational platforms used by Franks and coworkers (Franks 2005; Franks and Jaffe 2008) return horizontal 32×32 cm 2D fluorescence distributions sampled vertically every 6 to 24 cm, significantly different from the fluorescence data analyzed here that correspond to a time series recorded from a fixed depth at a rate of 2 Hz. Second, Franks's data were collected in stratified waters 250 to 450 m deep 10 km offshore of San Diego (California) where chlorophyll a concentrations were typically bounded between 0.1 and 0.6 µg l^{-1} (Franks and Jaffe 2008, Figure 2). In contrast, the data analyzed here were collected in the tidally mixed shallow coastal waters of the eastern English Channel where chlorophyll a concentrations ranged from 6 to 30 µg l^{-1}. It is then likely that the different spectral shapes returned by the analysis of those two data sets might also be related to their intrinsic differences. Unambiguously demixing apples and oranges in an intermittent context would then require further work through a thorough investigation of data sets collected following the different methods described above simultaneously in a range of marine environments.

8.4.2 FREQUENCY DISTRIBUTIONS AND MULTIFRACTALITY

As stressed above (Section 8.1.2) and as implicitly seen from the sets of equations defining most of the multifractal functions described in Section 8.2—that is, Equations (8.3), (8.21), and (8.23)—multifractal sets and signals will return a power-law behavior in cumulative frequency distributions (Chapter 5). However, a cumulative frequency distribution exhibiting a power-law behavior does not mean that the process being analyzed is multifractal. This is illustrated through the cumulative probability distribution function, $P(X \geq x) \propto x^{-\phi}$, and the probability distribution function, $P(X = x) \propto x^{-\mu}$, used to characterize intermittent distribution, when $1 < \mu \leq 3$ (see Section 5.1 and Equations 5.1 and 5.2). As discussed in Section 4.2.10.2 and Section 8.2.4, both nonintermittently and intermittently distributed quantities can be described by the qth-order structure functions. The function $\zeta(q)$ is linear for monofractal processes, that is, $\zeta(q) = qH$, where H is the Hurst exponent as defined in Chapter 4. In the case of multifractality, $\zeta(q)$ is nonlinear, concave, and takes the general form $\zeta(q) = Aq + Bq^{\alpha}$ (see Section 8.3.2.2 and Equation 8.72). For a Lévy distribution, the structure function exponent $\zeta(q)$ is defined as a bilinear model (Chechkin and Gonchar 2000; Nakao 2000):

$$\zeta(q) = q/\mu \qquad (8.84)$$

for $q < \mu$ (that is, $q < \phi + 1$), and

$$\zeta(q) = 1 \qquad (8.85)$$

for $q \geq \mu$ (that is, $q \geq \phi + 1$). In the special case of a Lévy flight, $\mu = 2$ (Shlesinger et al. 1996), leads to $\zeta(q) = q/2$ for $q < 2$, and $\zeta(q) = 1$ for $q \geq 2$. The cumulative probability distribution function estimated from a time series of the successive displacements of the calanoid copepod *Temora longicornis* (Figure 8.2A) clearly exhibits a power-law behavior with $\phi = 2.74$ (Figure 8.17A); that is, $\mu = 1.74$. The empirical function $\zeta(q)$ obtained from *T. longicornis* successive displacements is subsequently compared with the functions $\zeta(q)$ theoretically expected for a Brownian motion and a Lévy distribution characterized by $\mu = 1.74$ (Figure 8.17B). The nonlinearity of the empirical function clearly contrasts from the bilinear behavior related to Equations (8.84) and (8.85), showing that a power-law signature in frequency distributions does not imply multifractality. In contrast, multifractality is consistently related to a power-law signature in frequency distributions.

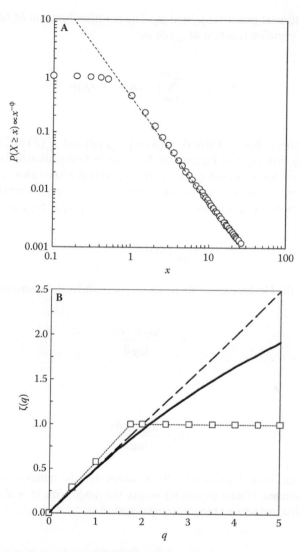

FIGURE 8.17 (A) The cumulative probability distribution function estimated from a time series of the successive displacements of the calanoid copepod *Temora longicornis* shown in Figure 8.2A. A power-law behavior with $\phi = 1.74$ (dashed line) is clearly visible for displacements ranging from 1 to 30 mm. (B) The structure function exponents expected for a Brownian motion $\zeta(q) = q/2$ (dashed line), a Lévy flight, $\zeta(q) = q/2$ for $q < 2$ and $\zeta(q) = 1$ for $q \geq 2$ (open squares), and the empirical $\zeta(q)$ estimated from a time series of the successive displacements of the copepod *Temora longicornis* shown in Figure 8.2A.

8.5 JOINT MULTIFRACTALS

8.5.1 JOINT MULTIFRACTAL MEASURES

Joint multifractals have been introduced by Meneveau et al. (1990) to assess the degree of correlation between two simultaneously recorded multifractal fields such as the intermittent fluxes ε_l and χ_l discussed above. To my knowledge, this approach has only subsequently been applied once to the joint analysis of crop yield and terrain slope (Kravchenko et al. 2000).

Consider two multifractal measures μ_1 and μ_2. The partition function $M_q(\delta)$ (Equation 8.5) can be rewritten as a joint partition function $M_{q_1 q_2}(\delta)$ as:

$$M_{q_1 q_2}(\delta) = \sum_{i=1}^{N(\delta)} [f_{1_i}(\delta)]^{q_1} [f_{2_i}(\delta)]^{q_2} \tag{8.86}$$

where $N(\delta)$ is the number of boxes of size δ, $f_{1_i}(\delta) = \sum_i^N \mu_1(\delta)$ and $f_{2_i}(\delta) = \sum_i^N \mu_2(\delta)$; see Equation (8.4). Note that when $q_1 = 0$ or $q_2 = 0$, Equation (8.86) reduces to Equation (8.5). As discussed for the single partition function, low values of q_1 and q_2 characterize the low values of the first and second measures, while at high q_1 and q_2 the joint partition function depends mostly on high values of μ_1 and μ_2. From Equation (8.11), the mass exponent of order (q_1, q_2), $\tau(q_1, q_2)$ comes from:

$$M_{q_1 q_2}(\delta) \propto \delta^{-\tau(q_1, q_2)} \tag{8.87}$$

From Section 8.2.2.1, the Hölder exponents α_1 and α_2 of multifractal measures μ_1 and μ_2 come as:

$$\alpha_1 = \frac{\log \mu_1(\delta)}{\log \delta} \tag{8.88}$$

and

$$\alpha_2 = \frac{\log \mu_2(\delta)}{\log \delta} \tag{8.89}$$

where δ is the box size. From Equation (8.15), it comes that the number of cells $N_{\alpha_1 \alpha_2}(\delta)$ with a singularity strength (that is, Hölder exponent) within the ranges $[\alpha_1, \alpha_1 + d\alpha_1]$ and $[\alpha_2, \alpha_2 + d\alpha_2]$ respectively for the first and second measure scales with δ as:

$$N_{\alpha_1 \alpha_2}(\delta) \propto \delta^{-f(\alpha_1, \alpha_2)} \tag{8.90}$$

where $f(\alpha_1, \alpha_2)$ is the joint singularity spectrum describing the abundance of cells with common α_1 and α_2 values. The joint singularity spectrum $f(\alpha_1, \alpha_2)$ and the joint mass exponent function $\tau(q_1, q_2)$ are connected via a double Legendre transform as:

$$\alpha_1(q_1, q_2) = \frac{d\tau(q_1, q_2)}{dq_1} \tag{8.91}$$

$$\alpha_2(q_1, q_2) = \frac{d\tau(q_1, q_2)}{dq_2} \tag{8.92}$$

$$f(\alpha_1, \alpha_2) = q_1 \alpha_1 + q_2 \alpha_2 - \tau(q_1, q_2) \tag{8.93}$$

As discussed in Section 8.2.2.1, the maximum value of the joint multifractal spectrum $f(\alpha_1, \alpha_2)$ is reached when $q_1 = 0$ and $q_2 = 0$, in which case $f(\alpha_1, \alpha_2)$ equals the box-counting dimension of the geometrical support of the measures μ_1 and μ_2.

Meneveau et al. (1990) first illustrated the applicability of joint multifractals to joint lognormal and binomial distributions, and simultaneous experimental distributions of turbulent shear and temperature gradients. A decade later, joint multifractals were used to assess the links between crop yield and topography, and allowed to differentiate yield distributions corresponding to field locations with high slopes and to make inferences about slope distributions that affect grain the most (Kravchenko et al. 2000). A more general procedure based on joint moments is proposed hereafter to study the joint properties of two multifractal measures.

8.5.2 The Generalized Correlation Functions

8.5.2.1 Definition

The joint correlation functions using structure functions introduced by Seuront and Schmitt (2005a, 2005b) are a continuation and development of the early study by Meneveau et al. (1990). Their approach provides estimates of the fractal dimension of mixed singularities, instead of the scale-invariant moment functions suggested hereafter. Another major difference between their joint multifractal formalism and the one provided here is that they did not normalize joint moments, as seen below, to provide joint correlations. As such, the joint correlation function introduced hereafter can be thought of as a more intuitive approach and as a high-order generalization of the standard correlation between two variables X and Y. Note that joint moments for scaling structure functions have also been proposed in the field of econophysics, to study correlations for multiple assets, in order to characterize their return distributions; see, for example, Muzy et al. (2001). However, the final objective of such a study is portfolio optimization, which is different from our analysis of the generalized correlation between two multifractal fields.

Instead of random variables X and Y, consider two stochastic processes (ΔX_τ) and (ΔY_τ) characterized by their pth- and qth-order structure functions as $\langle |(\Delta X_\tau|^p \rangle \approx \tau^{\zeta_X(p)}$ and $\langle |(\Delta Y_\tau|^q \rangle \approx \tau^{\zeta_Y(q)}$. The correlation between the two processes (ΔX_τ) and (ΔY_τ) then becomes a function of the scale and of the statistical orders of moment p and q, expressed by the generalized correlation functions (GCF hereafter) $c(p, q)$ as (Seuront and Schmitt 2005a):

$$c(p,q) = \frac{\langle [|(\Delta X_\tau)|^p |(\Delta Y_\tau)|^q] \rangle}{\langle |(\Delta X_\tau)|^p \rangle \langle |(\Delta Y_\tau)|^q \rangle} \mu \tau^{-r(p,q)} \tag{8.94}$$

where for more generality we take $p \neq q$ and $r(p, q) \geq 0$ (Seuront and Schmitt 2005a). The generalized correlation exponents (GCE hereafter) $r(p, q)$ are estimated as the slopes of the power-law trends of $c(p, q)$ vs. τ in a log-log plot, and expressed as

$$r(p,q) = \zeta_X(p) + \zeta_Y(q) - S(p,q) \tag{8.95}$$

where $\zeta_X(q)$ and $\zeta_Y(q)$ characterize the multiscaling properties of the single fluctuations $\langle |(\Delta X_\tau|^p \rangle$ and $\langle |(\Delta Y_\tau|^q \rangle$, and $S(p, q)$ characterizes the multiscaling properties of the joint fluctuations $\langle |(\Delta X_\tau|^p |(\Delta Y_\tau)|^q \rangle$. Both $c(p, q)$ and $r(p, q)$ are generalizations of the standard correlation function. In the special case $p = q = 1$, Equation (8.94) indeed recovers the standard expression of the correlation coefficient between (ΔX_τ) and (ΔY_τ). GCF and GCE hence express the correlation between $|(\Delta X_\tau)|^q$ and $|(\Delta Y_\tau)|^q$, together with their scale and moment dependence. Note that at a given scale τ, if the fluctuations of the stochastic processes (ΔX_τ) and (ΔY_τ) are independent, $r(p, q) = 0$ for any combination of p and q. Increasing values of $r(p, q)$ would thus characterize increasing dependence between $|(\Delta X_\tau)|^p$ and $|(\Delta Y_\tau)|^q$. Note that while independence implies uncorrelation, uncorrelation does not imply independence. Uncorrelation corresponds to the relation $r(1, 1) = 0$ and implies independence only in special cases such as for Gaussian processes. In the general case, this is no longer true: Independence between the stochastic processes means that $r(p, q) = 0$ whatever the values of p and q. Figure 8.18A

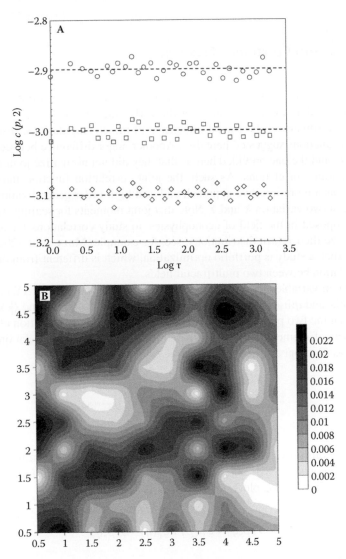

FIGURE 8.18 (A) The generalized correlation functions $c(p, q)$ vs. the time scale τ in log-log plots, for the grid-generated turbulent velocity time series and the *in vivo* fluorescence time series shown in Figure 8.14A,D, respectively. The function $c(p, q)$ shown here has been estimated for a constant value of the statistical order of moment q of velocity fluctuations ($q = 2$), and various values of the statistical order of moment p of *in vivo* fluorescence (that is, $p = 1$, 2, and 3, from bottom to top). The slopes of the linear regression estimated over the scaling ranges (dashed lines) provide estimates of the generalized correlation exponents $r(p, q)$ (B). (Modified from Seuront and Schmitt, 2005a.)

shows the GCF, $c(p, q)$, plotted in log-log plots vs. τ, for the grid-generated turbulent velocity time series and the *in vivo* fluorescence time series, respectively (shown in Figure 8.14A,D). As these time series have been independently sampled, they represent an archetypical example of two independent multifractal processes. The very low values taken by the functions $c(p, q)$ indicate the absence of any correlation between the turbulent velocity and fluorescence fluctuations, $|(\Delta V_\tau)|^p$ and $|(\Delta F_\tau)|^q$. This is confirmed and specified by the related values of the function $r(p, q)$, which remain statistically undistinguishable from zero, whatever the combinations of p and q values (Figure 8.18B); see Box 8.2.

The main advantages of this technique are that (1) it makes no assumptions about the spectrum or the probability distribution of either data set, (2) it takes into account the multiscale intrinsic properties of intermittent processes, and (3) it allows testing for the phenomenology responsible for the high intensity (rare and unexpected) fluctuations observed in intermittent distributions, considering their potential association with both high- and low-intensity fluctuations characterized by high and low orders of moment. The generalized correlation functions and exponents thus provide a general framework, as they express the correlation between the fields (ΔX_τ) and

---

BOX 8.2   GENERALIZED CORRELATION FUNCTIONS AND EXPONENTS IN SPECIAL CASES

The function $c(p, q)$ and the related scaling exponent $r(p, q)$ can be used as an analysis tool to study the couplings between two multifractal fields x ($x = \Delta X_l$) and y ($y = \Delta Y_l$). To provide some basis for discussion and interpretation of experimental results, some limit cases are considered.

If x and y are independent $r(p, q) = 0$. On the other hand, in case of perfect proportionality $x = Ky$, where K is a constant, or for random proportionality $x = \kappa y$, where κ is a random variable independent on y, it is readily seen that

$$r(p, q) = \zeta_Y(p) + \zeta_Y(q) - \zeta_Y(p + q) \tag{8.B2.1}$$

In particular, one may note that $r(p, q) > 0$ due to the convexity of the scaling functions $\zeta(p)$. This relation can be directly tested to verify the proportionality hypothesis. Furthermore, the shape of the surface obtained is symmetric in the p-q plane. In this specific case, the function $r(p, q)$ has the desirable advantage to reduce considerably the number of data points—that is, $r(p, q)$ values—needed to understand the relationship between the fields x and y.

Another very simple situation occurs when $x = Ky^b$ with $b > 0$ and K constant, or when $x = \kappa y^b$ with κ random and independent of y, then

$$r(p, q) = \zeta_Y(bp) + \zeta_Y(q) - \zeta_Y(bp + q) \tag{8.B2.2}$$

$r(p, q)$ in Equation (8.B2.2) is still positive but no longer symmetric in the p-q plane but in the bp-q plane. The value of b may be first estimated as the positive value such that

$$r(p, 0) = r(0, bp) \tag{8.B2.3}$$

Using the values of b, this can be tested by verifying that $r(p/b, q)$ is indeed symmetric in the p-q plane. More generally speaking, the more $r(p, q)$ is positive, the more the $x = \Delta X_l$ and $x = \Delta Y_l$ are dependent random variables.

(ΔY_τ), together with their scale dependence and their moment dependence. This indicates if high-intensity fluctuations of one field are highly correlated to low intensities of the other (or, in other words, the gradients of two fields are proportional or inversely proportional) and the scales over which these correlations occur.

8.5.2.2 Applications

The potential effect of varying turbulent forcing on the local structure of physical and biological parameters has been investigated using structure function analysis in Section 8.2.4.1 on the basis of 24 time series of temperature, salinity, and *in vivo* fluorescence (Figures 8.11 and 8.12). Those results demonstrated that temperature and salinity fluctuations remained similar under different turbulent and tidal forcings, while *in vivo* fluorescence (that is, phytoplankton biomass) clearly exhibited more intermittent fluctuations during ebb tide than flood tide and under conditions of low turbulence. However, this does not provide any information on the nature of biophysical couplings and their relation to turbulence and tide. This issue is investigated here using the generalized correlation functions and exponents introduced in Section 8.5.2.1.

The nature of the dependence between temperature and phytoplankton distributions has been assessed through the generalized correlation functions, $c(p, q)$ and the related generalized correlation exponents, $r(p, q)$, between temperature and fluorescence time series, for each of the 24 time series mentioned above. Figure 8.19 shows the GCF, $c(p, q)$, plotted in log-log plots versus the time scale τ, for simultaneously recorded temperature and fluorescence time series for ebb tides (Figure 8.19A) and flood tides (Figure 8.19B), as well as for temperature and fluorescence time series taken at different moments of the tidal cycle, and *a fortiori* independent (Figure 8.20A). Both the power-law behavior of the functions $c(p, q)$ over the whole range of available scales, and the positive values taken by the GCE, $r(p, q)$, indicate a form of dependence between temperature and fluorescence fluctuations. In addition, the values of the functions $c(p, q)$ are smaller during flood tide (Figure 8.19A) than during ebb tide (Figure 8.19B), suggesting a differential correlation between temperature and phytoplankton biomass fluctuations controlled by tidal processes. On the other hand, the weak values taken by the functions $c(p, q)$ estimated between independent temperature and fluorescence time series (Figure 8.20A) indicate a low correlation between temperature and phytoplankton biomass fluctuations, $\langle (\Delta T_\tau)^p \rangle$ and $\langle (\Delta F_\tau)^q \rangle$. This is confirmed by the related values of the functions $r(p, q)$, which remain close to zero, whatever the combinations of p and q values (Figure 8.20B).

These observations were refined comparing the functions $r(p, q)$ obtained between temperature and fluorescence time series in different tidal and turbulent conditions. Figure 8.21 shows the functions $r(p, q)$ obtained for all combinations of p and q values (between 0.5 and 5) with 0.1 increments for three levels of turbulence (10^{-4}, 10^{-5}, and 10^{-6} m^2·s^{-3}) during ebb and flood tides, respectively. In both case, the correlation between temperature and fluorescence fluctuations increases with increasing hydrodynamic conditions (Figure 8.21A,B,D,E) and is weaker, even nil, in low turbulent conditions (Figure 8.21C,F). On the other hand, the decorrelation observed between temperature and phytoplankton fluctuations during both ebb and flood tides under weak turbulent conditions suggests an increase in the biological contributions to the control of phytoplankton biomass distribution and confirms previous observations (cf. Figure 8.12). Phytoplankton fluctuations then appear independent from the temperature fluctuations under the lowest turbulence levels investigated here, that is, 5×10^{-7} m^2·s^{-3} (Figure 8.21C,F). This confirms the differential physical control suggested under strong turbulent conditions from the analysis of the shape of the function $\zeta_F(q)$, and its comparison with the function $\zeta_T(q)$ (see Figure 8.12).

More specifically, the overall shape of the functions $r(p, q)$ indicates that large phytoplankton fluctuations are associated, under strong enough turbulent conditions, to strong temperature gradients, and *vice versa*. This tendency seems to reflect, over a slightly wider range of scales, findings

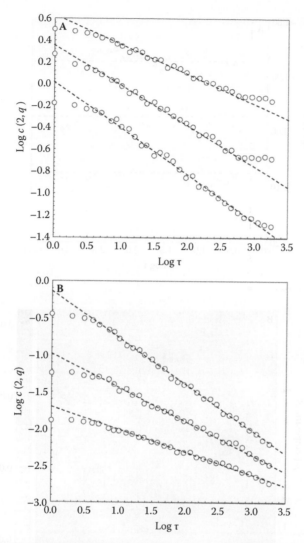

FIGURE 8.19 The generalized correlation function $c(p, q)$ vs. the time scale in log-log plots, for temperature and fluorescence time series simultaneously recorded during ebb tide (A) and flood tide (B). The function shown here have been estimated for a constant value of the statistical order of moment q of temperature fluctuations ($q = 2$), and various values of the statistical order of moment p of *in vivo* fluorescence (that is, $p = 1$, 2, and 3, from bottom to top). The slopes of the linear regression estimated over the scaling ranges (dashed lines) provide estimates of the generalized correlation exponents $r(p, q)$. (Modified from Seuront and Schmitt, 2005b.)

of Desiderio et al. (1993), who observed the occurrence of 0.1 to 0.2-meter-thick fluorescent layers just above local temperature gradients. As this suggests a proportionality relationship between temperature and phytoplankton fluctuations, Equation (8.B2.2) has been verified testing the validity of Equation (8.B2.3) over a wide range of b values (Box 8.2). Using b values ranging between 0.05 and 5 (with 0.05 increments), we then showed that Equation (8.B2.3) is verified for four of the six turbulence levels investigated during ebb tides, i.e., $\varepsilon = 10^{-4}$ m$^2 \cdot$s^{-3} with $b = 0.90$, $\varepsilon = 5 \times 10^{-5}$ m$^2 \cdot$s^{-3} with $b = 0.85$, $\varepsilon = 10^{-5}$ m$^2 \cdot$s^{-3} with $b = 0.80$, and $\varepsilon = 5 \times 10^{-6}$ m$^2 \cdot$s^{-3} with $b = 0.78$, and two of the six turbulence levels investigated during flood tides, that is, $\varepsilon = 10^{-4}$ m$^2 \cdot$s^{-3} with $b = 0.96$ and $\varepsilon = 5 \times 10^{-5}$ m$^2 \cdot$s^{-3}

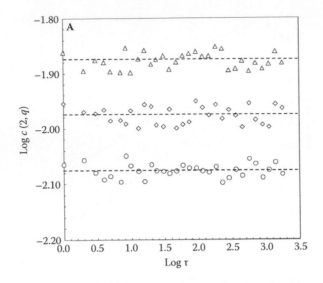

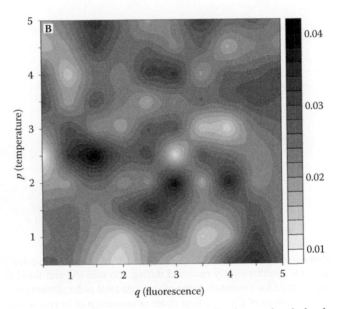

FIGURE 8.20 The generalized correlation function $c(p, q)$ vs. the time scale τ in log-log plots, for temperature and fluorescence time series simultaneously recorded at different moments of the tidal cycle, and *a priori* independent (A). The function $c(p, q)$ shown here have been estimated for a constant value of the statistical order of moment q of temperature fluctuations ($q = 2$), and various values of the statistical order of moment p of *in vivo* fluorescence (that is, $p = 1, 2,$ and 3, from bottom to top). The slopes of the linear regression estimated over the scaling ranges (dashed lines) provide estimates of the generalized correlation exponents $r(p, q)$ shown in (B). (Modified from Seuront and Schmitt, 2005b.)

with $b = 0.91$. Figure 8.22 shows the relation $r(p, 0)$ vs. $r(0, bp)$ corresponding to the functions $r(p, q)$ shown in Figure 8.21A,B with $b = 0.90$ and $b = 0.80$, respectively. The correlation shown between temperature and phytoplankton fluctuations under high turbulent conditions is then related to a power-law dependence relationship of the form $\Delta F_\tau \propto (\Delta T_\tau)^b$.

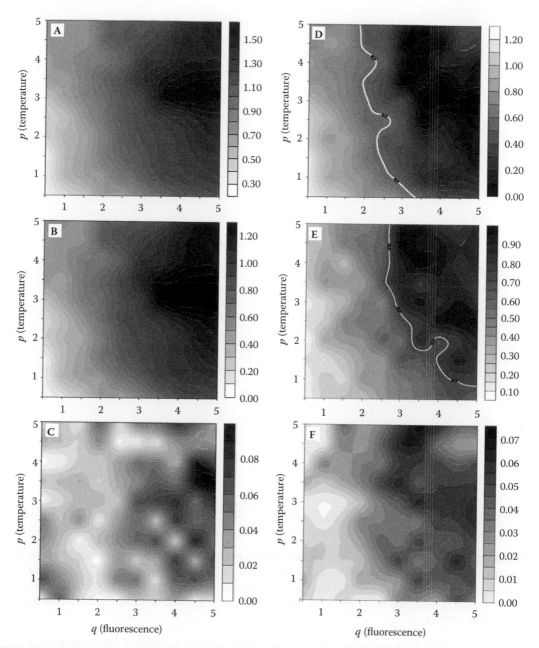

FIGURE 8.21 The generalized correlation exponents $r(p, q)$, shown as a function of both p and q, which characterize temperature and *in vivo* fluorescence fluctuations, respectively. The functions $r(p, q)$ correspond to three different levels of turbulence: $\varepsilon = 10^{-4}$ m²·s⁻³ (A,D), $\varepsilon = 10^{-5}$ m²·s⁻³ (B,E), and $\varepsilon = 10^{-6}$ m²·s⁻³ (E,F) investigated during ebb tide (left panel) and flood tide (right panel). (Modified from Seuront and Schmitt, 2005b.)

8.6 INTERMITTENCY AND MULTIFRACTALS: BIOLOGICAL AND ECOLOGICAL IMPLICATIONS

Intermittency can have implications in a wide range of biological and ecological processes in terrestrial and aquatic ecosystems. Based on the marine background of the author, this section explores a few areas of marine sciences that could be significantly affected by intermittency. Specifically regarding probabilistic arguments derived from the multifractal framework described

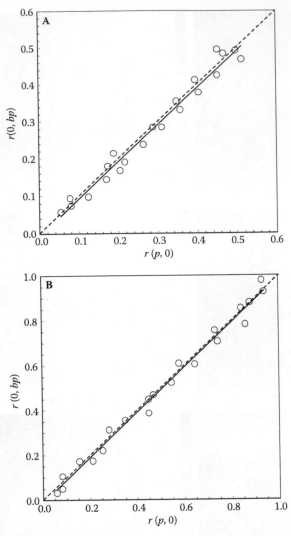

FIGURE 8.22 Plots of the generalized correlation exponents $r(p, 0)$ vs. $r(0, bp)$ for two levels of turbulence during ebb tide, $\varepsilon = 10^{-4}$ m²·s⁻³ with $b = 0.90$ (A) and $\varepsilon = 10^{-5}$ m²·s⁻³ with $b = 0.80$ (B). The slopes of the regression lines (continuous line) cannot be distinguished from the one of the relation $bp = q$ (dashed line). This shows the validity of Equation (8.B2.2), and the symmetry of the functions $r(p, q)$ in the bp-q plane. (Modified from Seuront and Schmitt, 2005b.)

in Section 8.2, this section demonstrates that taking into account turbulence intermittency can lead to substantial changes in the probability of the occurrence of rare events, and our subsequent understanding of their impact on marine life and salient processes in the ocean such as predator–prey (or male–female) encounter rates, nutrient fluxes around phytoplankton cells, physical coagulation of phytoplankton cells, and in the subsequent phytoplankton aggregate volume. It is stressed, however, that the probabilistic nature of the approach used hereafter does not hamper the generality of this section, as a similar approach can be equivalently implemented to assess the effect of intermittent fluctuations on terrestrial processes.

8.6.1 INTERMITTENCY, LOCAL DISSIPATION RATES, AND ZOOPLANKTON SWIMMING ABILITIES

Making use of the lognormal theory, Yamazaki et al. (2002) investigated the probability density function of locally averaged dissipation rate over a subdomain l within a parent domain L.

Considering that the minimum averaging scale l (that is, the smallest scale reachable by turbulent fluctuations) is 10 times the Kolmogorov length scale l_k ($l_k = (v^3/\varepsilon)^{1/4}$, where v is the kinematic viscosity), they focused on two typical conditions in the upper ocean: turbulence in the surface mixed layer and a turbulent patch in a seasonal thermocline. In both cases, the turbulence is characterized by the average dissipation rate ε (10^{-5} and 10^{-8} m$^2\cdot$s^{-3}, respectively), and the subsequent Kolmogorov length scale l_k (5.62×10^{-4} and 3.16×10^{-3} m, respectively), the largest scale L (10 and 1 m, respectively) and the smallest scale l ($l = 10l_k$) reachable by turbulent fluctuations (5.62×10^{-3} and 3.16×10^{-2} m, respectively). Assuming lognormality for the local dissipation rate ε_λ, they estimated the probability for a local value ε_λ to exceed the average value ε, that is, $\Pr(\varepsilon_\lambda \geq \varepsilon)$, as 0.248 and 0.328 for the mixed layer and the seasonal thermocline cases, respectively. In the multifractal framework, this probability is given by (see Equation 8.21):

$$\Pr(\varepsilon_\lambda \geq \lambda^\gamma) = \lambda^{-c(\gamma)} \tag{8.96}$$

γ is a given singularity (that is, an intermittency level), λ is the scale ratio (that is, $\lambda = L/l$), and $c(\gamma)$ is the codimension function defined as (see Equation 8.27):

$$c(\gamma) = C_1 \left(\frac{\gamma}{C_1\alpha'} + \frac{1}{\alpha} \right)^{\alpha'} \tag{8.97}$$

where C_1 and α are the parameters characterizing the log-Lévy distribution and α' is given by $1/\alpha + 1/\alpha' = 1$ (Section 8.2.3). Using values of C_1 and α used in the previous section ($C_1 = 0.16$ and $\alpha = 1.55$) and of λ estimated from above, the probability $\Pr(\varepsilon_\lambda \geq \varepsilon)$ became $\Pr(\varepsilon_\lambda \geq \varepsilon) = 0.634$ and $\Pr(\varepsilon_\lambda \geq \varepsilon) = 0.765$ for the mixed layer and the seasonal thermocline case, respectively (Seuront 2008).

Using root-mean-square turbulent velocity values estimated from dissipation rates observed in the seasonal thermocline and in a fjord, Yamazaki and Squires (1996) showed that nominal swimming speeds of several zooplankton species taken from a literature survey are larger than turbulent velocity fluctuations. They subsequently claimed than organism motion can be independent of the local turbulent flow field. Similar results have been recently obtained on the basis of behavioral observations of the calanoid copepod *Temora longicornis* (Seuront et al. 2004d). It is shown that root-mean-square turbulent velocity overcomes the swimming velocity of *T. longicornis* only for very high values of the dissipation rates ε, that is, $\varepsilon \geq 10^{-5}$ m$^2\cdot$s^{-3}. However, these statements are based on investigations of mean values of the dissipation rate ε, instead of the local values ε_λ. Using the multifractal framework, it is shown that these results can be quantitatively refined considering the probabilistic nature of ε_λ.

Under the isotropy assumption, the velocity difference ∂u between two points separated by a distance l can be estimated following the isotropic relation:

$$\varepsilon_\lambda \cong 7.5v \overline{\left(\frac{\partial u}{\partial z} \right)^2} \approx 7.5v \left(\frac{\delta u}{l} \right)^2 \tag{8.98}$$

where v is the kinematic viscosity, ca. 10^{-6} m$^2\cdot$s^{-1}. From values of l defined above, and values of ε_λ ranging between 1ε and 100ε, it can be seen that velocity differences ∂u are of the same order of magnitude that zooplankton swimming ability (regarded as bounded between 0.001 and 0.010 m$\cdot$s^{-1} for most mesozooplankton species) for ε_λ values up to 2ε and 70ε in the mixed layer and the seasonal thermocline, respectively (Figure 8.23). The probability described by Equation (8.96) is now rewritten as:

$$\Pr(\partial u < v_{zoo}) = \Pr(\varepsilon_\lambda \geq 2\varepsilon) \tag{8.99}$$

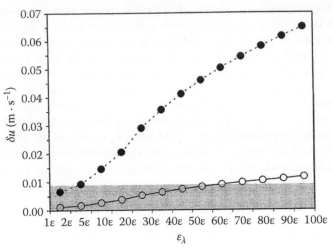

FIGURE 8.23 Estimates of the velocity difference δu between two points separated by a distance l in the surface mixed layer (black dots) and in a turbulent patch in the seasonal thermocline (open circles) for values of the local dissipation rate ε_λ ranging between ε and 100ε. The values of the scale l and of the average dissipation rate ε are 5.63×10^{-3} and 3.16×10^{-2} m, and 10^{-5} and 10^{-8} m^2 s^{-3} for the mixed layer and the seasonal thermocline cases, respectively. The average swimming speed of most mesozooplankton species, bounded between 0.001 and 0.010 m s^{-1}, is shown by the shaded area. (Modified from Seuront, 2008.)

in the surface mixed layer, and

$$\Pr(\partial u < v_{zoo}) = \Pr(\varepsilon_\lambda \geq 70\varepsilon) \qquad (8.100)$$

in a turbulent patch in the seasonal thermocline. This leads to $\Pr(\partial u < v_{zoo}) = 0.57$ for the mixed surface case, and $\Pr(\partial u < v_{zoo}) \rightarrow 1.00$ for the seasonal thermocline case. Thus, the surface mixed layer contains 57% of water volume whose shear can be overcome by mesozooplankton swimming. However, mesozooplankton swimming ability overcomes turbulent shear in almost 100% of the turbulent patches in the seasonal thermocline.

These simple arguments provide the first quantitative description of mesozooplankton swimming ability against turbulence, taking into account the probabilistic nature of intermittent turbulence. These results also confirm that organism motion can be independent of the local flow field in most areas of the world's oceans.

8.6.2 Intermittency, Local Dissipation Rates, and Biological Fluxes in the Ocean

The influence on turbulence intermittency on plankton processes will be considered through three major processes controlled by turbulent processes, and at the core of marine system functioning: nutrient fluxes around nonmotile phytoplankton cells, physical coagulation of phytoplankton cells, and predator–prey encounter rates.

In order to take into account the distribution of the local dissipation rate ε_λ in calculations, Equation (8.24) has been rewritten as (Seuront et al. 2001):

$$\langle \varepsilon_\lambda^q \rangle = \varepsilon^q \lambda^{K(q)} \qquad (8.101)$$

Equation (8.101) is used to evaluate the average of any polynomial function $f(\varepsilon_\lambda)$ of the intermittent field ε_l as:

$$f(\varepsilon_\lambda) = \sum_{q=0}^{N} a_q (\varepsilon_\lambda)^q \qquad (8.102)$$

where a_q are constants, and q the polynomial order of the function $f(\varepsilon_\lambda)$. The average of the function $f(\varepsilon_\lambda)$ in Equation (8.102) leads to:

$$f(\varepsilon_\lambda) = \sum_{q=0}^{N} a_q (\varepsilon_\lambda)^q \lambda^{K(q)} \tag{8.103}$$

8.6.2.1 Intermittency, Turbulence, and Nutrient Fluxes toward Phytoplankton Cells

Following Karp-Boss et al. (1996), the increased rate of nutrient flux due to turbulence around nonmotile phytoplankton cells of radius r(m) can be directly estimated using the Sherwood number, S, as:

$$S_1 = 1 + 0.29 \left[\frac{r}{D} \left(\frac{\varepsilon_0}{v} \right)^{1/2} \right]^{1/2} \tag{8.104}$$

for $r < l_B$, and

$$S_2 = 0.55 \left[\frac{r}{D} \left(\frac{\varepsilon_0}{v} \right)^{1/2} \right]^{1/3} \tag{8.105}$$

for $r > l_B$. D is the diffusivity ($D = 10^{-9}$ m²·s⁻¹), ε_0 the mean turbulent energy dissipation rate (m²·s⁻³), v the kinematic viscosity ($v = 10^{-6}$ m²·s⁻¹), and l_b the Batchelor microscale, the scale of the smallest variations in the ambient nutrient field. One may note here that the Batchelor microscale l_B is smaller than the Kolmogorov microscale l_K (that is, the scale of the smallest turbulent eddies) following $l_K = l_B (D/v)^{1/2}$.

Now the turbulent kinetic energy dissipation rate ε is expressed as an intermittent variable ε_l characterized by the scaling moment function $K_\varepsilon(q)$ (Section 8.2.4.2) and by its mean $\varepsilon_0 = \langle \varepsilon_\lambda \rangle$. Consider the average Sherwood numbers $\bar{S}_1$ and $\bar{S}_2$ associated with the intermittent (multifractal) variable ε_λ and defined as $\bar{S}_1 = \langle S_1 (\varepsilon_\lambda) \rangle$ and $\bar{S}_2 = \langle S_2 (\varepsilon_\lambda) \rangle$, when $r < l_B$ and $r > l_B$, respectively. Using Equation (8.101), Equations (8.104) and (8.105) can be rewritten as:

$$\bar{S}_1 = 1 + 0.29 k_1 \varepsilon_0^{1/4} \lambda^{K_\varepsilon(1/4)} \tag{8.106}$$

for $r < l_B$, and

$$\bar{S}_2 = 0.55 k_2 \varepsilon_0^{1/6} \lambda^{K_\varepsilon(1/6)} \tag{8.107}$$

for $r > l_B$, where k_1 and k_2 are constants. Because $\lambda \gg 1$, $K(1/4) < 0$ and $K(1/6) < 0$, therefore $\lambda^{K_\varepsilon(1/4)}$ and $\lambda^{K_\varepsilon(1/6)}$ act as dampening factors in Equations (8.106) and (8.107), yielding $(\bar{S}_1/S_1) < 1$ and $(\bar{S}_2/S_2) < 1$. This shows that using a mean value of the turbulent kinetic energy dissipation rate ε_0 instead of the intermittent distribution ε_λ leads to overestimating the turbulence contribution to the rates of nutrient fluxes around phytoplankton cells, whatever their size may be.

In order to quantify this difference, one needs to estimate the intermittent dampening factors $(\bar{S}_1/S_1)$ and $(\bar{S}_2/S_2)$ due to $\lambda^{K_\varepsilon(1/4)}$ and $\lambda^{K_\varepsilon(1/6)}$ in Equation (8.106) and Equation (8.107) using realistic values of λ, $K_\varepsilon(1/4)$ and $K_\varepsilon(1/6)$. The inertial subrange scale ratio λ is increasing with increasing intensities of turbulence and can be reasonably regarded as ranging between 10^2 and 10^5 (see, for example, Gregg [1990]). The scale-invariant moment exponents $K_\varepsilon(1/4)$ and $K_\varepsilon(1/6)$ have been estimated from high-resolution shear vertical profiles recorded in tidally mixed coastal waters following Seuront et al. (1999) as $K(1/4) = -0.053 \pm 0.005$ and $K(1/6) = -0.042 \pm 0.004$ (Seuront and Yamazaki, unpublished data). The resulting dampening factors $(\bar{S}_1/S_1)$ and $(\bar{S}_2/S_2)$ range between 1.06 and 1.19, and between 1.21 and 1.62, when $r < l_B$ and $r > l_B$, respectively. This leads to a consideration of a decrease in the rate of nutrient fluxes around phytoplankton cells due to microscale turbulence intermittency ranging between 6.26% and 19.07% for phytoplankton cells smaller than the Batchelor microscale l_B, and between 21.22% and 61.78% for phytoplankton cells larger than the Batchelor microscale l_B. The inertial subrange scale ratio λ increases with increasing intensities of turbulence. The decrease in the rate of nutrient fluxes around phytoplankton cells of any size is thus higher when the intensity of turbulence is high.

8.6.2.2 Intermittency, Turbulence, and Physical Coagulation of Phytoplankton Cells

Theoretical analyses of particle coagulation processes predict that aggregate formation depends on the probability of particle collision and on the efficiency with which two particles that collide and stick together afterwards (McCave 1990; Kiørboe 1997). The former is a function of particle concentration, size, and the mechanism by which particles are brought into contact—for example, Brownian motion, shear, or the differential settlement of particles (see Figure 3.28). The latter, which is not studied here, depends mainly on the physicochemical properties of the particle surface and may vary with the particle type.

Consider a monospecific phytoplankton cells suspension characterized by a cell radius r(m) and cell concentration C(cells·m^{-3}). Because all particles are of the same size and density and settle with the same velocity, and because encounters due to Brownian motion are insignificant for particles > 1 μm (McCave 1990), the only mechanism that may increase the relative velocity between phytoplankton cells and thus to bring them to collide is due to turbulent shear and can be expressed as (Kiørboe 1997):

$$E_1 = 10.4 r^3 C^2 (\varepsilon_0 / \nu)^{1/2} \tag{8.108}$$

where E_1 is the encounter rate due to turbulence (encounter·s^{-1}), ε_0 the mean turbulent energy dissipation rate (m^2·s^{-3}), and ν the kinematic viscosity ($\nu = 10^{-6}$ m^2·s^{-1}). As previously done in Section 8.6.2.1, the turbulent kinetic energy dissipation rate is an intermittent variable ε_l characterized by the scaling moment function $K_\varepsilon(q)$ defined above, and by its mean $\varepsilon_0 = \langle \varepsilon_\lambda \rangle$. Here, E_1 and $\bar{E}_1$ are regarded as estimates of average encounter rates, that is, $\bar{E}_1 = E(\varepsilon_0)$ and $\bar{E}_1 = \langle E(\varepsilon_\lambda) \rangle$ when the turbulent energy dissipation rates are regarded as homogeneous and intermittent variables, respectively. In this case, Equation (8.108) is rewritten as:

$$\bar{E}_1 = 10.4 r^3 C^2 \nu^{-1/2} \varepsilon_0^{1/2} \lambda^{K_\varepsilon(1/2)} \tag{8.109}$$

This finally yields:

$$\bar{E}_1 = E_1 \lambda^{K_\varepsilon(1/2)} \tag{8.110}$$

and as defined above, $\lambda \gg 1$ and $K_\varepsilon(1/2) < 0$, $\lambda^{K(1/2)}$ thus acts as a restraining factor, therefore $\bar{E}_1 < E_1$. Using the values of λ proposed in Section 8.6.2.1 (that is, $\lambda \in [10^2, 10^5]$) and $K_\varepsilon(1/2) = -0.063 \pm 0.005$ (Seuront and Yamazaki, unpublished data), factor $\lambda^{K_\varepsilon(1/2)}$ ranges between 0.48 and 0.75. Now, the intermittent nature of microscale turbulence leads to a decrease of its contribution to the physical coagulation of phytoplankton cells by 25% to 48%. This decrease is higher when the turbulence levels are high; that is, the inertial subrange scale ratio is large.

Considering the role played by large-particle aggregates in the vertical flux of organic matter in the ocean (Jackson and Bird 1998), the effects of this decrease in encounters due to intermittency on the growth in particle size is considered hereafter. The average solid volume of aggregates increases according to (Kiørboe et al. 1990):

$$V_t = V_0 e^{\alpha[7.8\phi(\varepsilon_0/\nu)^{1/2}/\pi]t}$$

(8.111)

where V_0 and V_t are the average volume of aggregates at time 0 and t, α is the stickiness coefficient (that is, $\alpha \in [0, 1]$), and ϕ is the volume-concentration of cells [$\phi = (4/3)\pi r^3 C_0$]. Introducing the precise statistical distributions of turbulent dissipation rate ε_λ in Equation (8.111) instead of its average value ε_0 leads to a decrease in the aggregate average volumes ranging between 22% and 41% in low and high hydrodynamic conditions, respectively.

8.6.2.3 Intermittency, Turbulence, and Encounter Rates in the Plankton

Following the seminal theory of Rothschild and Osborn (1988), Kiørboe and Saiz (1995) demonstrated that the encounter rate E (encounters s^{-1}) between plankton predators and preys can be expressed as the sum of the encounter rate due to organism behavior and the encounter rate due to microscale turbulence. The former is a function of particle concentration, swimming speed of predator and prey, and perceptive distance of the predator. The latter, which I explore in more detail here, is expressed as:

$$E_2 = C\pi R^2 w$$

(8.112)

where E_2 is the encounter rate due to microscale turbulence, C is the number of preys per unit volume (preys·m^{-3}), R is the perceptive distance of the predator (m), and w (m s^{-1}) is the root-mean-square turbulent velocity enhancing the relative motion between predator and prey. The rms turbulent velocity w is directly related to the intensity of turbulence, characterized by a mean value of the turbulent kinetic energy dissipation rate ε_0 (m^2 s^{-3}) following (Rothschild and Osborn 1988):

$$w = 1.9(\varepsilon_0 d)^{1/3}$$

(8.113)

where d is the separation distance between predator and prey when an encounter takes place, that is, $d = R$, for example, (Visser and MacKenzie 1998). Finally, inserting the expression for the rms turbulent velocity w (Equation 8.113) into Equation (8.112) yields:

$$E_2 = 1.9C\pi R^{7/3}\varepsilon_0^{1/3}$$

(8.114)

Taking an intermittent variable ε_λ characterized by the scaling moment function $K_\varepsilon(q)$ defined above, and by its mean, $\langle \varepsilon_\lambda \rangle = \varepsilon_0$. Here, E_2 and $\bar{E}_2$ are the estimates of average encounter rates due to turbulence; that is, $E_2 = E(\varepsilon_0)$ and $\bar{E}_2 = \langle E(\varepsilon_0) \rangle$ when the turbulent energy dissipation rates are regarded as homogeneous and intermittent (multifractal) variables, respectively. Equation (8.114) can be rewritten as:

$$\bar{E}_2 = 1.9C\pi R^{7/3}\varepsilon_0^{1/3}\lambda^{K_\varepsilon(1/3)}$$

(8.115)

Now $\bar{E}_2 = E_2\lambda^{K_\varepsilon(1/3)}$, and as defined above, $\lambda \gg 1$ and $K_\varepsilon(1/3) < 0$, and $\lambda^{K_\varepsilon(1/3)}$ acts as a restraining factor; therefore $\bar{E}_2 < E_2$. Considering values of the inertial subrange scale ratio λ bounded between

10^2 and 10^5, and $K(1/3) = -0.060 \pm 0.003$ (Seuront and Yamazaki, unpublished data), the factor $\lambda^{K_\varepsilon(1/3)}$ ranges between 0.50 and 0.76. Taking into account the intermittent structure of turbulent kinetic energy dissipation rate ε_λ instead of an average value, ε_0 then decreases the contribution of microscale turbulence to the predator–prey encounter rate of 25% to 50% for ε_0 values ranging from 10^{-10} to 10^{-2} m^2 s^{-3}.

9 Conclusion

As suggested by the titles of several seminal earlier works, such as *The Fractal Geometry of Nature* (Mandelbrot 1983) and *Fractals Everywhere* (Barnsley 1993), one could expect to find fractal and multifractal properties everywhere. However, evidence for scaling in Nature is not as ubiquitous as many writers would have us believe. Recent studies (Malcai et al. 1997; Avnir et al. 1998) show that most reports of fractal scaling accepted for publication in the *Physical Reviews* and *Physical Review Letters* from 1990 to 1996 span only one-half to two decades, especially for self-similar fractals and with remarkably few exceptions. In particular, on the basis of their extensive literature survey and additional experiments, these authors raised the possibility that the cutoffs are due to intrinsic properties of the measured systems rather than the specific experimental/sampling conditions and apparatus. This is the reason why it is stressed here that one must (1) ensure the reality and the quality of the scaling using appropriate, statistically sound procedures (see Chapter 7), (2) avoid extrapolating structural information without reliable *a priori* information that the data continue to scale beyond the actual measurement spatial or temporal scales, and (3) consider the absence of scaling as a valuable source of information.

One of the major issues ecologists have to deal with is not only to state the observed fractal dimension for the distribution of moving organisms, or multifractal spectrum for the distribution of a given species, but also to explain, or try to explain, from a phenomenological point of view where these structures come from. Some hypotheses could be proposed to explain the origin of fractality or multifractality, and probably raise more questions than provide answers.

Because many environmental parameters display statistical self-similarity over a certain range of scales, the observed biotic patterns could reflect the distribution of some abiotic factors presenting a template upon which organisms or ecological systems operate. This can be true in certain conditions for phytoplankton biomass (see arrows in Figures 8.11 and 8.12) when the multifractal behaviors of temperature and phytoplankton biomass cannot be distinguished, but it is definitely not the rule. In particular, while the multifractal distribution of temperature remains the same whatever the intensity of turbulence, the distribution of phytoplankton is more and more intermittent (that is, nonlinear) when the turbulence intensity decreases (Figures 8.11 and 8.12). Phytoplankton cells then cannot be regarded as purely passive scalars. They are more intermittently distributed in weak turbulent conditions (that is, characterized by a few dense patches over a wide range of low-density patches), suggesting the prevalence of biological activity such as aggregation processes. Alternatively, they are more homogeneously distributed in high turbulent conditions, suggesting a physical disruption of phytoplankton patches by turbulence or a smoothing of their distribution by an increased grazing pressure via the turbulence-induced increase in predator-prey encounters. It is nevertheless likely that the differences in turbulence intensity also affect key processes such as nutrient availability, light history, and infection, which in turn might likely affect the distribution of phytoplankton cells.

A full understanding of the origin of phytoplankton patchiness is still far from being achieved (Vilar et al. 2003). It is nevertheless stressed here that the knowledge of the multifractal distributions of simultaneously recorded relevant parameters such as temperature, salinity, light, turbulent shear, nutrient concentration, bacteria, and phytoplankton and zooplankton biomass could be the first step to infer their phenomenological links. We know that fractal patterns are often generated by processes operating in transition zones such as the marine intertidal flat (Bradbury et al. 1984; Pennycuick and Kline 1986) and that superposition of several environmental gradients acting on different scales could result in rather complex spatial pattern of biota (Azovsky et al. 1998). In that way, a multifractal distribution can intrinsically be regarded as a pattern characterized by multiple gradients and transition zones (see, for example, Figure 3.20A, Figure 5.23, Figure 5.26, and Figure 5.28). The multifractal nature of turbulent flow and purely passive scalars (temperature and salinity) could thus be the starting point of a cascade process leading to the observed multifractal distributions of nutrients (Seuront et al. 2002), phytoplankton (Seuront et al. 1996a, 1996b, 1999; Lovejoy et al. 2001; Seuront 2005b) and zooplankton biomass (Pascual et al. 1995; Seuront and Lagadeuc 2001).

Alternatively, the properties of habitats are not just a function of the landscape and seascape patterns but also the ability of the organisms to perceive and use this pattern. It is then likely that different organisms would have different ranges of fractality and multifractality, as different organisms should have different scales of perception of their environment, exhibiting patterns of different types upon one and the same template. From the discussion in Section 8.1.5, it is likely that the multifractal distribution of microphytobenthos biomass would not be used in the same way by different predators, such as the deposit-feeding amphipod *Corophium* sp. or the snail *Littorina* sp. As suggested by Seuront and Spilmont (2002) for microphytobenthos biomass, the observed spatial (multifractal) pattern could be the result of the dynamic balance between endogenous (growth, migration, and death) and exogenous (tides, hydrodynamism, sediment quality, interspecific and intraspecific competition for nutrient, grazing) processes. Theoretical studies have also suggested that internal properties of individuals and populations interact to produce space-time structure, even in homogeneous environments (Deutschman et al. 1993; Bascompte and Solé 1995). This is also consistent with the empirical results presented in Sections 3.2.3.2 and 3.2.4.2 showing the similarity of the fractal dimensions of the spatial patterns of *Corophium arenarium* and their microphytobenthic preys. Potentially, environmental complexity interacts with biotic processes and influences spatial patterns (Roughgarden 1974; Pascual and Caswell 1997).

Finally, it may seem natural that any developing system tends to increase its internal structure as long as its size permits (see Section 2.1), with any increase in size adding new structural elements. Similar concepts of fractal-like ecosystem organization have been proposed (O'Neill et al. 1986; Kolasa 1989; Holling 1992). However, why and how biological communities realize this type of organization is still a puzzle, and remains a matter for further investigation.

References

Abbott, M.R., Powell, T.M. and Richerson, P.J. (1982) The relationship of environmental variability to the spatial patterns of phytoplankton biomass in Lake Tahoe. *J. Plankton Res.*, 4, 927–941.

Abdalla, S.H.M. and Boddy, L. (1996) Effect of soil and litter type on outgrowth patterns of mycelia systems of *Phanerochaete velutina*. *FEMS Microbiol. Ecol.*, 20, 195–204.

Abdusalam, H.A. (2000) A probabilistic approach to the site percolation problems using fractal tree. *Chaos Solitons Fractals*, 11, 1407–1410.

Abraham, E.R. (2001) The fractal branching of an arborescent sponge. *Mar. Biol.*, 138, 503–510.

Ahl, C. and Niemeyer, J. (1989) The fractal dimension of the pore volume inside soils. *Z. Pflanzenernährungu. Bodenkunde*, 152, 457–458.

Aho, A., Hopcroft, J.E. and Ullman, J.D. (1974) *Design and Analysis of Computer Algorithms*. Addison-Wesley, Reading, MA.

Alados, C.L., Emlen, J.M., Wchocki, B. and Freeman, D.C. (1998) Instability of development and fractal architecture in dryland plants as an index of grazing pressure. *J. Arid Environ.*, 38, 63–76.

———. (1999) Characterization of branch complexity by fractal analyses. *Intl. J. Plant Sci.*, 160, 147–155.

Alados, C.L., Escós, J.M. and Emlen, J.M. (1996) Fractal structure of sequential behaviour patterns: An indicator of stress. *Anim. Behav.*, 51, 437–443.

Alados, C.L. and Huffman, M.A. (2000) Fractal long-range correlations in behavioural sequences of wild chimpanzees: A non-invasive analytical tool for the evaluation of Health. *Ethology*, 106, 105–116.

Alados, C.L., Pueyo, Y., Navas, D., Cabezudo, B., Gonzalez, A. and Freeman, D.C. (2005) Fractal analysis of plant spatial patterns: A monitoring tool for vegetation transition shifts. *Biodiv. Conserv.*, 14, 1453–1468.

Alekseev, V.M. and Yakobson, M.V. (1981) Symbolic dynamics and hyperbolic dynamic systems. *Phys. Rep.*, 75, 287–325.

Alldredge, A.L. and Gotschalk, C. (1988) *In situ* settling behavior of marine snow. *Limnol. Oceanogr.*, 33, 339–351.

———. (1990) The relative contribution of marine snow of different origins to biological processes in coastal waters. *Cont. Shelf Res.*, 10, 41–58.

Allen, T.F.H. and Starr, T.B. (1982) *Hierarchy: Perspectives for Ecological Complexity*. University of Chicago Press, Chicago.

Alligood, K.T., Sauer, T.D. and Yorke, J.A. (1996) *Chaos*. Springer, New York.

Allmon, W.D., Rosenberg, G., Portell, R.W. and Schindler, K.S. (1993) Diversity of Atlantic coastline plain mollusks since the Pliocene. *Science*, 260, 1626–1629.

Almagor, H. (1985) Nucleotide distribution and the recognition of coding regions in DNA sequences: An information theory approach. *J. Theor. Biol.*, 117, 127–136.

Almirantis, Y. and Provata, A. (1999) Long- and short-range correlations in genome organization. *J. Stat. Phys.*, 97, 233–262.

Altmann, S. 1965. Sociobiology of rhesus monkeys. II: Stochastics of social communication. *J. Theor. Biol.*, 8, 490–522.

Alvarez-Llamoza, O., Cosenza, M.G. and Ponce, G.A. (2008) Critical behavior of the Lyapunov exponent in type-III intermittency. *Chaos Solitons Fractals*, 36, 150–156.

Alvarez-Ramirez, J. (2002) Characteristic time scales in the American dollar–Mexican peso exchange currency market. *Physica A*, 309, 157–170.

Anderson, A.N. and McBratney, A.B. (1995) Soil aggregates as mass fractals. *Aust. J. Soil Res.*, 33, 757–778.

Anderson, A.N., McBratney, A.B. and Crawford, J. (1998). Applications of fractals to soil studies. *Adv. Agron.*, 63, 2–78.

Anderson, A.N., McBratney, A.B. and Fitzpatrick, E.A. (1996) Soil mass, surface and spectral fractal dimensions estimated from thin section photographs. *Soil. Sci. Soc. Am. J.*, 60, 962–969.

Anderson, D. and Leach, M. (2004) Harvesting and redistributing renewable energy: On the role of gas and electricity grids to overcome intermittency through the generation and storage of hydrogen. *Energy Pol.*, 32, 1603–1614.

Andreadis, I. and Sertelis, A. (2002) Evidence of a random multifractal turbulent structure in the Dow Jones Industrial Average. *Chaos Solitons Fractals*, 13, 1309–1315.

Antonietti, M. (2003) Nanostructured materials—Self-organization of functional polymers. *Nat. Mater.*, 2, 9–10.

Antonietti, M. and Förster, S. (2003) Vesicles and liposomes: A self-assembly principle beyond lipids. *Adv. Mater.*, 15, 1323–1333.

Antonietti, M. and Tauer, K. (2003) 90 years of polymer latexes and heterophase polymerization: More vital than ever. *Macromol. Chem. Phys.*, 204, 207–219.

Appleby, S. (1996) Multifractal characterization of the distribution pattern of the human population. *Geograph. Anal.*, 28, 147–160.

Aqvist, J. and Tapia, O. (1987) Surface fractality as a guide for studying protein–protein interactions. *J. Mol. Graph.*, 5, 30–34.

Argyris, J., Ciubotariu, C. and Maluttis, H.G. (2001) Fractal space, cosmic strings and spontaneous symmetry breaking. *Chaos Solitons Fractals*, 12, 1–48.

Arnéodo, A., Bacry, E. and Muzy, J. (1995) The thermodynamics of fractals revisited with wavelets. *Physica A*, 213, 232–275.

Ascioti, F.A., Beltrami, E., Carrol, T.O. and Wirrick, C. (1993) Is there chaos in plankton dynamics? *J. Plankton Res.*, 15, 603–617.

Asmus, P. (2003) How California hopes to manage the intermittency of wind power. *Electr. J.*, 16, 48–53.

Atkinson, R.P.D., Rhodes, C.J., Macdonald, D.W. and Anderson, R.M. (2002) Scale-free dynamics in the movement patterns of jackals. *Oikos*, 98, 134–140.

Ausloos, M. and Ivanova, K. (2002) Multifractal nature of stock exchange prices. *Comput. Phys. Comm.*, 147, 582–585.

Austin, D., Bowen, W.D. and McMillan, J.I. (2004) Intraspecific variation in movement patterns: modelling individual behaviour in a large marine predator. *Oikos*, 105, 15–30.

Avnir, D., Biham, O., Lidar, D. and Malcai, O. (1998) Is the geometry of nature fractal? *Science*, 279, 39–40.

Axtell, R.L. (2001) Zipf distribution of U.S. firm sizes. *Science*, 293, 1818–1820.

Azam, F. and Long, R.A. (2001) Sea snow microcosms. *Nature*, 414, 495–498.

Azovsky, A.I., Chertropod, M.V., Kucheruk, N.V., Rybnikov, P.V. and Sapozhnikov, F.V. (2000) Fractal properties of spatial distribution of intertidal benthic communities. *Mar. Biol.*, 136, 581–590.

Azovsky, A.I., Obrido, S.V., Burkovsky, I.V. and Stoljarov, A.P. (1998) Community structure of transitive zones at the complex environmental gradients (the example of macrobenthos of Chernaja estuary, Kandalaksha, the White Sea). *Oceanology*, 38, 371–379.

Baar, J., Comini, B., Elferink, M.O. and Kuyper, T.W. (1997) Performance of four ectomycorrhizal fungi on organic and inorganic nitrogen sources. *Mycol. Res.*, 101, 523–529.

Bak, P. and Chen, K. (1991) Self-organized criticality. *Sci. Am.*, 26–33.

Bak, P., Chen, K. and Creutz, M. (1989) Self-organized criticality in the game of life. *Nature*, 342, 780–782.

Bak, P. and Sneppen, K. (1993) Punctuated equilibrium and criticality in a simple model of evolution. *Phys. Rev. Lett.*, 24, 4083–4086.

Bak, P. and Tang, C. (1989) Earthquakes as self-organized critical phenomenon. *J. Geophys. Res.*, 95, 15635–15637.

Bak, P., Tang, C. and Wiesenfield, K. (1987) Self-organized criticality: An explanation of 1/*f* noise. *Phys. Rev. Lett.*, 59, 381–384.

———. (1988) Self-organized criticality. *Phys. Rev. A*, 38, 364–374.

Baker, G.L. and Gollub, J.P. (1990) *Chaotic Dynamics: An Introduction*. Cambridge University Press, Cambridge.

Baker, M.A. and Gibson, C.H. (1987) Sampling turbulence in the stratified ocean: Statistical consequences of strong intermittency. *J. Phys. Oceanogr.*, 17, 1817–1836.

Barales, V. and Trees, C.C. (1987) Spatial variability of the ocean color field in CZCS imagery. *Adv. Space Res.*, 7, 95–100.

Barat, P., Das, N.Kr., Ghose, D. and Sinha, B. (1999) Fractal pattern in hydrothermal emission. *Physica A*, 262, 9–15.

Barata, C., Baird, D.J., Medina, M., Albalat, A. and Soares, A.M.V.M. (2002) Determining the ecotoxicological mode of action of toxic chemicals in meiobenthic marine organisms: Stage specific short tests with *Tisbe battagliai*. *Mar. Ecol. Prog. Ser.*, 230, 183–194.

Barata, C., Calbet, A., Saiz, E., Ortiz, L. and Bayona, J. (2005) Predicting single and mixture toxicity of petrogenic polycyclic aromatic hydrocarbons to the copepod *Oithona davisae*. *Environ. Toxicol. Chem.*, 24, 210–217.

Barnsley, M.F. (1993, 2000) *Fractals Everywhere*. Morgan Kauffmann, London.

Barranguet, C. (1997) The role of microphytobenthic primary production in a Mediterranean mussel culture area. *Est. Coast. Shelf Sci.*, 44, 753–765.

Bartoli, F., Philippy, R., Doirisse, M., Niquet, S., Dubuit, M. (1991) Structure and self-similarity in silty and sandy soils: The fractal approach. *J. Soil Sci.*, 42, 167–185.

Bartumeus, F., Peters, F., Pueyo, S., Marrasé, C. and Catalan, J. (2003) Helical Lévy walks: Adjusting searching statistics to resource availability in microzooplankton. *Proc. Natl. Acad. Sci. USA*, 100, 12771–12775.

Bascompte, J. and Solé, R.V. (1995) Rethinking complexity: Modelling spatiotemporal dynamics in ecology. *Trends Ecol. Evol.*, 10, 361–366.

Bascompte, J. and Vilà, C. (1997) Fractals and search paths in mammals. *Landscape Ecology*, 12, 213–221.

Basillais, E. (1997) Coral surfaces and fractal dimensions: A new method. *C. R. Acad. Sci. Paris Life Sci.*, 320, 653–657.

Bassingthwaighte, J.B. (1988) Physiological heterogeneity: Fractals link determinism and randomness in structures and functions. *News Physiol. Sci.*, 3, 5–10.

Bassingthwaighte, J.B. and Beyer, R.P. (1991) Fractal correlation in heterogeneous systems. *Physica D*, 53, 71–84.

Bassingthwaighte, J.B. and Raymond, G.M. (1995) Evaluation of the dispersional analysis method for fractal time series. *Ann. Biomed. Eng.*, 23, 491–505.

Batchelor, G.K. and Townsend, A.A. (1949) The nature of turbulent motion at large wave numbers. *Proc. R. Soc. A*, 199, 238–250.

Batty, M. (1995) New ways of looking at cities. *Nature*, 377, 574.

Batty, M. and Longley, P. (1994) *Fractal Cities: A Geometry of Form and Function.* Academic Press, San Diego.

Baumert, H., Sündermann, J. and Simpson, J., Eds. (2005) *Marine Turbulences: Theories, Observations and Models.* Cambridge University Press, Cambridge.

Beecher, M.D. (1989) Signalling systems for individual recognition: An information theory approach. *Anim. Behav.*, 48, 248–261.

Bejar, J., Ugo, F., Elio, G. and Stefano, M. (1995) Low dimensional chaos is present in radon time variations. *J. Environ. Radioactivity*, 28, 73–89.

Benhamou, S. (1990) An analysis of movements of the wood mouse *Apodemus sylvaticus* in its home range. *Behav. Proc.*, 22, 235–250.

———. (2004) How to reliably estimate the tortuosity of an animal's path: Straightness, sinuosity, or fractal dimension? *J. Theor. Biol.*, 229, 209–220.

Benincà, E., Huisman, J., Heerkloss, R., Jöhnk, K.D., Branco, P., Van Nes, E.H., Scheffer, M. and Ellner, S.P. (2008) Chaos in a long-term experiment with a plankton community. *Nature*, 451, 822–826.

Bennett, A.F. and Denman, K.L. (1985). Phytoplankton patchiness: Inferences from particles statistics. *J. Mar. Res.*, 43, 307–335.

Berntson, G.M. (1994) Root systems and fractals: How reliable are calculations of fractal dimension? *Ann. Bot.*, 73, 281–284.

Berntson, G.M., Lynch J.P. and Snapp, S. (1998) Fractal geometry and plant root systems: current perspectives and future applications. In *Fractals in Soil Science*, Baveye, P., Parlange, J.Y. and Stewart, B.A., Eds., Lewis, New York, 113–152.

Berry, M.V. and Lewis, Z.V. (1980) On the Weierstrass-Mandelbrot fractal function. *Proc. R. Soc. London A*, 370, 459–484.

Berryman, A. and Milstein, J. (1989) Are ecological systems chaotic—And if not, why not? *Trends Ecol. Evol.*, 4, 26–28.

Bertness, M.D. (1999) *The Ecology of Atlantic Shorelines.* Sinauer, Sunderland, MA.

Bertrand, S., Bertrand, A., Guevara-Carrasco, R. and Gerlotto, F. (2007) Scale-invariant movements of fishermen: The same foraging strategy as natural predators. *Ecol. Appl.*, 17, 331–337.

Biesinger, Z. and Haefner, J.W. (2005) Proximate cues for predator searching: A quantitative analysis of hunger and encounter rate in the ladybird beetle, *Coccinella septempunctata. Anim. Behav.*, 69, 235–244.

Biham, O., Malcai, O., Levy, M. and Solomon, S. (1998) Generic emergence of power law distributions and Lévy-stable intermittent fluctuations in discrete logistic systems. *Phys. Rev. E*, 58, 1352–1358.

Billiones, R.G., Tackx, M.L. and Daro, M.H. (1999) The geometric features, shape factors and fractal dimensions of suspended particulate matter in the Scheldt estuary (Belgium). *Est. Coast. Shelf Sci.*, 48, 293–305.

Billock, V.A., de Guzman, G.C. and Scott Kelso, J.A. (2001) Fractal time and 1/f spectra in dynamic images and human vision. *Physica D*, 148, 136–146.

Bird, N., Cruz Díaz, M., Saa, A. and Tarquis, A.M. (2006) Fractal and multifractal analysis of pore-scale images of soil. *J. Hydrol.*, 322, 211–219.

Biven, L., Nazarenko, S.V. and Newell, A.C. (2001) Breakdown of wave turbulence and the onset of intermittency. *Phys. Lett. A*, 280, 28–32.

Bjørnsen, P.K. and Nielsen, T.G. (1991) Decimeter scale heterogeneity in the plankton during a pycnocline bloom of *Gyrodinuim aureolum. Mar. Ecol. Prog. Ser.*, 73, 263–267.

Blanchard, D. and Bourget, E. (1999) Scales of coastal heterogeneity: Influence on intertidal community structure. *Mar. Ecol. Prog. Ser.*, 179, 163–173.

Block, B.A., Booth, D.T. and Carey, F.G. (1992) Depth and temperature of the blue marlin, *Makaira nigricans*, observed by acoustic telemetry. *Mar. Biol.*, 114, 175–183.

Bloomfield, P. (2000). Fourier analysis of time series: An introduction. Wiley Interscience, New York.

Blumenshine, S.C., Vadeboncoeur, Y., Lodge, D.M., Cottingham, K.L. and Knight, S.E. (1997) Benthic-pelagic links: Responses of benthos to water-column nutrient enrichment. *J. N. Am. Benthol. Soc.*, 16, 466–479.

Blundell, K.M. and Rawlings, S. (1999) The inevitable youthfulness of known high-redshift radio galaxies. *Nature*, 399, 330–332.

Boddy, L., Wells, J.M., Culshaw, C. and Donnelly, D.P. (1999) Fractal analysis in studies of mycelium in soil. *Geoderma*, 88, 301–328.

Bohle-Carbonel, M. (1992) Pitfalls in sampling, comments on reliability and suggestions for simulation. *Cont. Shelf Res.*, 12, 3–24.

Bohr, T. (1998) *Dynamical Systems Approach to Turbulence*. Cambridge University Press, Cambridge.

Bolton, R.G. and Boddy, L. (1993) Characterisation of the spatial aspects of foraging mycelial cord systems using fractal geometry. *Mycol. Res.*, 97, 762–768.

Bonn, D., Kellay, H., Prochnow, M., Ben-Djemiaa, K. and Meunier, J. (1998) Delayed fracture of an inhomogeneous soft solid. *Science*, 280, 256–267.

Borda-de-Água, L., Hubbell, S.P. and McAllister, M. (2002) Species-area curves, diversity indices, and species abundance distributions: A multifractal analysis. *Am. Nat.*, 159, 138–155.

Boström, C. and Bonsdorff, E. (2000) Zoobenthic community establishment and habitat complexity: The importance of seagrass shoot-density, morphology and physical disturbance for faunal recruitment. *Mar. Ecol. Prog. Ser.*, 205, 123–138.

Bouruet-Aubertot, P., Sommeria, J., Le Cann, B. and Koudella, C.R. (2004) Intermittency of vertical density gradients at fine scale and link with mixing processes. *Deep-Sea Res. II*, 51, 2919–2941.

Boxshall, G.A. (1998) Preface. *Philos. Trans. R. Soc. Lond. B*, 353, 669–670.

Bradbury, J.W. and Vehrencamp, S.L. (1998) *Principles of Animal Communication*. Sinauer, New York.

Bradbury, R.H., Reichelt, R.E. and Green, D.G. (1984) Fractals in ecology: Methods and interpretation. *Mar. Ecol. Prog. Ser.*, 14, 295–296.

Brakensiek, D.L., Rawls, W.J., Logsdon, SD. and Edwards, W.M. (1992) Fractal description of macroporosity. *Soil Sci. Sot. Am. I.*, 56, 1721–1723.

Breitfuss, M.J. (1982) Defining the characteristics of burrows to better estimate abundance of the grapsid crab, *Helograpsus haswellianus* (Decapoda, Grapsidae), on East Australian salt marsh. *Crustaceana*, 76, 499–507.

Bremer, L.G.B. and Walstra, P. (1989) Theoretical and experimental study of the fractal nature of the structure of casein gels. *J. Chem. Soc. Faraday Trans. I.*, 85, 3359.

Breslin, M.C. and Belward, J.A. (1999) Fractal dimensions for rainfall time series. *Math. Comput. Simulation*, 48, 437–446.

Brewer, M.C. (1996) Daphnia Swimming Behavior and Its Role in Predator–Prey Interactions. Ph.D. thesis, University of Wisconsin–Milwaukee.

Briggs, K. (1991) A precise calculation of the Feigenbaum constants. *Math. Comput.*, 57, 435–439.

———. (1997) Feigenbaum Scaling in Discrete Dynamical Systems. Ph.D. thesis. University of Melbourne, Melbourne, Australia.

Brillouin, L. (1962) *Science and Information Theory*. Academic Press, London.

Brown, C.T., Liebovitch, L.S. and Glendon, R. (2007) Lévy flights in Dobe Ju/'hoansi foraging patterns. *Hum. Ecol.*, 35, 129–138.

Brown, J.H. and West, G.B. (2000) *Scaling in Biology*. Oxford University Press, Oxford.

Brown, W. and Ball, R. (1985) Computer-simulation of chemically limited aggregation. *J. Physiol.*, 18, 517–521.

Bruno, R., Carbone, V., Sorriso-Valvo, L., Pietropaolo, E. and Bavassano, B. (2005) Intermittency in plasma turbulence. In *Multiscale Coupling of Sun-Earth Processes*, Liu, A.T.Y., Kamide, Y. and Consolini, G., Eds., Elsevier, Amsterdam, 9–28.

Bruno, R., Carbone, V., Veltri, P., Pietropaolo, E. and Bavassano, B. (2001) Identifying intermittency events in the solar wind. *Planetary Space Sci.*, 49, 1201–1210.

Brusca, R.C. and Brusca, G.J. (1990) *Invertebrates*, Sinauer, Sunderland, MA.

Buczaki, S. (2002) *Fauna Britannica*. Hamlyn, London.

Buldyrev, S.V., Goldberger, A.L., Havlin, S., Peng, C.K. and Stanley, H.E. (1994) Fractals in biology and medicine: From DNA to the heartbeat. In *Fractal in Science*, Bunde, A. and Havlin, S., Eds., Springer-Verlag, Berlin, 49–87.

Bullock, P., Fedoroff, N., Jongerius, A., Stoops, G. and Tursina, T. (1985) *Handbook for Soil Thin Section Description*. Wayne Research, Albrighton, Wolverhampton.

Bundy, M.H., Gross, T.F., Coughlin, D.J. and Strickler, J.R. (1993) Quantifying copepod searching efficiency using swimming pattern and perceptive ability. *Bull. Mar. Sci.*, 53, 15–28.

Burda, Z., Johnston, D., Jurkiewicz, J., Kaminski, M., Nowak, M.A., Papp, G. and Zahed, I. (2002) Wealth condensation in pareto macroeconomies. *Phys. Rev. E*, 65, 1–4.

Burgess, T.M. and Webster, R. (1980) Optimal interpolation and isarithmic mapping of soil properties. I: The semivariogram and punctual kriging. II: Block kriging. *J. Soil Sci.*, 31, 315–342.

Burlando, B., Cattaneo-Vietti, R., Parodi, R. and Scardi, M. (1991) Emerging fractal properties in gorgonian forms (*Cnidaria: Octocorallia*). *Growth, Dev. Aging*, 55, 161–168.

Burlinson, F.C. and Lawrence, A.J. (2006) Development and validation of a behavioural assay to measure the tolerance of *Hediste diversicolor* to copper. *Environ. Poll.*, 145, 274–278.

Burrough, P.A. (1981) Fractal dimensions of landscape and other environmental data. *Nature*, 294, 240–242.

———. (1983a) Multiscale sources of spatial variation in soil. I: The application of fractal concepts to nested levels of soil variation. *J. Soil Sci.*, 34, 577–597.

———. (1983b) Multiscale sources of spatial variation in soil. II: A non-Brownian fractal model and its application in soil survey. *J. Soil Sci.*, 34, 599–620.

———. (1986) *Principles of Geographical Systems for Land Resources Assessment*. Clarendon, Oxford.

Burrus, C.S. and Parks, T.W. (1985) *DFT/FFT and Convolution Algorithms*. Wiley, New York.

Caccia, D.C., Percival, D., Cannon, M.J., Raymond, G. and Bassingthwaighte, J.B. (1997) Analyzing exact fractal time series: Evaluating dispersional analysis and rescaled range methods. *Physica* A, 246, 609–632.

Cachot, J., Geffart, O., Augagneur, S., Lacroix, S., Le Menach, K., Peluhet, L., Couteau, J., Denier, X., Devier, M.H., Pottier, D. and Budzinski, H. (2006) Evidence of genotoxicity related to high PAH content of sediments in the upper part of the Seine estuary (Normandy, France). *Aquat. Toxicol.*, 79, 257–267.

Cahalan, R.F., Oreopoulos, L., Wen, G., Marshak, A., Tsay, S.C. and DeFelice, T. (2001) Cloud characterization and clear-sky correction from Landsat-7. *Remote Sensing Environ.*, 78, 83–98.

Cain, M.L. (1990) Models of clonal growth in *Solidago altissima*. *J. Ecol.*, 78, 27–46.

Cain, M.L. (1994) Consequences of foraging in clonal plant species. *Ecology*, 75, 933–944.

Calbet, A., Saiz, E. and Barata, C. (2007) Lethal and sublethal effects of naphthalene and 1,2-dimethylnaphthalene on the marine copepod *Paracartia grani*. *Mar. Biol.*, 151, 195–204.

Çambel, A.B. (1993) *Applied Chaos Theory: A Paradigm for Complexity*. Academic Press, San Diego.

Campbell, B.M. and Griffen, D.J.G. (1966). The Australian sesarminae (crustacea: brachyura): genera *Helice*, *Helograpsus* nov., *Cyclograpsus* and *Paragrapsus*. *Memoirs Queensland Mus.*, 14, 127–74.

Campbell, J.B. (1978) Spatial variation of sand content and pH within single contiguous delineations of two soil mapping units. *Soil Sci. Soc. Am. J.*, 42, 460–464.

Campo Bagatin, A., Martinez, V.J. and Paredes, S. (2002) Multifractal fits to the observed main asteroid distribution. *Icarus*, 157, 549–553.

Cannon, M.J., Percival, D.B., Caccia, D.C., Raymond, G.M. and Bassingthwaighte, J.B. (1997) Evaluating scaled window variance methods for estimating the Hurst coefficient of time series. *Physica A*, 241, 606–626.

Cantero, J.J., Leon, R., Cisneros, J.M. and Cantero, A. (1998) Habitat structure and vegetation relationships in central Argentina salt marsh landscapes. *Plant Ecology*, 137, 79–100.

Cantor, G. (1883) Uber unendliche, lineare Punktmannigfaltigkeiten. *Mathematische Annalen*, 21, 545–591.

Carathéodory, C. (1914) Über das lineare mass von punktmengeneine verallgemeinerung das längenbegriffs: Nach. *Ges. Wiss. Göttingen*, 406–426.

Carlson, J.M. and Langer, J.S. (1989) Properties of earthquakes generated by fault dynamics. *Phys. Rev. Lett.*, 62, 2632–2635.

Carr, J.R. and Benzer, W.B. (1991) On the practice of estimating fractal dimension. *Math. Geol.*, 23, 945–958.

Cartamil, D.P. (2003) Fine-Scale Movement Patterns and Habitat Preferences of Ocean Sunfish (*Mola mola*) off the Coast of Southern California. Master's thesis, California State University, Long Beach.

Cartamil, D.P. and Lowe, C.G. (2004) Diel movement patterns of ocean sunfish *Mola mola* off southern California. *Mar. Ecol. Prog. Ser.*, 266, 245–253.

Cassie, R.M. (1959) An experimental study of factors inducing aggregation in marine plankton. *New Zealand J. Sci.*, 2, 339–365.

Cassie, R.M. (1963) Microdistribution in the plankton. *Oceanogr. Mar. Biol. Ann. Rev.*, 1, 223–252.

Castaing, B. and Dubrulle, B. (1995). Fully developed turbulence: A unifying point of view. *J. Phys. II France*, 5, 895–899.

Chapman, S.C., Hnat, B. and Rowlands, G. (2005) Intermittency and self-similarity in "natural parameters" in solar wind turbulence. In *Multiscale Coupling of Sun-Earth Processes*, Liu, A.T.Y., Kamide, Y. and Consolini, G., Eds., Elsevier, Amsterdam, 329–341.

Charney, J.G. (1971) Geostrophic turbulence. *J. Atm. Sci.*, 28, 1087–1095.

Chatfield, C. (1996, 2003) *The Analysis of Time Series: An Introduction.* CRC Press, Boca Raton.

Chatfield, C. and Lemon, R. (1970) Analysing sequences of behavioural events. *J. Theor. Biol.*, 29, 427–445.

Chaudhari, A., Sanders, C.C. and Lee, S.L. (2002) Effects of surface roughness on diffusion limited reactions. *Chem. Phys. Lett.*, 351, 341–348.

Chechkin, A.V. and Gonchar, V.Yu. (2000) Self and spurious multi-affinity of ordinary Lévy motion, and pseudo-Gaussian relations. *Chaos Solitons Fractals*, 11, 2379–2390.

Chen, K. and Bak, P. (1989) Is the universe operating at a self-organized critical state? *Phys. Lett. A*, 140, 299–302.

Chen, Z. Ivanov, P.Ch., Hu, K. and Stanley, H.E. (2002) Effect of nonstationarities on detrended fluctuation analysis. *Phys. Rev. E*, 65, 041107.

Chesson, P., Gebauer, R.L.E., Schwinning, S., Huntly, N., Wiegand, K., Ernest, M.S.K., Sher, A., Novoplantsky, A. and Weltzin, J.F. (2004) Resource pulses, species interactions, and diversity maintenance in arid and semi-arid environments. *Oecologia*, 141, 236–253.

Chhabra, A. and Jensen, R.V. (1989) Direct determination of the $f(\alpha)$ singularity spectrum. *Phys. Rev. Lett.*, 62, 1327–1330.

Chu, P.C. (2004) Multi-fractal thermal characteristics of the southwestern GIN sea upper layer. *Chaos Solitons Fractals*, 19, 275–284.

Chun, H.C., Giménez, D. and Woon, S.W. (2008) Morphology, lacunarity and entropy of intra-aggregate pores: Aggregate size and soil management effects. *Geoderma*, 146, 83–93.

Claps, P. and Olivetto, G. (1994) Fractal structure, entropy and energy dissipation in river networks. *Hydrology J. Indian Assoc. Hydrologists*, 17, 38–51.

Clarke, R.H. and Clarke, M.F. (1999) The social organization of a sexually dimorphic honeyeater: The crescent honeyeater *Phylidonyris pyrrhoptera*, at Wilsons Promontory, Victoria. *Australian J. Ecol.*, 24, 644–654.

Cody, M.L. (1971) Finch flocks in the Mohave Desert. *Theor. Pop. Biol.*, 2, 142–158.

Cole, B.J. (1995) Fractal time in animal behaviour: The movement activity of *Drosophila*. *Anim. Behav.*, 50, 1317–1324.

Coleman, D.C. and Crossley Jr., D.A. (2004) *Fundamentals of Soil Ecology.* Academic Press, San Diego.

Collinge, S.K. and Palmer, T.M. (2002) The influences of patch shape and boundary contrast on insect response to fragmentation in California grasslands. *Landscape Ecol.*, 17, 647–656.

Commito, J.A. and Rusignuolo, B.R. (2000) Structural complexity in mussel beds: The fractal geometry of surface topography. *J. Exp. Mar. Biol. Ecol.*, 255, 133–152.

Coplen, T.B. and Krouse, H.R. (1998) Sulphur isotope data consistency improved. *Nature*, 392, 32.

Corbit, J.D. and Garbary, D.J. (1995) Fractal dimension as a quantitative measure of complexity in plant development. *Proc. R. Soc. London B*, 262, 1–6.

Correig, A.M., Urquizú, M. and Vila, J. (1997) Aftershock series of event February 18, 1996: An interpretation in terms of self-organized criticality. *J. Geophys. Res.*, 102, 27404–27420.

Corrsin, S. (1951) On the spectrum of isotropic temperature in an isotropic turbulence. *J. Appl. Phys.*, 22, 469–473.

Coughlin, D.J., Strickler, J.R. and Sanderson, B. (1992) Swimming and search behaviour in clownfish, *Amphiprion perideraion*, larvae. *Anim. Behav.*, 44, 427–440.

Cousins, S.H. (1988) Fundamental components in ecology and evolution: Hierarchy, concepts and descriptions. In *Ecodynamics: Contribution to Theoretical Ecology*, Wolf, W., Soeden, C.S. and Drepper, F.R., Eds., Springer-Verlag, New York, 60–68.

Cowles, T.J. and Remillard, J.F. (1983) Effects of exposure to sublethal concentrations of crude oil on the copepod *Centropages hamatus*. 1: Feeding and egg production. *Mar. Biol.*, 78, 45–51.

Crawford, J.W., Ritz, K. and I.M. Young (1993a) Quantification of fungal morphology, gaseous transport and microbial dynamics in soil: An integrated framework utilizing fractal geometry. *Geoderma*, 56, 15–172.

Crawford, J.W., Sleeman, B.D. and Young, I.M. (1993b) On the relation between number–size distributions and the fractal dimensions of aggregates. *J. Soil Sci.*, 44, 555–565.

Crawford, J.W. and Young, I.M.1990. A multiple scaled fractal tree. *J. Theor. Biol.*, 145, 199–206.

Crist, T.O., Guertin, D.S., Wiens, J.A. and Milne, B.T. (1992) Animal movement in heterogeneous landscapes: An experiment with *Elodes* beetles in shortgrass prairie. *Funct. Ecol.*, 6, 536–544.

Crist, T.O. and Wiens, J.A. (1994) Scale effects of vegetation on forager movement and seed harvesting by ants. *Oikos*, 69: 37–46.

Critten, D.L. (1993) A review of light transmission into greenhouse crops. *Acta Hortic.*, 32, 9–31.

Cuddington, K. and Yodzis, P. (2002) Predator-prey dynamics and movement in fractal environments. *Am. Nat.*, 160, 119–134.

Currie, W.J.S. and Roff, J.C. (2006) Plankton are not passive tracers: Plankton in a turbulent environment. *J. Geophys. Res.*, 11, C05S07, doi:10.1029/2005JC002967.

Cushing, D.H. (1962) Patchiness. *Rapp. Cons. Intl. Explor. Mer*, 153, 152–163.

Czirók, A., Mategna, R.N., Havlin, S. and Stanley, H.E. (1995) Correlations in binary sequences and a generalized Zipf analysis. *Phys. Rev. E*, 52, 446–452.

Daan, R. (1989) Factors controlling the summer development of the copepod populations in the Southern Bight of the North Sea. *Neth. J. Sea Res.*, 23, 305–322.

Dam, H.G. and Peterson, W.T. (1993) Seasonal contrasts in the diel vertical distribution feeding behavior and grazing impact of the copepod *Temora longicornis* in Long Island Sound, *J. Mar. Res.*, 51, 561–594.

Davenport, J., Pugh, P.J.A. and McKechnie, J. (1996) Mixed fractals and anisotropy in subantarctic marine macroalgae from South Georgia: Implications for epifaunal biomass and abundance. *Mar. Ecol. Prog. Ser.*, 136, 245–255.

Davis, F.W. (1993) Introduction to spatial statistics. In *Patch Dynamics*, Levin, S.A., Powell, T.M. and Steele, J.H., Eds. Springer-Verlag, Berlin, 16–26.

Davis, H.T. (1989) On the fractal character of the porosity of natural sandstone. *Europhys. Lett.*, 8, 629–632.

De Bartolo, S.G., Veltri, M. and Primavera, L. (2006) Estimated generalized dimensions of river networks. *J. Hydrol.*, 322, 181–191.

de Boer, J., Flyvbjerg, H., Jackson, A.D. and Wettig, T. (1994) Simple model of self-organized biological evolution. *Phys. Rev. Lett.*, 73, 906–909.

DeCola, L. (1989) Fractal analysis of a classified landsat scene. *Photogrammetric Eng. Remote Sensing*, 55, 601–610.

deJong, S. and Burrough, P.A. (1996) Fractal approach to the classification of Mediterranean vegetation types in remotely sensed images. *Photogrammetric Eng. Remote Sensing*, 61, 1041–1053.

Dellino, P. and Liotino, G. (2002) The fractal and multifractal dimension of volcanic ash particles contour: A test study on the utility and volcanological relevance. *J. Volcanol. Geoth. Res.*, 113, 1–18.

Denman, K.L. and Abbott, M.A. (1988) Time evolution of surface chlorophyll patterns from cross-spectrum analysis of satellite color images. *J. Geophys. Res.*, 93, 6789–6798.

———. (1994) Time scales of pattern evolution from cross-spectrum analysis of advanced very high resolution radiometer and coastal zone color scanner imagery. *J. Geophys. Res.*, 99, 7433–7442.

Denman, K.L., Okubo, A. and Platt, T. (1977) The chlorophyll fluctuation spectrum in the sea. *Limnol. Oceanogr.*, 22, 1033–1038.

Denman, K.L. and Platt, T. (1976) The variance spectrum of phytoplankton in a turbulent ocean. *J. Mar. Res.*, 34, 593–601.

Desiderio, R.A., Cowles, T.J. and Moum, J.N. (1993) Microstructure profiles of laser-induced chlorophyll fluorescence spectra: evaluation of backscatter and forward-scatter fiber-optic sensors. *J. Atm. Ocean. Tech.*, 10, 209–224.

Despland, E. (2003) Fractal index captures the role of vegetation clumping in locust swarming. *Funct. Ecol.*, 17, 315–322.

Deutschman, D.H., Bradshaw, G.A., Childress, W.M., Daly, K., Grunbaum, D., Pascual, M., Schumaker, N.H. and Wu, J. (1993) Mechanisms of patch formation. In *Patch dynamics*, Levin, S., Powell, T. and Steele, J., Eds. Springer, Berlin, 184–209.

Dewey, T.G. and Datta, M.M. (1989) Determination of the fractal dimension of membrane protein aggregates using fluorescence energy transfer. *Biophys. J.*, 56, 415–420.

Dicke, M. and Burrough, P.A. (1988) Using fractal dimensions for characterizing tortuosity of animal trails. *Physiol. Entomol.*, 13, 393–398.

Dietler, G. and Zhang, Y.C. (1992) Fractal aspects of the Swiss landscape. *Physica A*, 191, 213–219.

Dingle, H. (1969) A statistical and information analysis of aggressive communication in the mantis shrimp *Gonodactylus bredini*. *Anim. Behav.*, 17, 561–575.

Diodati, P., Marchesoni, F. and Piazza, S. (1991) Acoustic emission from volcanic rocks: An example of self-organized criticality. *Phys. Rev. Lett.*, 67, 2239–2242.

Doall, M.H., Colin, S.P., Strickler, J.R. and Yen, J. (1998) Locating a mate in 3D: The case of *Temora longicornis*. *Philos. Trans. R. Soc. Lond. B*, 353, 681–689.

Donnelly, D.P. and Boddy, L. (1997a) Development of mycelial systems of *Stropharia caerulea* and *Phanerochaete Íelutina* on soil: Effect of temperature and water potential. *Mycol. Res.*, 101, 705–713.

———. (1997b) Resource acquisition by the mycelial-cord-former *Stropharia caerulea*: Effect of resource quantity and quality. *FEMS Microbiol. Ecol.*, 23, 195–205.

———. (1998) Developmental and morphological responses of mycelia systems of *Stropharia caerulea* and *Phanerochaete Íelutina* to soil nutrient enrichment. *New Phytol.*, 138, 519–531.

Donnelly, D.P., Wilkins, M.F. and Boddy, L. (1995) An integrated image analysis approach for determining bio-mass, radial extent and box-count fractal dimension of macroscopic mycelia systems. *Binary*, 7, 19–28.

Doval, M.D., Morono, A., Pazos, Y., Lopez, A., Madrinan, M., Cabanas, J.M. and Maneiro, J. (2006) Monitoring dissolved aromatic hydrocarbons in Rias Baixas embayments (NW Spain) after Prestige oil spills: Relationship with hydrography. *Est. Coast. Shelf Sci.*, 67, 205–218.

Dowling, N.A., Hall, S.J. and Mitchell, J.G. (2000) Foraging kinematics of barramundi during early stages of development. *J. Fish Biol.*, 57, 337–353.

Drake, J.B. and Weishampel, J.F. (2000) Multifractal analysis of canopy height measures in a longleaf pine savanna. *Forest Ecol. Management*, 128, 121–127.

———. (2001) Simulating vertical and horizontal multifractal patterns of a longleaf pine savanna. *Ecol. Model.*, 145, 129–142.

Dremin, I.M., Ivanov, O.V. and Nechitailo, V.A. (2004) An introduction to wavelets. In *Handbook of Scaling Methods in Aquatic Ecology: Measurement, Analysis, Simulation*, Seuront, L. and Strutton, P.G., Eds. CRC Press, Boca Raton, 279–296.

Dubrulle, B. (1994) Intermittency in fully developed turbulence: Log-Poisson statistics and generalized scale covariance. *Phys. Rev. Lett.*, 73, 959–962.

———. (2006) Anomalous scaling and generic structure function in turbulence. *J. Phys. II France*, 6, 1825–1840.

Dugundgi, J. (1966) *Topology*. McGraw-Hill, New York.

Eadie, J.M. and Keast, A. (1984) Resource heterogeneity and fish species diversity in lakes. *Can. J. Zool.*, 62, 1689–1695.

Ebeling, W. (1997) Prediction and entropy of nonlinear dynamical systems and symbolic sequences with LRO. *Physica D*, 109, 42–52.

Ebeling, W. and Nicolis, G. (1991) Entropy of symbolic sequences: The role of correlations. *Europhys. Lett.*, 14, 191–196.

———. (1992) Word frequency and entropy of symbolic sequences: A dynamical perspective. *Chaos Solutions Fractals*, 2, 635–650.

Ebeling, W. and Pöschel, T. (1995) Entropy, transinformation and word distribution of information-carrying sequences. *Intl. J. Bifurcation and Chaos*, 5, 51–61.

Edwards, A.M., Phillips, R.A., Watkins, N.W., Freeman, M.P., Murphy, E.J., Afanasyev, V., Buldyrev, S.V., da Luz, M.G.E., Raposo, E.P., Stanley, H.E. and Viswanathan, G.M. (2007) Revisiting Lévy flight search patterns of wandering albatrosses, bumblebees and deer. *Nature*, 449, 1044–1049.

Eghball, B., Settimi, J.R., Maranville, J.W. and Parkhurst, A.M. (1993) Fractal analysis for morphological description of corn roots under nitrogen stress. *Agron. J.*, 85, 287–289.

Eke, A., Hermán, P., Bassingthwaighte, J.B., Raymond, G.M., Percival, D.B., Cannon, M., I Balla and Ikrényi, C. (2000) Physiological time series: Distinguishing fractal noises from motions. *Eur. J. Physiol.*, 439, 403–415.

Eke, A., Hermán, P., Kocsis, L. and Kozak, L.R. (2002) Fractal characterization of complexity in temporal physiological signals. *Physiol. Meas.*, 23, 1–38.

Elber, R. (1989) Fractal analysis of protein. In *The Fractal Approach to Heterogeneous Chemistry: Surfaces, Colloids, Polymers*, Avnir, D., Ed. Wiley & Sons, Chichester, 345–361.

Ellner, S. (1992) Detecting low dimensional chaos in population dynamic data: A critical review. In *Chaos and Insect Ecology*, Logan, J. and Hain, F., Eds. University Press of Virginia, Blacksburg, 63–91.

Ellner, S., Gallant, A.R., McCarey, D. and Nychka, D. (1991) Convergence rates and data requirements for Jacobian-based estimates of Lyapunov exponents from data. *Phys. Lett.*, 153, 357–363.

Ellner, S. and Turchin, P. (1995) Chaos in a noisy world: New methods and evidence from time series analysis. *Am. Nat.*, 145, 343–375.

Enquist, B.J., Brown, J.H. and West, G.B. (1998) Allometric scaling of plant energetics and population density. *Nature*, 395, 163–165.

Enquist, B.J., West, G.B., Charnov, E.L. and Brown, J.H. (1999) Allometric scaling of production and life-history variation in vascular plants. *Nature*, 401, 907–911.

Erlandson, J. and Kostylev, V. (1995) Trail following, speed and the fractal dimension of movement in a marine prosobranch, *Littorina littorea*, during a mating and a non-mating season. *Mar. Biol.*, 122, 87–94.

Eshel, A. (1998) On the fractal dimensions of a root system. *Plant Cell Environ.*, 21, 247–251.

Estrada, M. and Berdalet, E. (1997) Phytoplankton in a turbulent world. *Sci. Mar.*, 61, 125–140.

Etzenhouser, M.J., Owens, M.K., Spalinger, D.E. and Murden, S.B. (1998) Foraging behaviour of browsing ruminants in a heterogeneous landscape. *Landscape Ecol.*, 13, 55–64.

Evangelou, S.N. and Economou, E.N. (1990) Eigenvector statistics and multifractal scaling of band random matrices. *Phys. Lett. A*, 151, 345–348.

Evertsz, C.J.G. and Mandelbrot, B.B. (1992) Multifractal measures. In *Chaos and Fractals: New Frontiers of Science*, Peitgen, H.O., Hartmut, J. and Saupe, D., Eds. Springer-Verlag, New York, 922–954.

Falconer, K.J. (1985) *The Geometry of Fractal Sets*. Cambridge University Press, Cambridge.

———. (1993) *Fractal Geometry. Mathematical Foundations and Applications*. Wiley, Chichester.

Faloutsos, M., Faloutsos, P. and Faloutsos, C. (1999) On power-law relationships of the Internet topology. *Comput. Commun. Rev.*, 29, 251–262.

Farmer, J.D., Ott, E. and Yorke, J.A. (1983) The dimension of chaotic attractors. *Physica D*, 7, 153–180.

Farmer, J.D. and Sidorowich, J.J. (1987) Predicting chaotic time series. *Phys. Rev. Lett.*, 59, 845–848.

———. (1989) Exploiting chaos to predict the future and reduce noise. In *Evolution, Learning and Cognition*, Lee, Y.C., Ed. World Scientific Press, New York, 277–304.

Fauchald, P., Erikstad, K.E. and Systad, G.H. (2000) Scale-dependent predatory prey interactions: The hierarchical spatial distribution of sea birds and prey. *Ecology*, 81, 773–783.

Faure, P., Neumeister, H., Faber, D. and Korn, H. (2003) Symbolic analysis of swimming trajectories reveals scale invariance and provides a model for fish locomotion. *Fractals*, 11, 233–243.

Feder, H.J.S. and Feder, J. (1991) Experiments and simulations modeling earthquakes. In *Spontaneous Formation of Space-time Structures and Criticality*, Riste, T. and Sherrington, D., Eds. Kluwer Academic Publishers, Dordrecht, 107–111.

Feder, J. (1988) *Fractals*. Plenum Press, New York.

Feder, J., Jossang, T. and Rosenqvist, E. (1984) Scaling behavior and cluster fractal dimension determined by light scattering from aggregation proteins. *Phys. Rev. Lett.*, 53, 1403–1406.

Federer, H. (1969) *Geometric Measure Theory*. Springer, New York.

Feigenbaum, M.J. (1979) The universal metric properties of nonlinear transformations. *J. Stat. Phys.*, 21, 669–706.

Feller, W. (1968, 1971) *An Introduction to Probability Theory and Its Applications*. Wiley, New York.

Ferguson, S.H., Rettie, W.J. and Messier, F. (1998b) Fractal measure of female caribou movements. *Rangifer*, 10, 139–147.

Ferguson, S.H., Taylor, M.K., Born, E.W. and Messier, F. (1998a) Fractals, sea ice landscape and spatial patterns of polar bears. *J. Biogeography*, 25, 1081–1092.

Fergusson, I.K., Compagno, L.J.V. and Marks, M.A. (2000) Predation by white sharks, *Carcharodon carcharias* (Chondrichthyes: Lamnidae) upon chelonians, with new record from the Mediterranean Sea and a first record on the ocean sunfish *Mola mola* (Osteichthyes: Molidae) as stomach contents. *Environ. Biol. Fish*, 58, 447–453.

Fernandes, M.B., Sicres, M.A., Boireau, A and Tronczynski, J. (1997) Polyaromatic hydrocarbon (PAH) in the Seine River and its estuary. *Mar. Pollut. Bull.*, 34, 857–867.

Ferrer i Cancho, R. (2005a) Zipf's law from a communicative phase transition. *Eur. Phys. J. B*, 47, 449–457.

———. (2005b). The variation of Zipf's law in human language. *Eur. Phys. J. B*, 44, 249–257.

———. (2005c). Decoding least effort and scaling in signal frequency distributions. *Physica A*, 345, 275–284.

Ferrer i Cancho, R. and Solé, R.V. (2001). Two regimes in the frequency of words and the origin of complex lexicons: Zipf's law revisited. *J. Quant. Linguistics*, 8, 165–173.

Ficken, M.S., Hailman, E.D. and Hailman, J.P. (1994) The chickadee call system of the Mexican chickadee. *Condor*, 96, 70–82.

Fisher, K.E., Wiebe, P.H. and Malamud, B.D. (2004) Fractal characterization of local hydrographic and biological scales of patchiness on Georges Bank. In *Handbook of Scaling Methods in Aquatic Ecology: Measurement, Analysis, Simulation*, Seuront, L. and Strutton, P.G., Eds. CRC Press, Boca Raton, 297–319.

Fitter, A.H. and Stickland, T.R. (1992) Fractal characterization of root system architecture. *Functional Ecol.*, 6, 632–635.

Fleury, V. (1997) Branched fractal patterns in non-equilibrium electrochemical deposition from oscillatory nucleation and growth. *Nature*, 390, 145–148.

Flyvbjerg, H., Sneppen, K. and Bak, P. (1993) Mean field theory for a simple model of evolution. *Phys. Rev. Lett.*, 71, 4087–4090.

Focardi, S., Marcellini, P. and Montanaro, P. (1996) Do ungulates exhibit a food density threshold? A field study of optimal foraging and movement patterns. *J. Anim. Ecol.*, 65, 606–620.

Fourountan-pour, K., Dutilleul, P. and Smith, D.L. (1999) Soybean canopy development as affected by population density and intercropping with corn: Fractal analysis in comparison with other quantitative approaches. *Crop Sci.*, 39, 1784–1991.

———. (2000) Effects of population density and intercropping with soybean on fractal dimension of corn plant skeletal images. *J. Agron. Crop Sci.*, 184, 89–100.

————. (2001) Inclusion of the fractal dimension of leafless plant structure in the Beer-Lambert law. *Agron. J.*, 93, 333–338.

Fougere, P.F. (1985) On the accuracy of spectrum analysis of red noise processes using maximum entropy and periodogram methods: Simulation studies and application to geographical data. *J. Geograph. Res.*, 90(A5), 4355–4366.

Frankhauser, P. (1994) *La Fractalité des Structures Urbaines*. Collection Villes, Anthropos, Paris.

Franks, P.J.S. (2005) Plankton patchiness, turbulent transport and spatial spectra. *Mar. Ecol. Prog. Ser.*, 294, 295–309.

Franks, P.J.S. and Jaffe, J.S. (2008) Microscale variability in the distributions of large fluorescent particles observed in situ with a planar laser imaging fluorometer. *J. Mar. Syst.*, 69, 254–270.

Fraser-Brunner, A. (1951) The ocean sunfishes (Family Molidae). *Bull. Br. Mus. Zool.*, 1, 87–121.

Frette, V., Christensen, K., Malthe-Sørenssen, A., Feder, J., Jøssang, T. and Meakin, P. (1996) Dynamics in a pile of rice. *Nature*, 379, 49–51.

Friesen, W.I. and Mikula, R.J. (1987) Fractal dimensions of coal particles. *J. Colloid Interf.*, 120, 263–271.

Frisch, U. (1996) *Turbulence: The Legacy of A.N. Kolmogorov*. Cambridge University Press, Cambridge.

Fritz, H., Said, S. and Weimerkirch, H. (2003) Scale-independent hierarchical adjustments of movement patterns in a long-range foraging seabird. *Proc. R. Soc. London B*, 270, 1143–1148.

Frontier, S. (1977) Réflexions pour une théorie des écosystèmes. *Bull. Ecol.*, 8, 445–464.

————. (1985) Diversity and structure in aquatic ecosystems. *Oceanogr. Mar. Biol. Ann. Rev.*, 23, 253–312.

————. (1987) Applications of fractal theory to ecology. In *Developments in Numerical Ecology*, Legendre, P. and Legendre, L., Eds. Springer Verlag, Berlin, 335–378.

————. (1994) Species diversity as a fractal properties of biomass. In *Fractals in the Natural and Applied Sciences* (A-41), Nowak, M.M., Ed. Elsevier, North-Holland, 119–127.

Frontier, S. and Bour, W. (1976) Note sur une collection de Chaetognathes récoltée au dessus du talus continental près de Nosy Be (Madagascar) *Cah. ORSTOM, sér. Océanogr.*, 14, 267–272.

Fujikawa, H. (1994) Diversity of the growth patterns of *Bacillus subtilis* colonies on agar plates. *FEMS Microbiol. Ecol.*, 13, 159–168.

Fujikawa, H. and Matsushita, M. (1989) Fractal growth of *Bacillus subtilis* on agar plates. *J. Phys. Soc. Jpn.*, 58, 3875–3878.

————. (1991) Bacterial growth in the concentration field of nutrient. *J. Phys. Soc. Jpn.*, 60, 88–94.

Gajbhiye, S.N., Mustafa, S., Mehta, P. and Nair, V.R. (1995) Assessment of biological characteristics on coastal environment of Murud (Maharashtra) during the oil spill (17 May 1993). *Indian J. Mar. Sci.*, 24, 196–202.

Garcia, F., Carrère, P., Soussana, J.F. and Baumont, R. (2005) Characterisation by fractal analysis of foraging paths of ewes grazing heterogeneous swards. *Appl. Anim. Behav. Sci.*, 93, 19–37.

Garcia-Pelayo, R. and Morley, P.D. (1993) Scaling law for pulsar glitches. *Europhys. Lett.*, 23, 185–188.

Gardner, R.H., O'Neill, R.V., Turner, M.G. and Dale, V.H. (1989) Quantifying scale-dependent effects of animal movement with simple percolation models. *Landscape Ecol.*, 3, 217–227.

Gargett, A.E. (1997) "Theories" and techniques for observing turbulence in the euphotic zone. *Sci. Mar.*, 61, 25–45.

Gasol, J.M. and del Giorgio, P.A. (2000) Using flow cytometry for counting natural planktonic bacteria and understanding the structure of planktonic bacterial communities. *Sci. Mar.*, 64, 197–224.

Gautestad, A.O. and Mysterud, I. (1993) Physical and biological mechanisms in animal movement processes. *J. Appl. Ecol.*, 30, 523–535.

Ghilardi, P., Kaikai, A. and Menduni, G. (1993) Self-similar heterogeneity in granular porous media at the representative elementary volume scale. *Water Resour. Res.*, 29, 1205–1215.

Gibson, C.H. (1991) Kolmogorov similarity hypotheses for scalar fields: Sampling intermittent turbulent mixing in the ocean and galaxy. *Proc. R. Soc. Lond. A*, 434, 149–164.

Gilbert, W.J. (1984) A cube-filling Hilbert curve. *Math. Intelligencer*, 6, 78–88.

Giménez, D., Allmaras, R.R., Huggins, D.R. and Nater, E.A. (1998) Mass, surface and fragmentation fractal dimensions of soil fragments produced by tillage. *Geoderma*, 86, 261–278.

Giménez, D., Karmon, J.L., Posadas, A. and Shaw, R.K. (2002) Fractal dimensions of mass estimated from intact and eroded soil aggregates. *Geoderma*, 64, 165–172.

Glazman, R.E. (1991) Fractal nature of surface geometry in a developed sea. In *Nonlinear Variability in Geophysics*, Schertzer, D. and Lovejoy, S., Eds., Kluwer, Dordrecht, 217–226.

Gleick, J. (1987) *Chaos*. Mandarin, London.

Glud, R., Kühl, M., Wenzhöfer, F. and Rysgaard, S. (2002) Benthic diatoms of a high Arctic fjord (Young Sound, NE Greenland): Importance for ecosystem primary production. *Mar. Ecol. Prog. Ser.*, 238, 15–29.

Godfrey, H.C.J. and Blythe, S.P. (1991) Complex dynamics in multispecied communities. *Philos. Trans. R. Soc. Ser. B*, 330, 221–233.

Gollub, J.P. and Benson, S.V. (1980) Many routes to turbulent convection. *J. Fluid Mech.*, 100, 449–470.

Goodchild, M.F. (1980) Fractals and the accuracy of geographical measures. *Math. Geogr.*, 12, 85–98.

Gouyet, J.F. (1992) *Physique et Structures Fractales*. Masson, Paris.

———. (1996) *Physics of Fractal Structures*. Springer-Verlag, New York.

Gower, J.F.R, Denman, K.L. and Holyer, R.J. (1980) Phytoplankton patchiness indicates the fluctuation spectrum of mesoscale oceanic structure. *Nature*, 288, 157–159.

Graben, P.B. Saddy, J.D., Schlesewsky, M. and Kurths, J. (2000) Symbolic dynamics of event-related brain potentials. *Phys. Rev. E*, 62, 5518–5540.

Grant, H.L., Stewart, R.W. and Moillet, A. (1962) Turbulence spectra from a tidal channel. *J. Fluid Mech.*, 12, 241–263.

Grasman, J., Brascamp, J.W., Van Leeuwen, J.L. and Van Putten, B. (2003) The multifractal structure of arterial trees. *J. Theor. Biol.*, 220, 75–82.

Grassberger, P. (1983) Generalized dimensions of strange attractors. *Phys. Lett.*, 97, 227–230.

Grassberger, P. and Procaccia, I. (1983) Characterization of strange attractors. *Phys. Rev. Lett.*, 50, 346–349.

Gregg, M.C. (1990) Uncertainties and limitations in measuring ε and χ_T. *J. Atmos. Ocean. Tech.*, 16, 1483–1490.

Guarini, J.M., Blanchard, G.F. and Gros, P. (2000) Quantification of the microphytobenthic primary production in European intertidal mudflats: A modelling approach. *Cont. Shelf Res.*, 20, 1771–1788.

Guegan, J.F., Lek, S. and Oberdoff, T. (1998) Energy availability and habitat heterogeneity predict global riverine fish diversity. *Nature*, 391, 382–384.

Gunnarsson, B. (1992) Fractal dimension of plants and body size distribution in spiders. *Functional Ecol.*, 6, 636–641.

Gurvich, A.S. (1960) Experimental research on frequency spectra of atmospheric turbulence. *Izv. Akad. Nauk SSSR, Geofiz. Ser.*, 1042–1055.

Gurvich, A.S. and Yaglom, A.M. (1967) Breakdown of eddies and probability distributions for small-scale turbulence. *Phys. Fluids*, 10, 59–65.

Haeckel, E. (1891) Plakton studien. *J. Zeitschriftfuer Naturwis.*, 25, 232–336.

Hailman, J.P. (1994) Constrained permutation in "chick-a-dee"-like calls of the black-lored tit, *Parus xanthogenys*. *Bioacoustics*, 6, 33–50.

Hailman, J.P. and Ficken, M.S. (1986) Combinatorial animal communication with computable syntax: "Chick-a-dee" calling qualifies as "language" by structural linguistics. *Anim. Behav.*, 34, 1899–1901.

Hailman, J.P., Ficken, M.S. and Ficken, R.W. (1985) The "chick-a-dee" calls of *Parus atricapillus*: A recombinant system of animal communication compared with written English. *Semiotica*, 56, 191–224.

———. (1987) Constraints on the structure of combinatorial "chick-a-dee" calls. *Ethology*, 75, 62–80.

Haldane, J.B.S. (1928) *On Being the Right Size*. Oxford University Press, London.

Haldane, J.B.S. and Spurway, H. (1954) A statistical analysis of communication in *Apis mellifera* and a comparison with communication in other animals. *Insectes Sociaux*, 1, 247–283.

Halley, J.M., Hartley, S., Kallimanis, A.S., Kunin, W.E., Lennon, J.J. and Sgardelis, S.P. (2004) Uses and abuses of fractal methodology in ecology. *Ecol. Lett.*, 7, 254–271.

Halsey, T.C., Jensen, M.H., Kadanoff, L.P., Procaccia, I. and Shraiman, B.I. (1986) Fractal measure and their singularities: The characterization of strange sets. *Phys. Rev. A*, 33, 443–453.

Hamazaki, T. (1996) Effects of patch shape on the number of organisms. *Landscape Ecol.*, 11, 299–306.

Hamburger, D., Biham, O. and Avnir, D. (1996) Apparent fractality emerging from models of random distributions. *Phys. Rev. E*, 53, 3342–3358.

Hamilton, J.D. (1994) *Time Series Analysis*. Princeton University Press, Princeton.

Hamilton, S.K., Melack, J.M., Goodchild, M.F. and Lewis, W.M. (1992) Estimation of the fractal dimension of terrain from lake size distributions. In *Lowland Floodplain Rivers: Geomorphological Perspectives*, Carling, P.A. and Petts, G.E., Eds. Wiley, New York, 145–163.

Hardy, A.C. (1936) Observation of the uneven distribution of oceanic plankton. *Discovery Rep.*, 11, 511–538.

Harnos, A., Horvath, G., Lawrence, A.B. and Vattay, G. (2000) Scaling and intermittency in animal behaviour. *Physica A*, 286, 312–320.

Hassell, M.P., Lawton, J.H. and May, R.M. (1976) Patterns of dynamical behaviour in single species population models. *J. Anim. Ecol.*, 45, 471–486.

Hastings, A., Hom, C.L., Ellner, S., Turchin, P. and Godfray, H.C.J. (1993) Chaos in ecology: Is Mother Nature a strange attractor? *Annu. Rev. Ecol. Syst.*, 24, 1–33.

Hastings, A. and Powell, T.M. (1991) Chaos in a three species food chain. *Ecology*, 72, 896–903.

Hastings, H.M., Schneider, B.S., Monticciolo, R., vun Kannon, D. and del Monte, D. (1982) Time scales, persistence and patchiness. *BioSystems*, 15, 281–289.

Hastings, H.M., Schneider, B.S., Schreiber, M.A., Gorray, K., Maytal, G. and Maimon, J. (1992) Statistical geometry of pancreatic islets. *Proc. R. Soc. London B*, 250, 257–261.

Hastings, H.M. and Sugihara, G. (1993) *Fractals. A User's Guide for the Natural Sciences*. Oxford University Press, Oxford.

Haury, L.R., McGowan, J.A. and Wiebe, P.H. (1978) Patterns and processes in the time-space scales of plankton distributions. In *Spatial Pattern in Plankton Communities*, Steele, J.H., Ed. Plenum, New York, 277–327.

Hausdorff, F. (1919) Dimension und usseres Mass. *Math. Annalen*, 79, 157–179.

Hausdorff, J.M., Mitchell, S.L., Firtion, R., Peng, C.K., Cudkowicz, M.E., Wei, J.Y. and Goldberger, A.L. (1997) Altered fractal dynamics of gait: Reduced stride-interval correlations with aging and Huntington's disease. *J. Appl. Physiol.*, 82, 262–269.

Hausdorff, J.M., Peng, C.K., Ladin, Z., Wei, J.Y. and Goldberger, A.L. (1995) Is walking a random walk? Evidence for long-range correlations in stride interval of human gait. *J. Appl. Physiol.*, 78, 349–358.

Havlin, S. (1995) The distance between Zipf plots. *Physica A*, 216, 148–150.

Hawrot, R.Y. and Niemi, G.J. (1996) Effects of edge type and patch shape on avian communities in a mixed confer-hardwood forest. *Auk*, 113, 586–598.

Hazlett, B. and Bossert, W. (1965) A statistical analysis of the aggressive communications systems of some hermit crabs. *Anim. Behav.*, 13, 357–373.

He, F., Legendre, P. and Bellehumeur, C. (1994) Diversity pattern and spatial scale: A study of a tropical rain forest in Malaysia Environ. *Ecol. Stat.*, 1, 265–286.

Hee, E., Økland, R.H., Bratli, H., Dramstad, W.E., Engan, G., Pedersen, O. and Solstad, H. (2007) Regularity of species richness relationships to patch size and shape. *Ecography*, 30, 589–597.

Heegaard, E., Økland, R.H., Bratli, H., Dramstad, W.E., Engan, G., Pedersen, O. and Solstad, H. (2007) Regularity of species richness relationships to patch size and shape. *Ecography* 30, 589–597.

Held, G.A., Solina, D.H., Keane, D.T., Haag, W.J., Horn, P.M. and Grinstein, G. (1990) Experimental study of critical mass fluctuations in an evolving sand pile. *Phys. Rev. Lett.*, 65, 1120–1132.

Helmlinger, G., Endo, M., Ferrara, N., Hlatky, L. and Jain, R.K. (2000) Growth factors: Formation of endothelial cell networks. *Nature*, 405, 139–141.

Hénon, M. (1976) A two-dimensional mapping with a strange attractor. *Comm. Math. Phys.*, 50, 69–77.

Hentschel, H.G.E. and Procaccia, I. (1983) The infinite number of generalized dimensions of fractals and strange attractors. *Physica D*, 8, 435–444.

Herdan, G. 1953 *Small Particle Statistics*. Elsevier, New York.

Hidalgo, C., van Milligen, B.P. and Pedrosa, M.A. (2006) Intermittency and structures in edge plasma turbulence. *C. R. Phys.*, 7, 679–685.

Hilborn, R.C. (1994) *Chaos and Nonlinear Dynamics*. Oxford University Press, Oxford.

Hill, P.S. (1992) Reconciling aggregation theory with observed vertical fluxes following phytoplankton blooms. *J. Geophys. Res.*, 97, 2295–2308.

Hirsch, M.W and Smale, S. (1974) *Differential Equations, Dynamical Systems, and Linear Algebra*. Academic Press, New York.

Hoddle, M.S. (2003) The effect of prey species and environmental complexity on the functional response of *Franklinothrips orizabensis*: A test of the fractal foraging model. *Ecol. Entomol.*, 28, 309–318.

Holling, C.S. (1992) Cross-scale morphology, geometry, and dynamics of ecosystems. *Ecol. Monogr.*, 62, 447–502.

Horne, D.S. (1987) Determination of the fractal dimension using turbidimetric techniques. *Faraday Discuss. Chem. Soc.*, 83:259.

———. (1989a) Application of fractal concepts to the study of caseinate aggregation phenomena. *J. Dairy Res.*, 56, 535.

———. (1989b) Studies on the aggregation of casein micelles. In *Food Colloids*, Bee, R.D., Richmond, P. and Mingin, J., Eds. Oxford University Press, Oxford.

Horowitz, E. and Sahni, S. (1978) *Fundamentals of Computer Algorithms*. Computer Science Press, Rockville, MD.

Hovel, K.A., Fonseca, M.S., Myer, D.L., Kenworthy, W.J. and Whitfield, P.E. (2002) Effects of seagrass landscape structure, structural complexity and hydrodynamic regime on macrofaunal densities in North Carolina seagrass beds. *Mar. Ecol. Prog. Ser.*, 243, 11–24.

Hsieh, C.H., Glaser, S.M., Lucas, A.J. and Sugihara, G. (2005) Distinguishing random environmental fluctuations from ecological catastrophes for the North Pacific Ocean. *Nature*, 435, 336–340.

Hu, K., Ivanov, P.Ch., Chen, Z., Carpena, P. and Stanley, H.E. (2001) Effect of trends on detrended Fluctuation analysis. *Phys. Rev. E*, 64, 011114.

Huang, J. and Turcotte, D.L. (1989) Fractal mapping of digitized images: Application to the topography of Arizona and comparisons with synthetic images. *J. Geophys. Res.*, 94, 7491–7495.

Huisman, J., Pham Ti, N.N., Karl, D.M. and Sommeijer, B. (2006) Reduced mixing generates oscillations and chaos in the oceanic deep chlorophyll maximum. *Nature*, 439, 322–325.

Huisman, J. and Weissing, F.J. (1999) Biodiversity of plankton by species oscillations and chaos. *Nature*, 402, 407–410.

Hunt, J.R. (1982) Self-similar particle size distributions during coagulation: Theory and experimental verification, *J. Fluid Mech.*, 122, 303–309.

Hurewicz, W. and Wallman, H. (1941) *Dimension Theory*. Princeton University Press, Princeton.

Hurst, H.E. (1951). Long-term storage capacity of reservoirs. *Trans. Am. Soc. Civ. Eng.*, 116, 770–808.

Hurst, H.E., Black, R. and Sinaika, Y.M. (1965) *Long-Term Storage in Reservoirs: An Experimental Study*, Constable, London.

Hutchinson, E.H. and MacArthur, R.H. (1959) A theoretical ecological model of size distributions among species of animals. *Am. Nat.*, 93, 117.

Hutchinson, G.E. (1961) The paradox of the plankton. *Am. Nat.*, 95, 137–145.

Hutchinson, N., Davies, M.S., Ng, J.S.S. and Williams, G.A. (2007) Trail following behaviour in relation to pedal mucus production in the intertidal gastropod *Monodonta labio* (Linnaeus). *J. Exp. Mar. Biol. Ecol.*, 349, 313–322.

Ibanez, F. (1986) Le déterminisme du chaos. *J. Rech. Océanogr.*, 11, 66–69.

Inoue, E. (1952a) Turbulent fluctuations in temperature in the atmosphere and oceans, *J. Met. Soc. Japan*, 30, 289–295.

Inoue, E. (1952b) On the Lagrangian correlation coefficient for turbulent diffusion and its application to atmospheric diffusion phenomena, *Geophysical Res. Pap.*, 19, 397–412.

Intaglieta, M. and Breit, G.A. (1991) Chaos and microcircularity control. *Prog. Appl. Microcirc.*, 18, 22–32.

Isern-Fontanet, J., Turiel, A., García-Ladona, E. and Font, J. (2007) Microcanonical multifractal formalism: Application to the estimation of ocean surface velocities. *J. Geophys. Res.*, 112, C05024, doi:10.1029/2006JC003878.

Israeloff, N.E., Kagalenko, M. and Chan, K. (1995) Can Zipf distinguish language from noise in noncoding DNA? *Phys. Rev. Lett.*, 76, 1976–1979.

Iudin, D.I. and Gelashvily, D.B. (2003) Multifractality in ecological monitoring. *Nucl. Instr. Meth. Phys. Res. A*, 502, 799–801.

Ivanov, P.Ch., Nunes Amaral, L.A., Golberger, A.L., Havlin, S., Rosenblum, M.G., Struzik, Z.R. and Stanley, H.E. (1999) Multifractacilty in human heartbeat dynamics. *Nature*, 399, 461–465.

Ivanov, P.Ch., Nunes Amaral, L.A., Golberger, A.L. and Stanley, H.E. (1998) Stochastic feedback and the regulation of biological rhythms. *Europhys. Lett.*, 43, 363–368.

Jackson, G. and Bird, A. (1998) Aggregation in the marine environment. *Environ. Sci. Tech.*, 32, 2805–2814.

Jackson, G.A. (1990) A model of the formation of marine algal flocs by physical coagulation processes. *Deep-Sea Res.*, 37, 1197–1211.

Jaeger, M.H. and Nagel, S.R. (1992) Physics of the granular state. *Science*, 255, 1523–1526.

Janssen, M., Hust, M., Rhiel, E. and Krumbein, W.E. (1999) Vertical migration behaviour of diatom assemblages of Wadden Sea sediments (Dangast, Germany): A study using cryoscanning electron microscopy. *Intl. Microbiol.*, 2, 103–110.

Jaynes, E.T. (1957) Information theory and statistical mechanics. *Phys. Rev.*, 106, 620–630.

Jeffries, M. (1993) Invertebrate colonization of artificial pondweed of differing fractal dimension. *Oikos*, 67, 142–148.

Jenkins, G.M. and Watts, D.G. (1968) *Spectral Analysis*. Holden Day, London.

Jensen, R.V. (1992) Quantum chaos. *Nature*, 355, 311–318.

Jeong, G.D. and Rao, A.R. (1996) Chaos characteristics of tree ring series. *J. Hydrol.*, 182, 239–257.

Jiang, J. and Plotnick, R.E. (1998) Fractal analysis of the complexity of Unites States coastlines. *Math. Geol.*, 30, 535–546.

Jiang, Q. and Logan, B.E. (1991) Fractal dimensions of aggregates determined from steady-state size distributions. *Environ. Sci. Technol.*, 25, 2031–2038.

Jiménez, J. (1997) Oceanic turbulence at millimeter scales. *Sci. Mar.*, 61, 47–56.

Jiménez, J. (1998) Small scale intermittency in turbulence. *Eur. J. Mech. B/Fluids*, 17, 405–419.

———. (2000) Intermittency and cascades. *Sci. Mar.*, 409, 99–120.

———. (2006) Intermittency in turbulence. In *Encyclopedia of Mathematical Physics*, 144–151.

Jiménez, J. and Wray, A.A. (1994) Columnar vortices in isotropic turbulence. *Meccanica*, 29, 453–464.

Jiménez, J., Wray, A.A., Safman, P.G. and Rogallo, R.S. (1993) The structure of intense vorticity in isotropic turbulence. *J. Fluid Mech.*, 255, 65–90.

Johnson, A.R., Milne, B.T. and Wiens, J.A. (1992) Diffusion in fractal landscapes: Simulations and experimental studies of tenebrionid beetle movements. *Ecology*, 73, 1968–1983.

Johnson, G.D., Tempelman, A. and Patil, G.P. (1995) Fractal based methods in ecology: A review for analysis at multiple spatial scales. *Coenoses*, 10, 123–131.

Jones, C., Lonergan, G. and Mainwaring, D. (1996) Wavelet packet computation of the Hurst exponent *J. Phys. A: Gen. Phys.*, 29, 2509–2527.

Jonsson, P.R. and Johansson M. (1997) Swimming behaviour, patch exploitation and dispersal capacity of a marine benthic ciliate in flume flow. *J. Exp. Mar. Biol. Ecol.*, 215, 135–153.

Joshi, R.R. and Selvam, A.M. (1999) Identification of self-organized criticality in atmospheric low frequency variability. *Fractals*, 7, 421–425.

Journel, A.G. and Huijbregts, C.J. (1978) *Mining Geostatistics*. Academic Press, London.

Jullien, R. and Kolb, M. (1984) Hierarchical model for chemically limited cluster-cluster aggregation. *J. Physiol.*, A17, 639–643.

Jullien, R., Kolb, M. and Botet, R. (1984) Aggregation by kinetic clustering of clusters in dimension *d*>2. *J. Phys. Lett.*, 45, L211–L216.

Kaandorp, J.A. (1991) Modelling growth forms of the sponge *Haliclona oculata* (Porifera, Demospongiae) using fractal techniques. *Mar. Biol.*, 110, 203–215.

Kaandorp, J.A. and de Kluijver, M.J. (1992) Verification of fractal growth models of the sponge *Haliclona oculata* (Porifera) with transplantation experiments. *Mar. Biol.*, 113, 133–143.

Kafetsopoulos, E., Gouskos, S. and Evangelou, S.N. (1997) 1/*f* noise and multifractal fluctuations in rat behavior. *Nonlinear Anal. Theor. Meth. Appl.*, 30, 2007–2013.

Kampichler, C. and Hauser, M. (1993) Roughness of soil pore surface and its effects on available habitat space of microarthropods. *Geoderma*, 56, 223–232.

Kanter, I. and Kessler, D.A. (1995) Markov processes: Linguistics and Zipf's law. *Phys. Rev. Lett.*, 74, 4559–4562.

Kantha, L.H. and Clayson, C.A. (2000) *Small Scale Processes in Geophysical Fluid Flows*. International Geophysics Series, Vol. 67. Academic Press, San Diego.

Kantz, H. and Schreiber, T. (2004) *Nonlinear Time Series Analysis*. Cambridge University Press, Cambridge.

Kaplan, D. and Glass, L. (1995) *Understanding Nonlinear Dynamics*. Springer-Verlag, New York.

Kapur, J.N. and Kesavan, H.K. (1992) Entropy optimization principles and their applications. In *Entropy and Energy Dissipation in Water Resources*, Singh, V.P. and Fiorentino, M., Eds. Kluwer Academic, Dordrecht.

Kareiva, P. (1990) Population dynamics in spatially complex environments: Theory and data. *Philos. Trans. R. Soc. London B*, 330, 175–190.

Kareiva, P.M. and Shigesada, N. (1983) Analyzing insect movement as a correlated random walk. *Oecologia*, 56, 234–238.

Karp-Boss, L.E., Boss, E. and Jumars, P.A. (1996) Nutrient fluxes to planktonic osmotrophs in the presence of fluid motion. *Mar. Biol. Ann. Rev.*, 34, 71–107.

Katz, A.J. and Thompson, A.H. (1985) Fractal sandstones pores: Implication for conductivity and pore formation. *Phys. Rev. Lett.*, 54, 1325–1328.

Kauffman, S. (1993) *The Origins of Order*. Oxford University Press, New York.

———. (1995) *At Home in the Universe*. Penguin, London.

Kauffman, S.A. and Johnsen, S. (1991) Coevolution to the edge of chaos: Coupled fitness landscapes, poised states, and coevolutionary avalanches. *J. Theor. Biol.*, 149, 467–505.

Kawahara, K., Yamauchi, Y., Nakazono, Y. and Miyamoto, Y. (1989) Spectral analysis on low frequency fluctuation in respiratory rhythm in the decerebrate cat. *Biol. Cybern.*, 61, 265–270.

Kaye, B.H. (1989, 1994) *A Random Walk through Fractal Dimensions*. VCH, New York.

Keen, A.M. (1971) *Sea Shells of Tropical West America: Marine Mollusks from Baja California to Peru*. Stanford University Press, Stanford.

Keitt, T.H. and Marquet, P.A. (1996) The introduced Hawaiian avifauna reconsidered: Evidence for self-organized criticality? *J. Theor. Biol.*, 182, 161–167.

Kelaher, B.P. (2003) Changes in habitat complexity negatively affect diverse gastropod assemblages in coralline algal turf. *Oecologia*, 135, 431–441.

Kendall, B.E., Bjørnstad, O.N., Bascompte, J., Keitt, T.H. and Fagan, W.F. (2000) Dispersal, environmental correlation, and spatial synchrony in population dynamics. *Am. Nat.*, 155, 628–636.

Kendall, M. (1976) *Time-Series*. Charles Griffin, London.

Kenkel, N.C. and Walker, D.J. (1993) Fractals and ecology. *Abstr. Bot.*, 17, 53–70.

Kerr, J.T., Southwood, T.R.E. and Cihlar, J. (2001) Remotely sensed habitat diversity predicts butterfly species richness and community similarity in Canada. *Proc. Natl. Acad. Sci. USA*, 98, 11365–11370.

Khattri, K.N. (1995) Fractal description of seismicity of India and inferences regarding earthquake hazard. *Curr. Sci.*, 69, 361–366.

Khlebtsov, N.G. and Melnikov, A.G. (1994) Structure factor and exponent of scattering by polydisperse fractal colloidal aggregates. *J. Colloid Interf. Sci.*, 163, 145–151.

Kida, S. (1991) Log-stable distribution and intermittency of turbulence. *J. Phys. Soc. Jpn.*, 60, 5–8.

Kilps, J.R., Logan, B.E. and Alldredge, A.L. (1994) Fractal dimensions of marine snow aggregates determined from image analysis of *in situ* photographs. *Deep-Sea Res.*, 41, 1159–1169.

King, R.B., Weissman, L.J. and Bassingthwaighte, J.B. (1989) Fractal descriptions for spatial statistics. *Ann. Biomed. Eng.*, 18, 111–122.

Kiørboe, T. (1997) Small-scale turbulence, marine snow formation, and planktivorous feeding. *Sci. Mar.*, 61, 141–165.

———. (2008) *A Mechanistic Approach to Plankton Ecology.* Princeton University Press, Princeton.

Kiørboe, T., Andersen, K.P. and Dam, H. (1990) Coagulation efficiency and aggregate formation in marine phytoplankton. *Mar. Biol.*, 107, 235–245.

Kiørboe, T. and Hansen, J.L.S. (1993) Phytoplankton aggregate formation: Observations of patterns and mechanisms of cell sticking and the significance of exopolymeric material. *J. Plankton Res.*, 15, 993–1018.

Kiørboe, T., Hansen, J.L.S., Alldredge, A.L., Jackson, G.A., Passow, U., Dam, H.G., Drapeau, D.T., Waite, A. and Garcia, C.M. (1996) Sedimentation of phytoplankton during a diatom bloom: Rates and mechanisms. *J. Mar. Res.*, 54, 1123–1148.

Kiørboe, T., Lunsgaard, C., Olesen, M. and Hansen, J.L.S. (1994) Aggregation and sedimentation processes during a spring phytoplankton bloom: A field experiment to test coagulation theory. *J. Mar. Res.*, 52, 297–232.

Kiørboe, T. and Saiz, E. (1995) Planktivorous feeding in calm and turbulent environment, with emphasis on copepods. *Mar. Ecol. Prog. Ser.*, 122, 135–145.

Kiørboe, T., Tiselius, P., Mitchell-Innes, B., Hansen, J.L.S., Wisser, A.W. and Mari, X. (1998) Intensive aggregate formation with low vertical flux during an upwelling-induced diatom bloom. *Limnol. Oceanogr.*, 43, 104–116.

Klafter, J. and Blumen, A. (1984) Fractal behavior in trapping reaction. *J. Chem. Phys.*, 80, 875–877.

Kleindorfer, S., Lambert, S. and Paton, D. (2006). Ticks (*Ixodes* sp.) and blood parasites (*Haemoproteus* sp.) in New Holland honeyeaters (*Phylidonyris novaehollandiae*): Evidence for site specificity and fitness costs. *Emu*, 106, 113–118.

Klinkenberg, B. (1993) A review of methods used to determine the fractal dimension of linear features. *Math. Geol.*, 25, 1003–1026.

Koch, H. (1904) Sur une courbe continue sans tangente, obtenue par une construction géométrique élémentaire. *Arkiv Matematik*, 1, 681–704.

———. (1906) Une méthode géométrique pour l'étude de certaines questions de la théorie des courbes planes. *Mathematische Annalen*, 30, 145–174.

Kolasa, J. (1989) Ecological systems in hierarchical perspective: Breaks in community structure and other consequences. *Ecology*, 70, 36–47.

Kolasa, J. and Pickett, S.T.A. (1991) *Ecological Heterogeneity.* Springer-Verlag, New York.

Kolasa, J. and Rollo, D.C. (1991) The heterogeneity of heterogeneity: A glossary. In *Ecological Heterogeneity*, Kolasa, J. and Pickett, S.T.A., Eds. Springer-Verlag, New York, 1–23.

Kolb, M. and Jullien, R. (1984) Chemically limited versus diffusion limited aggregation. *J. Phys. Lett.*, 45, 977–981.

Kolmogorov, A.N. (1941) The local structure of turbulence in incompressible viscous fluid for very large Reynolds numbers. *Dokl. Akad. Nauk SSSR*, 30, 299–303.

———. (1962) A refinement of previous hypotheses concerning the local structure of turbulence in a viscous incompressible fluid at high Reynolds number. *J. Fluid Mech.*, 13, 82–85.

Kooi, B.W. and Kooijman, S.A.L.M. (2000) Invading species can stabilize simple trophic systems. *Ecol. Model.*, 133, 52–72.

Korcak, J. (1938) Deux types fondamentaux de distributions statistiques. *Bull. Inst. Intern. Stat.*, III, 295–299.

Korin, G. (1992) *Fractal Model in the Earth Sciences.* Elsevier, New York.

Koscielny-Bunde, E., Kantelhardt, J.W., Braun, P., Bunde, A. and Havlin, S. (2006) Long-term persistence and multifractality of river runoff records: Detrended fluctuation studies. *J. Hydrol.*, 322, 120–137.

Kostylev, V. and Erlandson, J. (2001) A fractal approach for detecting spatial hierarchy and structure on mussel beds. *Mar. Biol.*, 139, 497–506.

Kot, M., Sayler, G.S. and Schultz, T.W. (1992) Complex dynamics in a model of microbial system. *Bull. Math. Biol.*, 54, 619–648.

Kozac, E., Pachepsky, Y.A. Sokolowska, Z. and Stepniewski, W. (1996) A modified number-based method for estimating fragmentation fractal dimensions of soils. *Soil. Sci. Soc. Am. J.*, 60, 1291–1297.

Kraichnan, R.H. (1967). Inertial ranges in two-dimensional turbulence. *Physics of Fluids*, 10, 1417–1423.

Kramer, D.L. and McLaughlin, R.L. (2001) The behavioral ecology of intermittent locomotion. *Amer. Zool.*, 41, 137–153.

Kravchenko, A.N., Bullock, D.G. and Boast, C.W. (2000) Joint multifractal analysis of crop yield and terrain slope. *Agron. J.*, 92, 1279–1290.

Kreyszig, J. (1988) *Advanced Engineering Mathematics*. Wiley, New York.

Krummel, J.R., Gardner, R.H., Sugihara, G. and O'Neill, R.V. (1987) Landscape patterns in a disturbed environment. *Oikos*, 48, 321–324.

Kübler, J.E. and Dugeon, S.R. (1996) Temperature dependent change in the complexity of form of *Chondrus crispus* fronds. *J. Exp. Mar. Ecol. Ecol.*, 207, 15–24.

Labat, D., Mangin, A. and Ababou, R. (2002) Rainfall-runoff relations for karstic springs: Multifractal analysis. *J. Hydrol.*, 256, 176–195.

Laherrere, J. and Sornette, D. (1998) Stretched exponential distributions in nature and economy: "Fat tails" with characteristic scales. *Eur. Phys. J. B*, 2, 525–539.

Laidre, K.L., Heide-Jørgensen, M.P., Logsdon, M.L., Hobbs, R.C., Dietz, R. and VanBlaricom, G.R. (2004) Fractal analysis of narwhal space use patterns. *Zoology*, 107, 3–11.

Lam, N. and DeCola, M. (1993) Fractal measurement. In *Fractals in Geography*, Lam, N. and DeCola, L., Eds. Prentice Hall, New York, 23–55.

Lampert, W. (1987) Feeding and nutrition in *Daphnia*. In: Peters, R.H. and De Bernardi, R. (eds.), *Daphnia. Mem. del Inst. Ital. di Idrol.*, 143–192.

Landini, G. (2001) Evidence of linguistic structure in the Voynich manuscript using spectral analysis. *Cryptologia*, 25, 275–295.

Langdon, C.G. (1992) Life at the edge of chaos. In *Artificial Life II*, Langdon, C.G., Taylor, C., Farmer, J.D. and Rasmussen, S., Eds. Addison-Wesley, Redwood City, 41–91.

Laurie, H. and Perrier, E. (2007) Is species richness multifractal? Lessons from the Protea Atlas. *South African Journal of Botany*, 73, 297.

Le Comber, S.C., Seabloom, E.W. and Romañach, S.S. (2006) Burrow fractal dimension and foraging success in subterranean rodents: A simulation. *Behav. Ecol.*, 17, 188–195.

Lee, B. and Richards, F.M. (1971) The Interpretation of protein structures: Estimation of static accessibility. *Mol. Biol.*, 55, 379–400.

Lee, D.S. (1986) Seasonal, thermal, and zonal distribution of ocean sunfish, *Mola mola* (Linnaeus), off the North Carolina coast. *Brimleyana*, 12, 163–299.

Legendre, L. and Demers, S. (1984) Towards dynamic biological oceanography and limnology. *Can. J. Fish. Aquat. Sci.*, 41, 2–19.

Legendre, P. and Legendre, L. (2003) *Numerical Ecology*. Elsevier, New York.

Leising, A.W. and Franks, P.J.S. (2000) Copepod vertical distribution within a spatially variable food source: A simple foraging-strategy model. *J. Plankton Res.*, 22, 999–1024.

Lekan, J.F. and Wilson, R.E. (1978) Spatial variability of phytoplankton biomass in the surface waters of Long Island. *Est. Coast. Mar. Sci.*, 6, 239–251.

Lerman, M. (1986) *Marine Biology*. Benjamin/Cummings, Menlo Park, CA.

Lesieur, M. and Sadourny, R. (1981). Satellite sensed turbulent ocean structure. *Nature*, 294, 673.

Levandowsky, M., Klafter, J. and White, B.S. (1988) Swimming behavior and chemosensory responses in the protistan microzooplankton as a function of the hydrodynamic regime. *Bull. Mar. Sci.*, 43, 758–763.

Levin, S.A. (1992) The problem of patterns and scale in ecology. *Ecology*, 73, 1943–1967.

Levin, S.A., Powell, T.M. and Steele, J.H., Eds. (1993) *Patch Dynamics*. Springer-Verlag, Berlin.

Levinton, J.S. (2001) *Marine Biology: Function, Biodiversity, Ecology*. Oxford University Press, Oxford.

Lewis, M. and Rees, D.C. (1985) Fractal surfaces of proteins. *Science*, 230, 1163–1165.

Li, D.H. and Gangzarczyk, J.J. (1989) Fractal geometry of particle aggregates generated in water and wastewater treatment processes. *Environ. Sci. Technol.*, 23, 1385–1389.

Li, W. (1992) Random texts exhibit Zipf's-law-like word frequency distribution. *IEEE Trans. Inf. Theory*, 38, 1842–1845.

Li, X. and Logan, B.E. (1997a) Collision frequencies of fractal aggregates with small particles by differential sedimentation. *Environ. Sci. Technol.*, 31, 1229–1236.

———. (1997b) Collision frequencies between fractal aggregates and small particles in a turbulent sheared fluid. *Environ. Sci. Technol.*, 31, 1237–1242.

Li, X., Passow, U. and Logan, B.E. (1998) Fractal dimension of small (15–200 μm) particles in Eastern Pacific coastal waters. *Deep-Sea Res. I*, 45, 115–131.

Lin, M.Y., Lindsay, H.M., Weitz, D.A., Ball, R.C., Klein, R. and Meakin, P. (1989) Universality in colloid aggregation. *Nature*, 339, 360–362.

Little, S., Ellner, S., Pascual, M., Neubert, M., Kaplan, D., Sauer, T., Caswell, H. and Solow, A. (1996) Detecting non-linear dynamics in spatio-temporal systems, examples from ecological models. *Physica D*, 96, 321–333.

Livina, V., Kizner, Z., Braun, P., Molnar, T., Bunde, A. and Havlin, S. (2007) Temporal scaling comparison of real hydrological data and model runoff records. *J. Hydrol.*, 336, 186–198.

Loehle, C. (1983) The fractal dimension and ecology. *Sci. Tech.*, 6, 131–142.

———. (1990) Home range: A fractal approach. *Landscape Ecol.*, 5, 39–52.

Loehle, C. and Bai-Lian, L. (1996) Statistical properties of ecological and geologic fractals. *Ecol. Model.*, 85, 271–284.

Logan, B.E. and Alldredge, A.L. (1989) The increased potential for nutrient uptake by flocculating diatoms. *Mar. Biol.*, 101, 443–450.

Logan, B.E. and Wilkinson, D.B. (1990) Fractal geometry of marine snow and other biological aggregates. *Limnol. Oceanogr.*, 35, 130–136.

———. (1991) Fractal dimensions and porosities of *Zoogoloea ramigera* and *Saccharomyces cerevisiaea* aggregates. *Biotechnol. Bioengng.*, 38, 389–396.

Logan, B.E. and Kilps, J.R. (1995) Fractal dimensions of aggregates formed in different fluid mechanical environments. *Wat. Res.*, 29, 443–453.

Logan, J. and Allen, J.C. (1992) Nonlinear dynamics and chaos in insect populations. *Annu. Rev. Entomol.*, 37, 455–471.

Longley, P.A. and Batty, M. (1989) On the fractal measurement of geographical boundaries. *Geogr. Anal.*, 21, 47–67.

Longuet-Higgins, M.S. (1994) A fractal approach of breaking waves. *J. Phys. Oceanogr.*, 24, 1834–1838.

Lorenz, E.N. (1963) Deterministic nonperiodic flow. *J. Atm. Sci.*, 20, 130–141.

Lorenzen, C.J. (1967) Determination of chlorophyll and phaeopigments: Spectrometric equations. *Limnol. Oceanogr.*, 12, 343–346.

Lotka, A.J. (1926) The frequency distribution of scientific productivity. *J. Washington Acad. Sci.*, 16, 317–323.

Lovejoy, S. (1982) Area-perimeter relation for rain and cloud areas. *Science*, 216, 185–187

Lovejoy, S., Currie, W.J.S., Teissier, Y., Claereboudt, M.R., Bourget, J.C. and Schertzer, D. (2001) Universal multifractals and ocean patchiness: Phytoplankton, physical fields and coastal heterogeneity. *J. Plankton Res.*, 23, 117–141.

Lovejoy, S. and Mandelbrot, B.B. (1985) Fractal properties of rain, and a fractal model. *Tellus*, 37A, 209–232.

Lovejoy, S. and Schertzer, D. (2006) Multifractals, cloud radiances and rain. *J. Hydrol.*, 322, 59–88.

Lubchenco, J. (1978) Species diversity in a marine intertidal community: Importance of herbivore food preference and algal competitive abilities. *Am. Nat.*, 112, 23

———. (1983) *Littorina* and focus: Effects of herbivores, substratum heterogeneity, and plant escapes during succession. *Ecology*, 64, 1116–1123.

Luo, X. and Schramm, D.N. (1992) Fractals and cosmological large-scale structure. *Science*, 256, 513–515.

Lynch, J.P. and van Beem, J. (1993) Growth and architecture of seedling roots of common bean genotypes. *Crop Sci.*, 33, 1253–1257.

MacArthur, R.H. (1960) On the relative abundance of species. *Am. Nat.*, 94, 25–36.

———. (1965) Patterns of species diversity. *Biol. Rev.*, 40, 510–533.

MacArthur, R.H. and Wilson, E.O. (1967) *The Theory of Island Biogeography*. Princeton University Press, Princeton.

MacIntyre, H.L., Geider, R.J. and Miller, D.C. (1996) Microphytobenthos: The important role of the "secret garden" of unvegetated, shallow-water marine habitats. I: Distribution, abundance and primary production. *Estuaries*, 19, 186–201.

Mackas, D.L., Denman, K.L. and Abbott, M.R. (1985) Plankton patchiness: Biology in the physical vernacular. *Bull. Mar. Sci.*, 37, 652–674.

MacKay, D.M. (1972) Formal analysis of communicative processes. In *Non-verbal Communication*, Hinde, R.A., Ed. Cambridge University Press, Cambridge, 3–25.

Mäkikallio, T.H., Huikuri, H.V., Hintze, U., Videbæk, J. Mitrani, R.D., Castellanos, A., Myerburg, R.J. and Møller, M. (2001) Fractal analysis and time- and frequency-domain measures of heart rate variability as predictors of mortality in patients with heart failure, *Am. J. Cardiol.*, 87, 178–182.

Makse, H.A., Havlin, S. and Stanley, H.E. (1995) Modelling urban growth patterns. *Nature*, 377, 608–612.

Malamud, B.D. and Turcotte, D.L. (1999) Self-affine time series. I: Generation and analyses. *Adv. Geophys.*, 40, 1–90.

Malcai, O., Lidar, D.A. and Biham, O. (1997) Scaling range and cutoffs in empirical fractals. *Phys. Rev. E*, 56, 2817–2828.

Mandelbrot, B.B. (1953) Contribution à la théorie mathématique des communications. *Thèse Univ. Paris Publ. Inst. Stat. Univ. Paris* 2, 1–121.

———. (1967) How long Is the coast of Britain? Statistical self-similarity and fractional dimension. *Science*, 156, 636–638.

Mandelbrot, B.B. (1974) Intermittent turbulence in self-similar cascades: Divergence of high moments and dimension of the carrier. *J. Fluid Mech.*, 62, 305–330.

———. (1975) *Les Objects Fractals: Forme, Hasard et Dimension*. Flammarion, Paris.

Mandelbrot, B.B. (1976) Intermittent turbulence and fractal dimension: Kurtosis and the spectral exponent 5/3+. In: Teman, R. (Ed.), *Turbulence and Naviers Stokes Equations*. Lectures notes in Mathematics, 55, 121–145. Springer, New York.

———. (1977) *Fractals. Form, Chance, and Dimension*. Freeman, San Francisco.

———. (1983) *The Fractal Geometry of Nature*. Freeman, San Francisco.

———. (1985) Self-affine fractals and fractal dimension. *Phys. Scr.*, 32, 257–260.

———. (1986) Self-affine fractal sets. In *Fractals in Physics*, Pietronero, L. and Tosatti, E., Eds.. North Holland, Amsterdam.

Mandelbrot, B.B. and van Ness, J.W. (1968) Fractional Brownian motions, fractional noises and applications. *SIAM Rev.*, 10, 422–437.

Mandelbrot, B.B. and Wallis, J.R. (1969) Some long-run properties of geophysical records. *Water Resour. Res.*, 5, 321–340.

Manrubia, S.C. and Solé, R.V. (1996) Self-organized-criticality in rainforest dynamics. *Chaos Solitons Fractals*, 7, 523–541.

Mantegna, R.N., Buldyrev, S.V., Goldberger, A.L., Havlin, S., Peng, C.K., Simons, M. and Stanley, H.E. (1994) Linguistic features of noncoding DNA sequences. *Phys. Rev. Lett.*, 73, 3169–3172.

———. (1995) Systematic analysis of coding and noncoding DNA sequences using methods of statistical linguistics. *Phys. Rev. Lett.*, 52, 2939–2950.

Mantegna, R.N. and Stanley, H.E. (1995) Scaling behaviour in the dynamics of an economic index. *Nature*, 376, 46–49.

Mårell, A., Ball, J.P. and Hofgaard, A. (2002) Foraging and movement paths of female reindeer: Insights from fractal analysis, correlated random walks, and Lévy flights. *Can. J. Zool.*, 80, 854–865.

Margalef, R. (1957) La teoria de la informacion en ecologia. *Mem. Real Acad. Ciencias Artes Barcelona*, 32, 373–449.

María, G.A., Escós, J. and Alados, C.L. (2004) Complexity of behavioural sequences and their relation to stress conditions in chickens (*Gallus gallus domesticus*): A non-invasive technique to evaluate animal welfare. *Appl. Anim. Behav.*, 86, 93–104.

Marie, D., Partensky, F., Jacquet, S. and Vaulot, D. (1997) Enumeration and cell cycle analysis of natural populations of marine picoplankton by flow cytometry using a novel nucleic acid dye. *Appl. Environ. Microbiol.*, 63:186–193.

Marie, D., Partensky, F. and Vaulot, D. (1999) Enumeration of phytoplankton, bacteria and viruses in marine samples. *Curr. Protocols Cytometry*, 11.11.1–11.11.15.

Marsh, J. (1982) Aspects of the Ecology of Three Saltmarshes in the Derwent Region, and an Investigation into the Role of the Burrowing Crab *Helograpsus haswellianus* (Whitelegge, 1889). Honours thesis, University of Tasmania, 1–141.

Marsili, M. and Zhang, Y.C. (1998) Intercating individuals leading to Zipf's law. *Phys. Rev. Lett.*, 80, 2741–2744.

Martin, J.R. (2004) A portrait of locomotor behaviour in *Drosophila* determined by a video-tracking paradigm. *Behav. Proc.*, 67, 207–219.

Martín, M.Á. and Montero, E. (2002) Laser diffraction and multifractal analysis for the characterization of dry soil volume-size distributions. *Soil Till. Res.*, 64, 113–123.

Martinez, V.J., Paredes, S., Borgani, S. and Coles, P. (1995) Multiscaling properties of large-scale structure in the universe. *Science*, 269, 1245–1247.

Masugi, M. and Takuma, T. (2007) Multi-fractal analysis of IP-network traffic for assessing time variations in scaling properties. *Physica D*, 225, 119–126.

Matheron, G. (1971) *La théorie des variables régionalisées et ses applications*. Cahiers du Centre de Morphologie Mathématique de Fontainebleau. Fasc 5. ENSMP, Paris.

Matsushita, M. and Fujikawa, H. (1990) Diffusion-limited growth in bacterial colony formation. *Physica A*, 168, 498–506.

Matsuyama, T. and Matsushita, M. (1992) Self-similar colony morphogenesis by gram-negative rods as the experimental model of fractal growth by a cell population. *Appl. Environ. Microbiol.*, 58, 1227–1232.

———. (1993) Fractal morphogenesis by a bacterial cell population. *Crit. Rev. Microbiol.*, 19, 117–135.

Matsuyama, T., Sogawa, M. and Nakagawa, Y. (1989) Fractal spreading growth of *Serratia marcescens* which produces surface active exolipids. *FEMS Microbiol. Lett.*, 61, 243–246.

May, R.M. (1974) Biological populations with non-overlapping generations: Stable points, stable cycles and chaos. *Science*, 186, 645–647.

———. (1975) Biological populations obeying difference equations: stable points, stable cycles and chaos. *J. Theor. Biol.*, 51, 511–524.

———. (1976) Simple mathematical models with very complicated dynamics. *Nature*, 261, 459–467.

———. (1980) Nonlinear phenomena in ecology and epidemiology. *Ann. N.Y. Acad. Sci.*, 357, 267–281.

———. (1987) Chaos and the dynamics of biological populations. *Proc. R. Soc. London Ser. A*, 413, 27–44.

McCann, C. (1961) The sunfish, *Mola mola*, in New Zealand waters. *Res. Dominion Mus.*, 4, 7–20.

McCauley, J.L. (2001) Are galaxy distributions scale invariant? A perspective from dynamical systems theory. *Physica A*, 309, 183–213.

McCave, I.N. (1990) Size spectra and aggregation of suspended particles in the deep ocean. *Deep-Sea Res.*, 31, 329–352.

McCowan, B., Hanser, S.F. and Doyle, L.R. (1999) Quantitative tools for comparing animal communication systems: Information theory applied to bottlenose dolphin whistle repertoires. *Anim. Behav.*, 57, 409–419.

McCulloch, C.E. and Cain, M.L. (1989) Analyzing discrete movement data as a correlated random walk. *Ecology*, 70, 383–388.

McHardy, I. and Czerny, B. (1987) Fractal X-ray time variability and spectral invariance of the Seyfert galaxy NGC5506. *Nature*, 325, 696–698.

McMahon, T.A. and Bonner, J.T. (1983) *On Size and Life*. Scientific American Books, New York.

Meakin, P. (1983) Formation of fractal clusters and networks by irreversible diffusion-limited aggregation. *Phys. Rev. Lett.*, 51, 1119–1122.

———. (1986) A new model for biological pattern formation. *J. Theor. Biol.*, 118, 101–113.

———. (1988) Fractal aggregates. *Adv. Coll. Interf. Sci.*, 28, 249–331.

Meakin, P. and Family, F. (1987) Structure and dynamics of reaction-limited aggregation. *Phys. Rev. A*, 36, 5498–5501.

Medlin, L.K. (2006) The evolution of the diatoms and a report on the current status of their classification. In *Functioning of Microphytobenthos in Estuaries*, Kromkamp, J.C., de Brouwer, J., Blanchard G.F., Forster, R.M. and Créach, V., Eds. Academy of Arts and Sciences, Amsterdam, 3–8.

Meesmann, M., Boese, J. and Chialvo, D.R. (1993) Demonstration of $1/f$ fluctuations and white noise in the human heart rate by the variance-time curve: Implications for self-similarity. *Fractals*, 1, 312–320.

Meneveau, C. and Sreenivasan, K.R. (1991) Simple multifractal cascade model for fully developed turbulence. *Phys. Rev. Lett.*, 59, 1424–1427.

Meneveau, C., Sreenivasan, K.R., Kailasnath, P. and Fan, M.S. (1990) Joint multifractal measures: Theory and application to turbulence. *Phys. Rev. A*, 41, 894–913.

Mercik, S., Weron, K. and Siwy, Z. (1999) Statistical analysis of ionic current fluctuations in membrane channels. *Phys. Rev. E*, 60, 7343–7345.

Michell, J.F. (1999) *Who Wrote Shakespeare?* Thames & Hudson, Slovenia.

Mihail, J.D., Obert, M., Bruhn, J.N. and Taylor, S.J. (1995) Fractal geometry of diffuse mycelia and rhizomorphs of *Armillaria* species. *Mycol. Res.*, 99, 81–88.

Mihail, J.D., Obert, M., Taylor, S.J. and Bruhn, J.N. (1994) The fractal dimension of young colonies of *Macrophomina phaseolina* produced from microsclerotia. *Mycologia*, 86, 350–356.

Miles, A. and Sundbäck, K. (2000) Diel variations in microphytobenthic productivity in areas of different tidal amplitude. *Mar. Ecol. Prog. Ser.*, 205, 11–22.

Millán, H. and Orellana, R. (2001) Mass fractal dimensions of soil aggregates from different depths of a compacted vertisol. *Geoderma*, 101, 65–76.

Miller, D.J. and Lea, R.N. (1972) Guide to the coastal marine fishes of California. *Calif. Fish Game Bull.*, 157, 210.

Milne, B.T. (1988) Measuring the fractal geometry of landscapes. *Appl. Math. Computation*, 27, 67–79

———. (1991) Lessons from applying fractal models to landscape patterns. In *Quantitative Methods in Landscape Ecology: The Analysis and Interpretation of Landscape Heterogeneity*, Turner, M.G. and Gardner, R.H., Eds. Springer-Verlag, New York, 199–235.

Mistri, M. and Ceccherelli, V.U. (1993) Growth of the Mediterranean Gorgonian *Lophogorgia ceratophyta* (L., 1758). *Mar. Ecol.*, 14, 329–340.

Mitchell, J.G. and Furhman, J.A. (1989) Centimeter scale vertical heterogeneity in bacteria and chlorophyll *a*. *Mar. Ecol. Prog. Ser.*, 54, 141–148.

Mitchell, J.G. and Seuront, L. (2008) Towards a seascape typology. II: Zipf of one-dimensional patterns. *J. Mar. Sys.*, 69, 328–338.

Moktadir, Z., Kraft, M. and Wensick, H. (2008) Multifractal properties of Pyrex and silicon surfaces blasted with sharp particles. *Physica A*, 387, 2083–2090.

Monin, A.S. and Ozmidov, R. (1985) *Turbulence in the Ocean*. Reidel, Boston.

Monin, A.S. and Yaglom, A.M. (1975) *Statistical Fluid Mechanics: Mechanics of Turbulence*. MIT Press, Cambridge.

Montroll, E.W. and Badger, W.W. (1974) *Introduction to Quantitative Aspects of Social Phenomena*. Gordon and Breach Science, New York.

Montroll, E.W. and Shlesinger, M.F. (1982) On $1/f$ noise and other distributions with long tails. *Proc. Natl. Acad. Sci. USA*, 79, 3380–3383.

Morrisey, D., De Witt, T.H., Roper, D. and Williamson, R. (1999) Variation in the depth and morphology of the mud crab *Helice crassa* among different types of intertidal sediments in New Zealand. *Mar. Ecol. Prog. Ser.*, 182, 231–242.

Morse, D.R., Lawton, J.H., Dodson, M.M. and Williamson, M.H. (1985) Fractal dimension of vegetation and the distribution of arthropod body lengths. *Nature*, 314, 731–733.

Mouillot, D. and Viale, D. (2001) Satellite tracking of a fin whale (*Balaenoptera physalus*) in the northwestern Mediterranean Sea and fractal analysis of its trajectory. *Hydrobiology*, 452, 163–171.

Mountain, D.G. and Taylor, M.H. (1996) Fluorescence structure in the region of the tidal mixing front on the southern flank of Georges Bank. *Deep-Sea Res II*, 43, 1831–1853.

Muller, W. H. (1979) *Botany: A Functional Approach*. Macmillan, New York, 164.

Mulligan, R.F. (2004) Fractal analysis of highly volatile markets: An application to technology equities. *Q. Rev. Econ. Finance*, 44, 155–179.

Mullin, M.M., Stewart, E.F. and Fuglister, F.J. (1975) Ingestion by planktonic grazers as a function of concentration of food. *Limnol. Oceanogr.*, 20:259–262.

Mundt, M.D., Maguire, W.B. and Chase, R.R.P. (1991) Chaos in the sunspot cycle: Analysis and prediction. *J. Geophys. Res.*, 96, 1705–1716.

Muriel, S.B. and Grez, A.A. (2002) Effect of plant patch shape on the distribution and abundance of three lepidopteran species associated with *Brassica oleracea*. *Agric. For. Entomol.*, 4, 179–185.

Muzy, J.F., Sornette, D., Delour, J. and Arneodo, A. (2001) Multifractal returns and hierarchical portfolio theory. *Quant. Finance*, 1, 131–148.

Naeem, S. and Colwell, R.K. (1991) Ecological consequences of heterogeneity of consumable resources. In *Ecological Heterogeneity*, Kolasa, J. and Pickett, S.T.A., Eds. Springer-Verlag, New York, 224–255.

Nakao, H. (2000) Multi-scaling properties of truncated Lévy flights. *Phys. Lett. A*, 266, 282–289.

Nams, V.O. (1996) The V Fractal: A new estimator for fractal dimension of animal movement paths. *Landscape Ecol.*, 11, 289–297.

Nams, V.O. (2005) Using animal movement paths to measure a response to spatial scales. *Oecologia*, 143, 179–188.

Nams, V.O. and Bourgeois, M. (2004) Fractal analysis measures habitat use at different spatial scales: An example with American marten. *Can. J. Zool.*, 82, 1738–1747.

Newell, A.C., Nazarenko, S. and Biven, L. (2001) Wave turbulence and intermittency. *Physica D*, 152–153, 520–550.

Nielsen, K.L., Lynch, J., Jablokow, A.G. and Curtis, P.S. (1994) Carbon cost of root systems: An architectural approach. *Plant Soil*, 165, 161–169.

Nielsen, K.L., Lynch, J.P. and Weiss, H.N. (1997) Fractal geometry of bean root systems: Correlations between spatial and fractal dimension. *Am. J. Bot.*, 84, 26–33.

Nielsen, K.L., Miller, C.R., Beck, D. and Lynch, J.P. (1999) Fractal geometry of root systems: Field observations of contrasting genotypes of common bean (*Phaseolus vulgaris* L.) grown under different phosphorus regimes. *Plant Soil*, 206, 181–190.

Niklas, K.J. (1994) *Plant Allometry: The Scaling of Form and Process*. University of Chicago Press, Chicago.

Noever, D.A. (1993) Himalayan sandpiles. *Phys. Rev. E*, 47, 724–744.

Nomann, B.E. and Pennings, S.C. (1998) Fiddler crab-vegetation interactions in hypersaline habitats. *J. Exp. Mar. Biol. Ecol.*, 225, 53–68.

Nonnenmacher, T.F., Losa, G.A. and Weibel, E.R. (1994). *Fractals in Biology and Medicine*. Birkhäuser, Cambridge.

Normant, F. and Tricot, C. (1993) Fractal simplification of lines using convex hulls. *Geogr. Anal.*, 25, 118–129.

Nozdrin, Y. (1974) Influence of buoyancy forces on the spectra of turbulent processes in the ocean. *Oceanology*, 14, 255–277.

Nunes Amaral, L.A., Goldberger, A.L., Ivanov, P.Ch. and Stanley, H.E. (1998) Scale-independent measure and pathologic cardiac dynamics. *Phys. Rev. Lett.*, 81, 2388–2391.

Nychka, D., Ellner, S., Gallant, A.R. and McCarey, D. (1992) Finding chaos in noisy systems, *J. Royal Stat. Soc. B*, 54, 399–426.

Obert, M., Pfeifer, P. and Sernetz, M. (1990) Microbial growth patterns described by fractal geometry. *J. Bacteriol.*, 172, 1180–1185.

Obukhov, A.M. (1941) Spectral energy distribution in a turbulent flow. *Dokl. Akad. Nauk. SSSR*, 32, 22–24.

———. (1949) Structure of the temperature field in a turbulent flow. *Izv. Akad. Nauk. S.S.S.R., Geogr. Jeofiz.*, 13, 55.

———. (1962) Some specific features of atmospheric turbulence. *J. Fluid Mech.*, 13, 77–81.

Odum, E.P. (1971) *Fundamentals in Ecology*. Saunders, Philadelphia.

Okubo, A. (1980) *Diffusion and Ecological Problems: Mathematical Models*. Springer-Verlag, New York.

Olami, Z., Feder, K. and Christensen, K. (1992) Self-organized criticality in a continuous, nonconservative cellular automaton modeling earthquakes. *Phys. Rev. Lett.*, 68, 197–200.

Oleschko, K., Brambila, R., Brambila, F., Parrot, J.F. and López, P. (2000a) Fractal analysis of Teotihuacan, Mexico. *J. Archaeol. Sci.*, 27, 1007–1016.

Oleschko, K., Figueroa, S.B., Miranda, M.E., Vuelvas, M.A. and Solleiro, E.R. (2000b) Mass fractal dimensions and some selected physical properties of contrasting soils and sediments of Mexico. *Soil Tillage Res.*, 55, 43–61.

Oleschko, K., Fuentes, C., Brambilla, F. and Alvarez, R. (1997) Linear fractal analysis of three Mexican soils in different management systems. *Soil Technol.*, 10, 207–223.

Olsen, E.R., Ramsey, R.D. and Winn, D.S. (1993) A modified fractal dimension as a measure of landscape diversity. *Photogrammatic Eng. Remote Sens.*, 59, 1517–1520.

O'Neill, R.V., De Angelis, D.L., Waide, J.B. and Allen, T.F.H. (1986) A hierarchical concept of ecosystems. *Monogr. Popul. Biol.*, 23, 1–252.

Orrock, J.L., Danielson, B.J., Burns, M.J. and Levey, D.J. (2003) Spatial ecology of predator-prey interactions: Corridors and patch shape influence seed predation. *Ecology*, 84, 2589–2599.

Ott, E. (1993, 2002) *Chaos in Dynamical Systems*. Cambridge University Press, Cambridge.

Ott, F.S., Harris, R.P. and O'Hara, S.C.M. (1978) Acute and sublethal toxicity of naphthalene and three methylated derivatives to the estuarine copepod, *Eurytemora affinis*. *Mar. Environ. Res.*, 1, 49–58.

Ozier-Lafontaine, H., Lecompte, F. and Sillon, J.F. (1999) Fractal analysis of the root architecture of *Gliricidia sepium* for the spatial prediction of root branching, size and mass: Model development and evaluation in agroforestry. *Plant Soil*, 209, 167–180.

Ozik, J., Hunt, B.R. and Ott, E. (2005) Formation of multifractal population patterns from reproductive growth and local resettlement. *Phys. Rev. E*, 72, 1–15.

Paar, V., Cvitan, M., Ocelic, N. and Josipovic, M. (1997) Fractal dimension of coastline of the Coratian island Cres. *Acta Geogr. Croatica*, 32, 21–34.

Pachepsky, Y., Giménez, D. and Rawls, W.L. (2000a) Bibliography on applications of fractals in soil science. *Dev. Soil Sci.*, 27, 273–295.

———. (2000b) *Development in Soil Science*, Elsevier, Amsterdam.

Pachepsky, Y. and Timlin, D. (1998) Water transport in soils as in fractal media. *J. Hydrol.*, 204, 98–107.

Packard, N.H., Crutchfield, J.P., Farmer, J.D. and Shaw, R.S. (1980) Geometry from a time series. *Phys. Rev. Lett.*, 45, 712–716.

Paczuski, M., Maslov, S. and Bak, P. (1995) Avalanche dynamics in evolution, growth and depinning models. *Phys. Rev. E*, 53, 414–418.

Paladin, G. and Vulpiani, A. (1987) Anomalous scaling laws in multifractal objects. *Phys. Rep.*, 156, 147–225.

Palmer, M.W. (1988) Fractal geometry: A tool for describing spatial patterns of plant communities. *Plant Ecol.*, 75, 91–102.

Pareto, V. (1896) *Oeuvres Complètes*. Droz, Geneva.

Parisi, G. and Frisch, U. (1985) A multifractal model of intermittency. In *Turbulence and Predictability in Geophysical Fluid Dynamics and Climate Dynamics*, Ghil, M., Benzi, R. and Parisi, G., Eds. North-Holland, Amsterdam, 84–88.

Pascual, M., Ascitoti, F.A. and Caswell, H. (1995) Intermittency in the plankton: A multifractal analysis of zooplankton biomass variability. *J. Plankton Res.*, 17, 1209–1232.

Pascual, M. and Caswell, H. (1997) Environmental heterogeneity and biological pattern in a chaotic predator-prey system. *J. Theor. Biol.*, 185, 1–13.

Paterson, D.M. and Crawford, R.M. (1986) The structure of benthic diatoms assemblages: A preliminary account of the use and evaluation of low-temperature scanning electron microscopy. *J. Exp. Mar. Biol. Ecol.*, 96, 279–289.

Peitgen, H.O., Jurgens, H. and Saupe, D. (1992) *Chaos and Fractals: New Frontiers of Science*. Springer-Verlag, New York.

Peitgen, H.O. and Saupe, D. (1988) *The Science of Fractal Images*. Springer-Verlag, New York.

Peng, C.K., Buldyrev, S., Goldberger, A., Havlin, S., Sciortino, F., Simons, M. and Stanley, H.E. (1992) Long-range correlations in nucleotide sequences. *Nature*, 356, 168–171.

Peng, C.K., Buldyrev, S., Havlin, S., Simons, M., Stanley, H.E. and Goldberger, A.L. (1994) Mosaic organization of DNA nucleotides. *Phys. Rev. E*, 49, 1685–1689.

Peng, C.K., Mietus, J., Hausdorff, J., Havlin, S., Stanley, H.E. and Goldberger, A.L. (1993) Long-range anticorrelations and non-Gaussian behavior of the heartbeat. *Phys. Rev. Lett.* 70, 1343–1346.

Pennycuick, C.J. and Kline, N.C. (1986) Units of measurement for fractal extent, applied to the coastal distribution of bald eagle nests in the Aleutian Islands, Alaska. *Oecologia*, 68, 254–258.

Perfect, E. and Kay, B.D. (1991) Fractal theory applied to soil aggregation. *Soil Sci. Soc. Am.*, 55, 1552–1558.

Perfect, E., Kay, B.V. and Rasiah, V. (1994) Unbiased estimation of the fractal dimension of soil aggregate size distributions. *Soil Tillage Res.*, 31, 187–198.

Perfect, E., Rasiah, V. and Kay, B.V. (1992) Fractal dimensions of soil aggregate-size distributions calculated by number and mass. *Soil Sci. Soc. Am.*, 56, 1407–1409.

Perline, R. (1996) Zipf's law, the central limit theorem, and the random division of the unit interval. *Phys. Rev. E*, 54, 220–223.

Perrier, E., Rieu, M., Sposito, G. and de Marsily, G. (1996) A computer model of the water retention curve for soils with a fractal pore size distribution. *Water Resour. Res.*, 32, 3025–3031.

Perrin, J. (1906) La discontinuité de la matière. *Revue du mois*, 1, 323–344.

Perrin, J. (1913) *Les Atomes*. Alcan, Paris.

Pershing, A.J., Wiebe, P.H., Manning, J.P. and Copley, N.J. (2001) Evidence for vertical circulation cells in the well-mixed area of Georges Bank and their biological implications. *Deep-Sea Res. II*, 48, 283–310.

Peta, O., Hitier, B., Olivesi, R., Delesmont, R., Morel, M. and Loquet, N (1998) *Suivi régional des nutriments sur le littoral nord/Pas-de-Calais/Picardie*. Bilan de l'année 1997. IFREMER, Boulogne-sur-Mer.

Peters, E.E. (1994) *Fractal Market Analysis: Applying Chaos Theory to Investment and Economics*. J. Wiley and Sons, New York.

Peters, H.H. (1983) *The Ecological Implications of Body Size*. Cambridge University Press, Cambridge.

Peterson, D.L. and Clarke, V.T. (1998) *Ecological Scales: Theory and Applications*. Columbia University Press, New York.

Peterson, D.L. and Parker, V.T. (1998) *Ecological Scale*. Columbia University Press, New York.

Petren, K. and Case, T.J. (1998) Habitat structure determines competition intensity and invasion success in gecko lizards. *Proc. Natl. Acad. Sci. USA*, 95, 11739–11744.

Pettit, F.K. and Bowie, J.U. (1999) Protein surface roughness and small molecular binding sites. *J. Mol. Biol.*, 285, 1377–1382.

Pfeifer, P. and Avnir, D. (1983) Chemistry in non integer dimensions between two and three. I: Fractal theory of heterogeneous surfaces. *J. Chem. Phys.*, 79, 3558–3564.

Pfeifer, P. and Obert, M. (1989) Fractals: Basic concepts and terminology. In *The Fractal Approach to Heterogeneous Chemistry*, Avnir, D., Ed. Wiley, Chichester, 11–40.

Pfeifer, P., Welz, U. and Wippermann, H. (1985) Fractal surface dimension of proteins: Lysozyme. *Chem. Phys. Lett.*, 113, 535–540.

Philip, S.R. (1994) Structural and lithological controls on coastline profiles in Fife, eastern Britain. *Terra Res.*, 6, 251–254.

Phillips, J.D. (1985) Measuring complexity of environmental gradients. *Vegetatio*, 64, 95–102.

Phillips, M.L., Clark, W.R., Nusser, S.M., Sovada, M.A. and Greenwood, R.J. (2004) Analysis of predator movement in prairie landscapes with contrasting grassland composition. *J. Mammal.*, 85, 187–195.

Pianka, E.R. (1966) Latitudinal gradients in species diversity: A review of concepts. *Am. Nat.*, 100, 33.

————. (1974) Niche overlap and diffuse competition. *Proc. Natl. Acad. Sci. USA*, 71, 2141–2145.

————. (1988) *Evolutionary Ecology*. Harper Collins, New York.

Piao, M., Okamura, H., Luo, R. and Aoyama, I. (2000) Interactive effect of heavy metals and pesticide on *Daphnia magna*. *Jpn. J. Environ. Toxicol.*, 3, 23–32.

Pielou, E.C. (1966) Species-diversity and pattern-diversity in the study of ecological succession. *J. Theor. Biol.*, 10, 370–383.

————. (1975) *Ecological Diversity*. Wiley, New York.

Pikovsky, A.S. (1984) On the interaction of strange attractors. *Z. Phys. B: Cond. Matt.*, 55, 149–154.

Platt, N., Spiegel, E.A. and Tresser, C. (1993) On-off intermittency: A mechanism for bursting. *Phys. Rev. Lett.*, 70, 279–282.

Platt, T. (1972) Local phytoplankton abundance and turbulence. *Deep-Sea Res.*, 19, 183–187.

Platt, T. and Denman, K.L. (1975) Spectral analysis in ecology. *Ann. Rev. Ecol. Syst.*, 6, 189–210.

Platt, T., Harrison, W.G., Lewis, M.R., Li, W.K.W., Santhyendranath, S., Smith, R.E. and Vezina, A.F. (1989) Biological production of the oceans: The case for a consensus. *Mar. Ecol. Prog. Ser.*, 52, 77–88.

Polidori, L., Chorowicz, J.J. and Guillande, R. (1991) Description of terrain as a fractal surface, and application to digital elevation model quality assessment. *Photogrammatic Eng. Remote Sensing*, 57, 1329–1332.

Pomeau, L.R. and Manneville, P. (1980) Intermittent transition to turbulence in dissipative dynamical systems. *Comm. Math. Phys.*, 74, 189–197.

Pomeroy, L.R. (1980) Detritus and its role as a food source. In *Fundamentals of Aquatic Ecosystems*, Barnes, R.K. and Mann, K.H., Eds. Blackwell, Oxford, 84–102.

Pond, S. and Pickard, G.L. (1983) *Introductory Dynamical Oceanography*. Butterworth-Heineman, Oxford.

Pool, R. (1989) Is it chaos or is it just a noise? *Science*, 243, 310–313.

Pope, S.B. (2000) *Turbulent Flows*. Cambridge University Press, Cambridge.

Popova, E.E., Fasham, M.J.R., Osipov, A.V. and Ryabchenko, V.A. (1997) Chaotic behavior of an ocean eco-system model under seasonal external forcing. *J. Plankton Res.*, 19, 1495–1515.

Popper, K. (2002) *The Logic of Scientific Discovery*. Routledge, London.

Porter, J.R. and Semenov, M.A. (1999) Climate variability and crop yields in Europe. *Nature*, 400, 724.

Powell, T.M. and Okubo, A. (1994) Turbulence, diffusion and patchiness in the sea. *Philos. Trans. R. Soc. London B*, 343, 11–18.

Powell, T.M., Richerson, P.J., Dillon, T.M., Agee, B.A., Dozier, B.J., Godden, D.A. and Myrup, L.O. (1975) Spatial scales of current speed and phytoplankton biomass fluctuations in Lake Tahoe. *Science*, 189, 1088–1089.

Priestley, M.B. (1980) State dependent models: A general approach to nonlinear time series analysis. *J. Time Ser. Anal.*, 1, 47–71.

Provata, A. and Almirantis, Y. (2000) Fractal cantor patterns in the sequence structure of DNA. *Fractals*, 8, 15–27.

Puche, H. and Su, N. (2001) Application of fractal analysis for tunnel systems of subterranean termites (Isoptera: Rhinotermitidae) under laboratory conditions. *J. Mammal.*, 30, 545–549.

Pyke, J.H. (1984) Optimal foraging theory: A critical review. *Ann. Rev. Ecol. Syst.*, 15, 523–575.

Quenette, P.Y. and Desportes, J.P. (1992) Temporal and sequential structures of vigilance behaviour of wild boars (*Sus scrofa*). *J. Mammal.*, 73, 535–540.

Rahbek, C. and Graves, G.R. (2001) Multiscale assessment of patterns of avian species richness. *Proc. Natl. Acad. Sci. USA*, 98, 4534–4539.

Rainey, P.B. and Travisano, M. (1998), Adaptative radiation in a heterogeneous environment. *Nature*, 394, 69–72.

Ramos-Fernández, G., Mateos, J.L., Miramontes, O., Cocho, G., Larralde, H. and Ayala-Orozco, B. (2004) Lévy walk patterns in the foraging movements of spider monkeys (*Ateles geoffroyi*). *Behav. Ecol. Sociobiol.*, 55, 223–230.

Ramsden, J.J. and Vohradsky, J. (1998) Zipf-like behavior in procaryotic protein expression. *Phys. Rev. E*, 58, 7777–7780.

Ramsey, J.B. and Yuan, H.J. (1989) Bias and error bars in dimension calculations and their evaluations in some simple models. *Phys. Lett. A*, 134, 287–297.

Rand, D.A. (1994) Measuring and characterizing spatial patterns, dynamics and chaos in spatially extended dynamical systems and ecologies. *Philos. Trans. R. Soc. London Ser. A*, 348, 497–514.

Raper, J.A. and Amal, R. (1993) Measurement of aggregate fractal dimension using static light scattering. *Part. Syst. Charact.*, 10, 239–245.

Rasband, S.N. (1990) *Chaotic Dynamics of Nonlinear Systems*. Wiley, New York.

Rasiah, V., Kay, B.D. and Perfect, E. (1992) Evaluation of selected factors influencing aggregate fragmentation using fractal theory. *Can. J. Soil Sci.*, 72, 97–106.

———. (1993) New mass-based model for estimating fractal dimension of soil aggregates. *Soil Sci. Soc. Am. J.*, 57, 891–895.

Raup, D.M. (1982) Mass extinctions in the marine fossil record. *Science*, 1501–1503.

Rawls, W.J., Brakensiek, D.L. and Logsdon, S.D. (1993) Predicting saturated hydraulic conductivity utilizing fractal principles. *Soil Sci. Soc. Am. J.*, 57, 1193–1197.

Reiger, G. (1983) Mysterious *Mola mola. Sea Frontiers*, 29, 367–371.

Rényi, A. 1970. *Probability Theory*. North-Holland, Amsterdam.

Rex, K.D. and Malanson, G.P. (1990) The fractal shape of riparian forest patches. *Landscape Ecol.*, 4, 249–258.

Reynolds, A.M. (2006) Cooperative random Lévy flight searches and the flight patterns of honeybees. *Phys. Lett. A*, 354, 384–388.

Reynolds, A.M. and Frye, M.A. (2007) Free-odor tracking in Drosophila is consistent with an optimal intermittent scale-free search. *PlosONE*, 2, e354.

Reynolds, A.M., Smith, A.D., Reynolds, D.R., Carreck, N.L. and Osborne, J.L. (2007) Honeybees perform optimal scale-free searching flights when attempting to locate a food source. *J. Exp. Biol.*, 210, 3763–3770.

Riaux-Gobin, C. (1997) Microphytobenthos. In: Dauvin, J.-C. (Ed.), Les biocénoses marines et littorales françaises des côtes Atlantique, Manche et Mer du Nord: Synthèses, menaces et perspectives. Collection patrimoines naturels, vol. 28. M.N.H.N., Paris, 103–111.

Riaux-Gobin, C. and Bourgoin, P. (2002) Microphytobenthos biomass at Kerguelen's Land (Subantarctic Indian Ocean): Repartition and variability during austral summers. *J. Mar. Syst.*, 32, 295–306.

Richards, F.M. (1977) Areas, volumes, packing, and protein structure. *Ann. Rev. Biophys. Bioengin.*, 6, 151–176.

Richards, G.R. (2000) The fractal structure of exchange rates: Measurement and forecasting. *J. Intl. Financial Markets, Inst. Money*, 10, 163–180.

Richardson, L.F. (1922) *Weather Prediction by Numerical Processes*. Cambridge University Press, Cambridge.

Ricklefs, R.E. and Schluter, D. (1993) *Species Diversity in Ecological Communities: Historical and Geographical Perspectives*. University of Chicago Press, Chicago.

Ricotta, C. (2000) From theoretical ecology to statistical physics and back: Self-similar landscape metrics as a synthesis of ecological diversity and geometrical complexity. *Ecol. Model.*, 125, 245–253.

Ricotta, C., Pacini, A. and Avena, G. (2002) Parametric scaling from species to growth-form diversity: An interesting analogy with multifractal functions. *BioSystems*, 65, 179–186.

Riebesel, U. and Wolf-Gladrow, D.A. (1992) The relationship between physical aggregation of phytoplankton particle flux: A numerical model. *Deep-Sea Res. A*, 39, 1085–1102.

Riethmüller, R., Heineke, M., Kühl, H. and Keuker-Rüdiger, R. (2000) Chlorophyll *a* concentration as an index of sediment surface stabilisation by microphytobenthos? *Cont. Shelf Res.*, 20, 1351–1372.

Rieu, M. and Sposito, G. (1991) Fractal fragmentation, soil porosity, and soil water properties. *Soil Sci. Soc. Am, J.*, 55, 1231–1244.

Rigaut, J.P. (1991) *Fractals, Non-Integral Dimensions and Applications*. John Wiley and Sons, Chichester.

Rigon, R., Rinaldo, A. and Rodriguez-Iturbe, I. (1994) On landscape self-organization. *J. Geophys. Res.*, 99, 11971–11987.

Rinaldo, A., Maritan, A., Colaiori, F., Flammini, A., Rigon, R., Ignacio, I., Rodriguez-Iturbe, I. and Banavan, J.R. (1996) Thermodynamics of fractal river networks. *Phys. Rev. Lett.*, 76, 3364–3367.

Risch, S.J. and Carroll, C.R. (1986) Effects of seed predation by a tropical ant on competition among weeds. *Ecology*, 67, 1319–1327.

Ritchie, M.E. (1998) Scale-dependent foraging and patch choice in fractal environments. *Evol. Biol.*, 12, 309–330.

Ritz, K. and Crawford, J. (1990) Quantification of the fractal nature of colonies of *Trichoderma viride. Mycol. Res.*, 94, 1138–1141.

Robertson, M.C., Sammis, C.G., Sahimi, M. and Martin, A.J. (1995) Fractal analysis of three-dimensional spatial distributions of earthquakes with a percolation interpretation, *J. Geophys. Res.*, 100, 609–620.

Rodriguez-Iturbe, I., de Power, B.F., Sharifi, M.B. and Georgakakos, K.B. (1989) Chaos in rainfall. *Water Resour. Res.*, 25, 1667–1675.

Roemich, D. and McGowan, J.A. (1995) Climatic warming and the decline of zooplankton in the California current. *Science*, 267, 1324–1326.

Rogers, C.A. (1970) *Hausdorff Measures*. Cambridge University Press, Cambridge.

Romañach, S.S. and Le Comber, S.C. (2004) Measures of pocket gopher (*Thomomys bottae*) burrow geometry: Correlated of fractal dimension. *J. Zool.*, 262, 399–403.

Romañach, S.S., Reichman, O.J., Rogers, W.E. and Cameron, G.N. (2005) Effects of species, gender, age and habitat on pocket gopher foraging tunnel geometry. *J. Mammal.*, 86, 750–756.

Rosenzweig, M.L. (1995) *Species Diversity in Space and Time.* Cambridge University Press, Cambridge.

Rössler, O.E. (1976) An equation for continuous chaos. *Phys. Lett.*, 57A, 397–398.

Rothman, D.H., Grotzinger, J.P. and Flemings, P. (1994) Scaling in turbidite deposition. *J. Sed. Res. A*, 64, 355–359.

Rothschild, B.J. and Osborn, T.R. (1988) Small-scale turbulence and plankton contact rates. *J. Plankton Res.*, 10, 465–474.

Roughgarden, J. (1974) Population dynamics in a spatially varying environment: How population size tracks spatial variation in carrying capacity. *Am. Nat.*, 108, 649–664.

Rouse, H. and Ince, S. (1963) *History of Hydraulics.* Dover, New York.

Rowe, G. and Harvey, I. (1985) Information content in finite sequences: Communication between dragonfly larvae. *J. Theor. Biol.*, 116, 275–290.

Rubin, D.M. (1992) Use of forecasting signatures to help distinguishing periodicity, randomness and chaos in ripples and other spatial patterns. *Chaos*, 2, 525–535.

Ruelle, D. and Takens, F. (1971) On the nature of turbulence. *Communs. Math. Phys.*, 20, 167–192.

Russell, R.W., Hunt, G.L., Coyle, K.O. and Cooney, R.T. (1992) Foraging in a fractal environment: spatial patterns in a marine predator-prey system. *Landscape Ecol.*, 7, 195–209.

Saiz, E., Tiselius, P., Jonsson, P.R., Verity, P. and Paffenhöfer, G.A. (1993) Experimental records of the effects of food patchiness and predation on egg production of *Acartia tonsa. Limnol. Oceanogr.*, 38, 280–289.

Salas, E., Ozier-Lafontaine, H. and Nygren, P. (2004) A fractal root model applied for estimating the root biomass and architecture in two tropical legume tree species. *Ann. For. Sci.*, 61, 337–345.

Samain, J.F., Moal, J., Alayse Danet, A.M., Daniel, J.Y. and Le Coz, J.R. (1981) Model for rapid detection of sublethal effects of pollutants. 2: An *in situ* example: Anomalous metabolism of the hyponeustonic copepod *Anomaloeem patersoni* coinciding with an oil spill. *Mar. Biol.*, 64, 35–41.

Schaefer, D.W. (1989) Polymers, fractals, and ceramic materials. *Science*, 243, 1023–1027.

Schaefer, D.W., Martin, J.E., Wiltzius, P. and Cannell, D.S. (1984) Fractal geometry of colloidal aggregates. *Phys. Rev. Lett.*, 52, 2371–2374.

Schäfer, A. and Teyssen, T. (1987) Size, shape and orientation of grains in sands and sandstones: Image analysis applied to rock thin sections. *Sed. Geol.*, 52, 251–271.

Schapira, M., Buscot, M.J., Leterme, S.C., Pollet, T., Chapperon, C. and Seuront, L. (2009) Distribution of heterotrophic bacteria and virus-like particles along a salinity gradient in a hypersaline coastal lagoon. *Aquatic Microbial Ecology*, 54, 171–183.

Scheffer, M. (1991) Should we expect strange attractors behind plankton dynamics—And if so, should we bother? *J. Plankton Res.*, 13, 1291–1305.

Schertzer, D. and Lovejoy, S. (1983) The dimension and intermittency of atmospheric dynamics. In *Turbulent Shear Flows 4*, Launder, B., Ed. Springer-Verlag, Karlsruhe, 7–33.

———. (1987) Physically based rain and cloud modeling by anisotropic multiplicative turbulent cascades. *J. Geophys. Res.*, 92, 9693–9714.

Schertzer, D., Lovejoy, S., Schmitt, F., Chigirinskya, Y. and Marsan, D. (1998) Multifractal cascade dynamics and turbulent intermittency. *Fractals*, 5, 427–471.

Scheuring, I. (1991) The fractal nature of vegetation and the species-area relation. *Theor. Popul. Biol.*, 39, 170–177.

Scheuring, I. and Riedi, R.H. (1994) Application of multifractals to the analysis of vegetation pattern. *J. Vegetation Sci.*, 5, 489–496.

Schindell, D., Rind, D., Balachandran, N. and Lonergan. P. (1999) Solar cycle variability, ozone, and climate. *Science*, 285, 305–308.

Schinner, A. (2007) The Voynich manuscript: Evidence of the hoax hypothesis. *Cryptologia*, 31, 95–107.

Schluter D. and Ricklefs, R.E. (1993) Species diversity in ecological communities. Historical and geographical perspectives. University of Chicago Press, Chicago.

Schmid, B. and Harper, J.L. (1985) Clonal growth in grassland perennials. I: Density and pattern-dependent competition between plants with different growth forms. *J. Ecol.*, 73, 793–808.

Schmidt, P.W. (1989) Use of scattering to determine the fractal dimension. In *The Fractal Approach to Heterogeneous Chemistry: Surfaces, Colloids, Polymers*, Avnir, D., Ed. Wiley and Sons, Chichester, 67–78.

Schmidt-Nielsen, K. (1984) *Scaling, Why Animal Size Is So Important?* Cambridge University Press, Cambridge.

Schmitt, F.G. and Seuront, L. (2001) Multifractal random walk in copepod behavior. *Physica A*, 301, 375–396.

Schmittbuhl, J., Violette, J.P. and Roux, S. (1995) Reliability of self-affine measurements. *Phys. Rev. E*, 51, 131–147.

Schonauer, D. and Kreibig, U. (1985) Topography of samples with variably aggregated metal particles. *Surf. Sci.*, 156, 100–111.

Schopf, T.J.M. (1970) Taxonomic diversity gradients of ectoprocts and bivalves and their geologic implications. *Geol. Soc. Am. Bull.*, 81, 3765–3768.

Schroeder, M. (1991) *Fractals, Chaos, Power Laws*. Freeman, New York.

Schuler, J., Frank, J. Saenger, W. and Georgalis, Y. (1999) Thermally induced aggregation of human transference receptor studied by light-scattering techniques. *Biophys. J.*, 77, 1117–1125.

Schuster, F.L. and Levandowsky, M. (1996) Chemosensory responses of *Acanthamoeba castellanii*: Visual analysis of random movement and responses to chemical signals. *J. Eur. Microbiol.*, 43, 150–158.

Schwartz, F.J. and Lindquist, D.G. (1987) Observations on *Mola* basking behavior, parasites, echeneidid associations, and body-organ weight relationships. *J. Elisha Mitchell Sci. Soc.*, 103, 14–20.

Sergeev, V. and Williams, W.D. (1985) *Daphniopsis australis* nov.sp. (Crustacea: Cladocera), a further daphniid in Australian salt lakes. *Hydrobiology*, 120, 119–128

Serôdio, J. and Catarino, F. (2000) Modelling the primary productivity of intertidal microphytobenthos: Time scales of variability and effects of migratory rhythms. *Mar. Ecol. Prog. Ser.*, 192, 13–30.

Serôdio, J., da Silva, J.M. and Catarino, F. (1997) Non-destructive tracing of migratory rhythms of intertidal benthic microalgae using *in vivo* chl-*a* fluorescence. *J. Phycol.*, 33, 545–553.

Serra, J. (1968) Les structures gigognes: Morphologie mathematique et interpretation metallogenique. *Mineralium Deposita*, 3, 135–154.

Seuront, L. (1998) Fractals and multifractals: New tools to characterize space-time heterogeneity in marine ecology. *Océanis*, 24, 123–158.

———. (1999) Space-Time Heterogeneity in Pelagic Ecology: Implications on Carbon Fluxes. Ph.D. thesis, Université des Sciences et Technologies, Lille, France.

———. (2004) Small-scale turbulence in the plankton: Low-order deterministic chaos or high-order stochasticity? *Physica A*, 341, 495–525.

———. (2005a) Systemic Approach of Microscale Patterns and Processes in Marine Ecology. D.Sc. Thesis, University of Sciences and Technologies, Lille, France.

———. (2005b) Hydrodynamic and tidal controls of small-scale phytoplankton patchiness. *Mar. Ecol. Prog. Ser.*, 302, 93–101.

Seuront, L., Yamazaki, H. and Schmitt, F. (2005) Intermittency. In: Baumert, H., Sundermann, J. and Simpson, J. (Eds.), *Marine Turbulences: Theories, Observations and Models*. Cambridge University Press, Cambridge, 66–78.

———. (2005c) First record of the calanoid copepod *Acartia omorii* (Copepoda: Calanoida: Acartiidae) in the southern bight of the North Sea. *J. Plankton Res.*, 27, 1301–1306.

———. (2006) Effect of salinity on the swimming behaviour of the estuarine calanoid copepod *Eurytemora affinis*. *J. Plankton Res.*, 28, 805–813.

Seuront, L. (2008) Microscale complexity in the ocean: Turbulence, intermittency and plankton life. *Mathematical Modelling of Natural Phenomena*, 3, 1–41.

Seuront, L., Brewer, M. and Strickler, J.R. (2004a) Quantifying zooplankton swimming behavior: The question of scale. In *Handbook of Scaling Methods in Aquatic Ecology: Measurement, Analysis, Simulation*, Seuront, L. and Strutton, P.G., Eds. CRC Press, Boca Raton, 333–359.

Seuront, L. and Strutton, P.G. (2004b) *Handbook of Scaling Methods in Aquatic Ecology: Measurement, Analysis, Simulation*. CRC Press, Boca Raton.

Seuront, L. and Schmitt, F.G. (2004c) Eulerian and Lagrangian properties of biophysical intermittency in the ocean. Geophys. *Res. Lett.*, 31, L03306, doi:10.1029/2003GL018185.

Seuront, L., Devreyker, D. and Flamme, G. 2003. An ocean sunfish in the eastern English Channel. *Bull. Coord. Mammol.*, 10, 8.

Seuront, L., Duponchel, A.C. and Chapperon, C. (2007) Heavy-tailed distributions in the intermittent motion behaviour of the intertidal gastropod *Littorina littorea*. *Physica A*, 385, 573–582.

Seuront, L., Gentilhomme, V. and Lagadeuc, Y. (2002) Small-scale nutrient patches in tidally mixed coastal waters. *Mar. Ecol. Prog. Ser.*, 232, 29–44.

Seuront, L., Hwang, J.S., Tseng, L.C., Schmitt, F.G., Souissi, S., Shih, C.T. and Wong, C.K. (2004b) Individual variability in the swimming behavior of the tropical copepod *Oncaea venusta* (Copepoda: Poecilostomatoida). *Mar. Ecol. Prog. Ser.*, 283, 199–217.

Seuront, L. and Lagadeuc, Y. (1997) Characterisation of space-time variability in stratified and mixed coastal waters (Baie des Chaleurs, Québec, Canada): Application of fractal theory. *Mar. Ecol. Prog. Ser.*, 159, 81–95.

————. (1998) Spatio-temporal structure of tidally mixed coastal waters: Variability and heterogeneity. *J. Plankton Res.*, 20, 1387–1401.

————. (2001) Multiscale patchiness of the calanoid copepod *Temora longicornis* in a turbulent coastal sea. *J. Plankton Res.*, 23, 1137–1145.

Seuront, L. and Leterme, S.C. (2006) Microscale patchiness in microphytobenthos distributions: Evidence for a critical state. In *Functioning of Microphytobenthos in Estuaries*, Kromkamp, J.C., de Brouwer, J., Blanchard, G.F., Forster, R.M. and Créach, V., Eds. Academy of Arts and Sciences, Amsterdam, 165–183.

————. (2007) Increased zooplankton behavioral stress in response to short-term exposure to hydrocarbon contamination. *Open Oceanogr. J.*, 1, 1–7.

Seuront, L. and Mitchell, J.G. (2008) Towards a seascape typology. I: Zipf versus Pareto laws. *J. Mar. Sys.*, 69, 310–327.

Seuront, L. and Schmitt, F.G. (2005a) Multiscaling statistical procedures for the exploration of biophysical couplings in intermittent turbulence. I: Theory. *Deep-Sea Res. II*, 52, 1308–1324.

————. (2005b) Multiscaling statistical procedures for the exploration of biophysical couplings in intermittent turbulence. II: Applications. *Deep-Sea Res. II*, 52, 1325–1343.

Seuront, L., Schmitt, F.G., Brewer, M.C., Strickler, J.R. and Souissi, S. (2004d) From random walk to multifractal random walk in zooplankton swimming behavior. *Zool. Stud.*, 43, 8–19.

Seuront, L., Schmitt, F. and Lagadeuc, Y. (2001) Turbulence intermittency, small-scale phytoplankton patchiness and encounter rates in plankton: Where do we go from here? *Deep-Sea Res. I*, 48, 1199–1215.

Seuront, L., Schmitt, F., Lagadeuc, Y., Schertzer, D. and Lovejoy, S. (1999) Multifractal analysis as a tool to characterize multiscale inhomogeneous patterns: Example of phytoplankton distribution in turbulent coastal waters. *J. Plankton Res.*, 21, 877–922.

Seuront, L., Schmitt, F., Lagadeuc, Y., Schertzer, D., Lovejoy, S. and Frontier, S. (1996a) Multifractal structure of phytoplankton biomass and temperature in the ocean. *Geophys. Res. Lett.*, 23, 3591–3594.

Seuront, L., Schmitt, F., Schertzer, S., Lagadeuc, Y. and Lovejoy, S. (1996b) Multifractal intermittency of Eulerian and Lagrangian turbulence of ocean temperature and plankton fields. *Nonlinear Proc. Geophys.*, 3, 236–246.

Seuront, L. and Spilmont, N. (2002) Self-organized criticality in intertidal microphytobenthos patch patterns. *Physica A*, 313, 513–539.

Seuront, L., Vincent, D. and Mitchell, J.G. (2006) Biologically induced modification of seawater viscosity in the eastern English Channel during a *Phaeocystis globosa* spring bloom. *J. Mar. Sys.*, 61, 118–133.

Seuront, L., Yamazaki, H. and Souissi, S. (2004c). Hydrodynamic disturbance and zooplankton swimming behavior. *Zool. Stud.*, 43, 376–387.

Seymour, J.R., Mitchell, J.G., Pearson, L. and Waters, R.L. (2000) Heterogeneity in bacterioplankton abundance from 4.5 millimetre resolution sampling. *Aquat. Microb. Ecol.*, 22, 143–153.

Seymour, J.R., Mitchell, J.G. and Seuront, L. (2004) Microscale heterogeneity in the activity of coastal bacterioplankton communities. *Aquat. Microb. Ecol.*, 35, 1–16.

Seymour, J.R., Seuront, L., Doubell, M.J. and Mitchell, J.G. (2008) Microscale variability of bacteria and viruses during a *Phaeocystis globosa* bloom in the eastern English Channel. *Est. Coast. Shelf Sci.*, in press.

Seymour, J.R., Seuront L. and Mitchell, J.G. (2005) Microscale and small-scale temporal dynamics of a coastal planktonic microbial community. *Mar. Ecol. Prog. Ser.*, 300, 21–37.

————. (2007) Microscale gradients of planktonic microbial communities above the sediment surface in a mangrove estuary. *Est. Coast. Shelf Sci.*, 73, 651–666.

Seymour, J.R., Seuront, L., Doubell, M.J. and Mitchell, J.G. (2008) Mesoscale and microscale spatial variability of bacteria and viruses during a *Phaeocystis globosa* bloom in the Eastern English Channel. *Est. Coast. Shelf Sci.*, 80, 589–597.

Shannon, C.E. (1948) A mathematical theory of communication. *Bell Syst. Tech. J.*, 27, 379–423.

————. (1951) Prediction and entropy of printed English. *Bell Syst. Tech. J.*, 30, 50–64.

Shannon, C.E. and Weaver, W. (1963) *The Mathematical Theory of Communication*. University of Illinois Press, Urbana.

Shashack, M. and Brand, S. (1991) Relations among spatio-temporal heterogeneity, population abundance, and variability in a desert. In *Ecological Heterogeneity*, Kolasa, J. and Pickett, S.T.A., Eds. Springer-Verlag, New York, 202–223.

She, Z.S. and Waymire, E.C. (1994) Quantized energy cascade and log-Poisson statistics in fully developed turbulence. *Phys. Rev. Lett.*, 74, 262–265.

Shimizu, N., Ogino, C., Kawanishi, T. and Hayashi, Y. (2002) Fractal analysis of *Daphnia* motion for acute toxicity bioassay. *Environ. Toxicol.*, 17, 441–448.

Shlesinger, M.F. (1986) Lévy walks versus Lévy flights. In *On Growth and Form: Fractal and Non-Fractal Patterns in Physics*, Stanley, H.E. and Ostrowsky, N., Eds. Vol. 283. Nijhoff, Dordrecht, 279–283.

Shlesinger, M.F., Zaslavsky, G.M. and Frish, U. (Eds., 1996), *Lévy Flights and Related Topics in Physics*. Springer, Berlin.

Shlyakhter, I., Rozenoer, M., Dorsey, J. and Tellier, S. (2001) Reconstructing 3D tree models from instrumented photographs. *IEEE Comput. Graphics Appl.*, 21, 53–61.

Shorrocks, B., Marsters, J., Ward, I. and Evennett, P.J. (1991) The fractal dimension of lichens and the distribution of arthropod body lengths. *Functional Ecol.*, 5, 457–460.

Shraiman, B.I. and Siggia, E.D. (2000) Scalar turbulence. *Nature*, 405, 639–646.

Siegel, S. and Castellan, N.J. (1988) *Nonparametric Statistics*. McGraw-Hill, New York.

Siggia, E.D. (1991) Numerical study of small scale intermittency in three dimensional turbulence. *J. Fluid Mech.*, 107, 375–406.

Simonsen, I. (2003) Measuring anti-correlations in the Nordic electricity spot market by wavelets. *Physica A*, 322, 597–606.

Simonsen, I., Hansen, A., and Nes, O. (1998) Determination of the Hurst exponent by use of wavelet transforms. *Phys. Rev. E*, 58, 2779–2787.

Sims, D.W. and Southall, E.J. (2002) Occurrence of ocean sunfish, *Mola mola*, near fronts in the western English Channel. *J. Mar. Biol. Ass. UK*, 82, 927–928.

Sims, D.W., Southall, E.J., Humphries, N.E., Hays, G.C., Bradshaw, C.J.A., Pitchford, J.W., James, A., Ahmed, M.Z., Brierley, A.S, Hindell, M.A., Morritt, D., Musyl, M.K., Shepard, E.L.C., Wearmouth, V.J., Wilson, R.P., Witt, M.J. and Metcalfe, J.D. (2008) Scaling laws of marine predator search behavior. *Nature*, 451, 1098–1103.

Slater, P.J.B. (1973) Describing sequences of behavior. In *Perspectives in Ethology*, Bateson, P.P.G. and Klopfer, P.H., Eds. Vol. I, 131–153. Plenum, New York.

Smith, L.A. (1988) Intrinsic limits on dimension calculations. *Phys. Lett. A*, 133, 283–288.

———. (2007) *Chaos: A Very Short Introduction*. Oxford University Press, Oxford.

Smith, R.C., Zhang, X. and Michaelsen, J. (1988) Variability of pigment biomass in the California current system as determined by satellite imagery. 1: Spatial variability. *J. Geophys. Res.*, 93, 10863–10882.

Smith, T., Boto, K.G., Frusher, S. and Giddens, R. (1991) Keystone species and mangrove forest dynamics: The influence of burrowing by crabs on soil nutrient status and forest productivity. *Est. Coast. Shelf. Sci.*, 33, 419–432.

Snegirev, A.Yu., Makhviladze, G.M. and Roberts, J.P. (2001) The effects of particle coagulation and fractal structure on the optical properties and detection of smoke. *Fire Safety J.*, 36, 73–95.

Snover, M.L. and Commito, J.A. (1998) The fractal geometry of *Mytilus edulis* L. spatial distribution in a soft-bottom system. *J. Exp. Mar. Biol. Ecol.*, 223, 53–64.

Sokal, R.R. and Rohlf, F.J. (1995) *Biometry: The Principles and Practice of Statistics in Biological Research*. Freeman, San Francisco.

Solé, R.V. and Bascompte, J. (1995) Measuring chaos from spatial informations. *J. Theor. Biol.*, 175, 139–147.

Solé, R.V., Lopez, D., Ginovart, M. and Valls, J. (1992) Self-organized criticality in Monte Carlo simulated ecosystems. *Phys. Lett. A*, 172, 56–61.

Solé, R.V. and Manrubia, S.C. (1995a) Self-similarity in rain forests: Evidence for a critical state. *Phys. Rev. E*, 51, 6250–6253.

Solé, R.V. and Manrubia, S.C. (1995a) Are rainforest self-organized in a critical state? *J. Theor. Biol.*, 172, 31–40.

Solé, R.V. and Manrubia, S.C. (1995b) Self-similarity in rainforests: Evidence for a critical state. *Phys. Rev. E*, 51, 6250–6253.

Somfai, E., Czirók, A. and Vicsek, T. (1994a) Self-affine roughening in a model experiment on erosion in geomorphology. *J Physica A*, 205, 355–366.

———. (1994b) Power-law distribution of landslides in an experiment on the erosion of a granular pile. *J. Phys. a*, 27, 757–760.

Sorriso-Valvo, L., Carbone, V., Guiliani, P., Veltri, P., Bruno, R., Antoni, V. and Martines, E. (2001) Intermittency in plasma turbulence. *Planetary Space Sci.*, 49, 1177–1191.

Spek, L.Y. and van Noordwijk, M. (1994) Proximal root diameters as predictors of total root system size for fractal branching models. II: Numerical model. *Plant Soil*, 164, 119–127.

Sprott, J.C. (2003) *Chaos and Time-Series Analysis*. Oxford University Press, Oxford.

Stach, S., Cybo, J. and Chmiela, J. (2001) Fracture surface-fractal or multifractal? *Mater. Characterization*, 26, 163–167.

Stanley, H.E. (1995) Power laws and universality. *Nature*, 378, 554–557.

Stanley, H.E., Buldyrev, S.V., Goldberger, A.L., Haudorff, J.M., Havlin, S., Mietus, J., Peng, C.K., Sciortino, F. and Simons, M. (1992) Fractal landscapes in biological systems: Long-range correlations in DNA and interbeat heart intervals. *Physica A*, 191, 1–12.

Stanley, H.E., Buldyrev, S.V., Goldberger, A.L., Havlin, S., C.K. Peng and Simons, M. (1999) Scaling features of noncoding DNA. *Physica A*, 273, 1–18.

Stanley, H.E. and Meakin, P. (1988) Multifractal phenomena in physics and chemistry. *Nature*, 335, 405–409.

Stanley, L.D. (1981) Neogene mass extinction of western Atlantic mollusks. *Nature*, 293, 457–459.

———. (1986) Anatomy of a regional mass extinction: Plio-Pleistocene decimation of the western Atlantic bivalve fauna. *Palaios*, 1, 17–36.

Stauffer, D. (1979) Scaling theory of percolation cluster. *Phys. Rep.*, 54, 1–74.

Steele, J.H. and Henderson, E.W. (1992) A simple model for plankton patchiness. *J. Plankton Res.*, 14, 1397–1404.

Steinberg, J. and Conant, R. (1974) An information analysis of the inter-male behaviour of the grasshopper *Chortophaga viridifasciata*. *Anim. Behav.*, 22, 617–627.

Steneck, R.S. and Watling, L. (1982) Feeding capabilities and limitation of herbivorous molluscs: A functional group approach. *Mar. Biol.*, 68, 299–319.

Stephens, D.W. and Krebs, J.R. (1986) *Foraging Theory*. Princeton University Press, Princeton.

Steuer, R., Ebeling, W., Russell, D.F., Bahar, S., Neiman, A. and Moss, F. (2001) Entropy and local uncertainty of data from sensory neurons. *Phys. Rev. E*, 64, 061911.

Stewart, R.W. (1969) *Turbulence*. Cambridge University Press, Cambridge.

Strickler, J.R. (1985) Feeding currents in calanoid copepods: Two new hypotheses. *Symp. Soc. Exp. Biol.*, 39, 459–485.

Strogatz, S.H. (1994) *Nonlinear Dynamics and Chaos*. Perseus, Reading, MA.

Strutton, P.G., Mitchell, J.G. and Parslow, J.S. (1996) Non-linear analysis of chlorophyll a transects as a method of quantifying spatial structure. *J. Plankton Res.*, 18, 1717–1726.

———. (1997) Using non-linear analysis to compare the spatial structure of chlorophyll with passive tracers. *J. Plankton Res.*, 19, 1553–1564.

Sugihara, G. (1994) Nonlinear forecasting for the classification of natural time series. *Philos. Trans. R. Soc. London B*, 348, 477–495.

Sugihara, G., Grenfell, B. and May, R.M. (1990) Distinguishing error from chaos in ecological time series. *Philos. Trans. R. Soc. London B*, 330, 235–251.

Sugihara, G. and May, R.M. (1990a) Applications of fractals in ecology. *TREE*, 5, 76–86.

———. (1990b) Nonlinear forecasting as a way of distinguishing chaos from measurement error in time series. *Nature*, 344, 734–741.

Sugiyama, S. (1993) World view materialized in Teotihuacan, Mexico. *J. Latin Am. Antiquity*, 4, 103–129.

Sumbera, R. Burda, H, Chitaukali, W.N. and Kubovaa, J. (2003) Silvery mole-rats (*Heliophobius argenteocinereus*, Bathyergidae) change their burrow architecture seasonally. *Naturwissenschaften*, 90, 370–373.

Sumita, I. and Olson, P. (1999) A laboratory model for convection in Earth's core driven by a thermally heterogeneous mantle. *Science*, 286, 1547–1549.

Summerhayes, C.P. and Thorpe, S.A. (1996) *Oceanography: An Illustrated Guide*. Manson, London.

Sun, S.F. (2004) *Physical Chemistry of Macromolecules: Basic Principles and Issues*. Wiley, Chichester.

Suwa, N. and Takahashi, T. (1971) *Morphological and Morphometrical Analysis of Circulation in Hypertension and Kidney*. Urban and Scwarzenberg, Munich.

Svendsen, H. (1997) Physical oceanography and marine ecosystems: Some illustrative examples. *Sci. Mar.*, 61, 93–108.

Takayasu, H. (1997) *Fractals in the Physical Sciences*. Manchester University Press, Manchester.

Takens, F. (1981). Detecting strange attractors in turbulence. *Lect. Notes Math.*, 898, 366–381.

Tamburro, A.M. and Guantieri, V. (1991) Classical and fractal description of elastin structure. In *Ecological Physical Chemistry*, Rossi, C. and Tiezzi, E., Eds. Proc. Intl. Workshop, 8–12 November 1990, Sienna, Italy. Elsevier, Amsterdam, 391–400.

Tang, S., Ma, Y. and Sebastine, I.M. (2001) The fractal nature of *Escherishia coli* biological flocs. *Coll. Surf.*, 20, 211–218.

Taniguchi, H., Nakano, S. and Mutsunori, T. (2003) Influences of habitat complexity on the diversity and abundance of epiphytic invertebrates on plants. *Freshwater Biol.*, 48, 718–728.

Tate, N.J. (1998) Estimating the fractal dimension of synthetic topographic surfaces. *Comput. Geosciences*, 25, 325–332.

Tatsumi, J., Yamauchi, A. and Kono, Y. (1989) Fractal analysis of plant root systems. *Ann. Bot.*, 64, 499–503.

Taylor, G.I. (1938) The spectrum of turbulence. *Proc. R. Soc. London Ser. A*, 164, 476–490.

Telesca, L., Lapenna, V. and Vallianatos, F. (2002) Monofractal and multifractal approaches in investigating scaling properties in temporal patterns of the 1983–2000 seismicity in the western Corinth graben, Greece. *Phys. Earth Planetary Interiors*, 131, 63–79.

Tennekes, H. and Lumley, J.L. (1972) *A First Course in Turbulence*. MIT Press, Cambridge.

Theiler, J., Eubank, S., Longtin, A., Galdrikian, B. and Farmer, J.D. (1992) Testing for nonlinearity in time series: The method of surrogate data. *Physica D*, 58, 77–94.

Thompson, A.W. (1961) *On Growth and Form*. Cambridge University Press, Cambridge.

Tiño, P. (2002) Multifractal properties of Hao's geometric representations of DNA sequences. *Physica A*, 304, 480–494.

Tiselius, P. (1992) Behavior of *Acartia tonsa* in patchy food environments. *Limnol. Oceanogr.*, 37, 1640–1651.

Tiselius, P., Jonsson, P.R., Karrtveld, S., Olsen, E.M. and Jorstad, T. (1997) Effects of copepod foraging behavior on predation risk: An experimental study of the predatory copepod *Pareuchaeta norvegica* feeding on *Acartia clausi* and *A. tonsa* (Copepoda). *Limnol. Oceanogr.*, 42, 164–170.

Toledo, P.G., Novy, R.A., Davis, H.T. and Scriven, L.E. (1990) Hydraulic conductivity of porous media at low water content. *Soil Sci. Soc. Am. J.*, 54, 673–679.

Tong, H. and Lim, K.S. (1980) Threshold autogression, limit cycles and cyclical data- with discussion. *J. R. Stat. Soc. B*, 42, 245–292.

Torres, F.E., Russel, W.B. and Schowalter, W.R. (1990) Floc structure and growth kinetics for rapid shear coagulation of polystyrene colloids. *J. Coll. Interf. Sci.*, 142, 554–574.

Tricot, C. (1995) *Curves and Fractal Dimension*. Springer-Verlag, New York.

Troll, G. and Graben, P.B. (1998) Zipf's law is not a consequence of the central limit theorem. *Phys. Rev. E*, 57, 1347–1355.

Truffier, S., Hitier, B., Olivesi, R., Delesmont, R., Morel, M. and Loquet N (1997) *Suivi régional des nutriments sur le littoral nord/Pas-de-Calais/Picardie*. Bilan de l'année 1996. IFREMER, Boulogne-sur-Mer.

Tsonis, A.A., Elsner, J.B. and Georgakakos, K.P. (1993) Estimating the dimension of weather and climate attractors: important issues about the procedure and interpretation. *J. Atmos. Sci.*, 50, 2549–2555.

Tsuda, A. (1995) Fractal distribution of the oceanic copepod *Neocalanus cristatus* in the subarctic Pacific. *J. Oceanogr.*, 51, 261–266.

Turchin, P. (1991) Translating foraging movements in heterogeneous enviroments into the spatial distribution of foragers. *Ecology*, 72, 1253–1266.

Turchin, P. (1996) Fractal analysis of animal movements: A critique. *Ecology*, 77, 2086–2090.

———. (1998) *Quantitative Analysis of Movement*. Sinauer, Sunderland, MA.

Turchin, P., Odendaal, F.J. and Rausher, M.D. (1991) Quantifying insect movement in the field. *Environ. Entomol.*, 20, 955–963.

Turcott, R.G. and Teich, M.C. (1995) Interevent-interval and counting statistics of the human heartbeat for heart-failure and normal patients. *Ann. Biomed. Eng.*, 24, 269–293.

Turcotte, D.L. (1986) Fractals and fragmentation. *J. Geophys. Res.*, 91, 1921–1926.

———. (1989) Fractals in geology and geophysics. *Pure Appl. Geophys.*, 131, 171–196.

———. (1991) Fractals in geology: What they are and what are they good for? *GSA Today*, 1, 1–4.

———. (1992) *Fractals and Chaos in Geology and Geophysics*. Cambridge University Press, Cambridge.

Turiel, A. and Pérez-Vicente, C.J. (2003) Multifractal geometry in stock market time series. *Physica a*, 322, 629–649.

Turner, M.G. and Gardner, R.H. (1991) Quantitative methods in landscape ecology: An introduction. In *Quantitative Methods in Landscape Ecology*, Turner, M.G. and Gardner, R.H., Eds. Springer-Verlag, New York, 3–14.

Tyler, S.W. and Wheatcraft, S.W. (1989) Application of fractal mathematics to soil water retention estimation. *Soil Sci. Soc. Am. J.*, 53, 987–996.

———. (1990) Fractal processes in soil water retention. *Water Resour. Res.*, 26, 1047–1054.

———. (1992) Fractal scaling of soil particle-size distributions: Analysis and limitations. *Soil Sci. Soc. Am. J.*, 56, 362–369.

Uttieri, M., Nihongi, A., Mazzocchi, M.G., Strickler, J.R. and Zambianchi, E. (2007) Pre-copulatory swimming behaviour of *Leptodiaptomus ashlandi* (copepoda: calanoida): A fractal approach. *J. Plankton Res.*, 29, 17–26.

Uttieri, M., Zambianchi, E., Strickler, J.R. and Mazzocchi, M.G. (2005) Fractal characterization of three-dimensional zooplankton swimming trajectories. *Ecol. Model.*, 185, 51–63.

Valentine, J.W. (1989) How good was the fossil record? Clues from the California Pleistocene. *Paleobiology*, 15, 83–94.

van Beek, J., Roger, S. and Bassingthwaigthe, J. (1989) Regional myocardial flow heterogeneity explained with fractal networks. *Am. J. Physiol.*, 257, 1670–80.

van Hes, H.M. (1993) The spatial nature of soil variability and its implications for field studies. In *Patch Dynamics*, Levin, S.A., Powell, T.M. and Steele, J.H., Eds. Springer-Verlag, Berlin, 27–36.

Van Hees, W.W.S. (1994) A fractal model of vegetation complexity in Alaska. *Landscape Ecol.*, 9, 271–278.

Van Noordwijk, M. and Purnomosidhi, P. (1995) Root architecture in relation to tree-crop-soil interactions and shoot pruning in agroforestry. *Agrofor. Syst.*, 30, 161–173.

Van Noordwijk, M., Spek, L.Y and, De Willingen, P. (1994) Proximal root diameters as predictors of total root system size for fractal branching models. I: Theory. *Plant Soil*, 164, 107–118.

Vassilicos, J.C., Demos, A. and Tata, F. (1993) No evidence of chaos but some evidence of multifractals in the foreign exchange and stock markets. In *Applications of Fractals and Chaos. The Shape of Things*, Crilly, A.J., Eranshaw, R.A. and Jones, H., Eds. Springer-Verlag, Berlin, 249–265.

Vaulot, D. (1989) CytoPC: Processing software for flow cytometric data. *Signal Noise*, 2, 8.

Vétier, N., Banon, S., Chardot, V. and Hardy, J. (2003) Effect of temperature and aggregation rate on the fractal dimension of renneted casein aggregates. *J. Dairy Sci.*, 86, 2504–2507.

Vétier, N., Desobry-Banon, S., Ould Eleya, M.M. and Hardy, J. (1997) Effect of temperature and acidification rate on the fractal dimension of acidified casein aggregates. *J. Dairy Sci.*, 80, 3161–3166.

Vicsek, T. (1993) *Fractal Growth Phenomena*. Word Scientific, Singapore.

Vidale, J.E. and Earle, P.S. (2000) Fine-scale heterogeneity in the Earth's inner core. *Nature*, 404, 273–275.

Vidondo, B., Prairie, Y.T., Blanco, J.M. and Duarte, C.M. (1997) Some aspects of the analysis of size spectra in aquatic ecology. *Limnol. Oceanogr.*, 42, 184–192.

Vilar, J.M.G., Solé, R.V. and Rubí, J.M. (2003) On the origin of plankton patchiness. *Physica A*, 317, 239–246.

Visser, A.W. and MacKenzie, B.R. (1998) Turbulence-induced contact rates of plankton: The question of scale. *Mar. Ecol. Prog. Ser.*, 166, 307–310.

Viswanathan, G.M., Afanasyev, V., Buldyrev, S.V., Havlin, S., da Luz, M.G.E., Raposo, E.P. and Stanley, H.E. (2001) Lévy flights search patterns of biological organisms. *Physica a*, 295, 85–88.

Viswanathan, G.M., Afanasyev, V., Buldyrev, S.V., Murphy, E.J., Prince, P.A. and Stanley, H.E. (1996) Lévy flight search patterns of wandering albatrosses. *Nature*, 381, 413–415.

Viswanathan, G.M., Bartumeus, F., Buldyrev, S.V., Catalan, J., Fulco, U.L., Havlin, S., da Luz, M.G.E., Lyra, M.L., Raposo, E.P. and Stanley, H.E. (2002) Lévy flight random searches in biological phenomena. *Physica A*, 314, 208–213.

Voss, R. (1985) Random fractal forgeries. *NATO ASI Ser. F17*, 805–835.

Voss, R.F. (1988) Fractals in nature: From characterization to simulation. In *The Science of Fractal Images*, Peiten, H.O. and Saupe, D., Eds. Springer, New York, 21–70.

Voss, R.F. and Clark, J. (1975) 1/*f* noise in music and speech. *Nature*, 258, 317–318.

———. (1978) 1/*f* noise in music: Music from 1/*f* noise. *J. Acoust. Soc. Am.*, 63, 258–263.

Watanabe, M.S. (1996) Zipf's law in percolation. *Phys. Rev. E*, 53:4187–4190.

Waters, R.L. and Mitchell, J.G. (2002) Centimeter-scale spatial structure of estuarine *in vivo* fluorescence profiles. *Mar. Ecol. Prog. Ser.*, 237, 51–63.

Waters, R.L., Seymour, J.R. and Mitchell, J.G. (2003) Geostatistical characterization of centimetre-scale spatial structure of in vivo fluorescence. *Mar. Ecol. Prog. Ser.*, 251, 49–58.

Weber, L.H., El-Shayed, S.Z. and Hampton, I. (1986) The variance spectra of phytoplankton, krill and water temperature in the Antarctic Ocean south of Africa. *Deep-Sea Res.*, 33, 1327–1343.

Webster, R. and Butler, B. (1976) Soil classification and survey studies at Ginninderra. *Aust. J. Soil Res.*, 14, 1–24.

Wei, W.S. (2005) *Time Series Analysis: Univariate and Multivariate Methods*. Addison Wesley, Reading, MA.

Weitz, D.A. and Oliveria, M. (1984) Fractal structures formed by kinetic aggregation of aqueous gold colloids. *Phys. Rev. Lett.*, 52, 1433–1436.

Wells, J.M., Donnelly, D.P. and Boddy, L. (1997). Patch formation and developmental polarity in mycelial cord systems of *Phanerochaete velutina* on a nutrient-depleted soil. *New Phytol.*, 136, 653–665.

West, B.J. and Goldberger, A.L. (1987) Physiology in fractal dimensions. *Am. Sci.*, 75, 354–365.

West, B.J. and Shlesinger, M.F. (1989) On the ubiquity of 1/*f* noise. *Intl. J. Mod. Phys.*, B3, 795.

———. (1990) The noise in natural phenomena. *Am. Sci.*, 78, 40–45.

West, G.B., Brown, J.H. and Enquist, B.J. (1997) A general model for the origin of allometric scaling laws in biology. *Science*, 276, 122–126.

———. (1999) A general model for the structure and allometry of plant vascular systems. *Nature*, 400, 664–667.

Wheeler, A. (1969) *The fishes of the British Isles and Northwest Europe*. Macmillan, London.

Wiebe, P.H., Mountain, D.G., Stanton, T.K., Greene, C.H., Lough, G., Kaartvedt, S., Dawson, J. and Copley, N. (1996) Acoustical study of the spatial distribution of plankton on Georges Bank and the relationship between volume backscattering strength and the taxonomic composition of the plankton. *Deep-Sea Res II*, 43, 1971–2001.

Wiens, J.A. (1989) Spatial scaling in ecology. *Funct. Ecol.*, 3, 385–397.

Wiens, J.A. (1999) Ecological heterogeneity: An ontogeny of concepts and approaches. In *The Ecological Consequences of Environmental Heterogeneity*, Hutchings, M.J., John, E.A. and Stewart, A.J.A., Eds. Cambridge University Press, Cambridge, 9–31.

———. (2001) *The Landscape Context of Dispersal*. Oxford University Press, Oxford.

Wiens, J.A., Crist, T.O. and Milne, B.T. (1993) On quantifying insect movements. *Environ. Entomol.*, 22, 709–715.

Wiens, J.A., Crist, T.O., With, K.A. and Milne, B.T. (1995) Fractal patterns and insect movement in microlandsape mosaics. *Ecology*, 76, 663–666.

Wiens, J.A. and Milne, B.T. (1989) Scaling of landscapes in landscape ecology or landscape ecology in a beetle's perspective. *Landscape Ecol.*, 3, 87–96.

Wilcox, D.C. (1998) *Turbulence Modeling for CFD*. DCW Industries.

Williams, G.P. (1997) *Chaos Theory Tamed*. Joseph Henry Press, Washington.

Williamson, M.H. and Lawton, J.H. (1991) Measuring habitat structure with fractal geometry. In *Habitat Structure: the Physical Arrangement of Objects in Space*, Bell, S., McCoy, E.D. and Mushinsky, H.R., Eds. Chapman and Hall, London, 69–86.

Wilson, E.O. (1962) Chemical communication among workers of the fire ant *Solenopsis saevissima* (Fr. Smith). 2: An information analysis of the odour trail. *Anim. Behav.*, 10, 148–158.

Wilson, R.S. and Greaves, J.O.B. (1979) The evolution of the bugs system: Recent progress in the analysis of bio-behavioral data. In: Jacoff, F.S. (ed.), *Advances in Marine Environmental Research*. Proc. Symp. US EPA (EPA-600/9-79-035), 252–272.

With, K.A. (1994) Using fractal analysis to assess how species perceive landscape structure. *Landscape Ecol.*, 9, 25–36.

Witten, T.A. and Cates, M.E. (1986) Tenuous structures from disorderly growth processes. *Science*, 232, 1607–1612.

Wolf, A., Swift, J.B., Swinney, H.L. and Vastano, J.A. (1985) Determining Lyapunov exponents from a time series. *Physica*, 16D, 285–317.

Wu, K.K.S., Lahav, O. and Rees, M.J. (1999) The large-scale smoothness of the universe. *Nature*, 397, 225–230.

Wu, Q., Borkovec, M. and Sticher, H. (1993) On particle-size distributions in soils. *Soil Sci. Soc. Am. J.*, 57, 883–890.

Yaglom, A.M. (1966) The influence of fluctuations in energy dissipation on the shape of turbulent characteristics in the inertial interval. *Sov. Phys. Dokl.*, 11, 26–29.

Yamamoto, M., Nakahamam H. and Shima, K. (1986) Neuronal activities during paradoxical sleep. *Adv. Neurolog. Sci.*, 30, 1010–1022.

Yamazaki, H. (1990) Breakage models: Lognormality and intermittency. *J. Fluid Mech.*, 79, 159–165.

Yamazaki, H., Mackas, D. and Denman, K. (2002) Coupling small scale physical processes with biology. In *The Sea: Biological-Physical interaction in the Ocean*, Robinson, A.R., McCarthy, J.J. and Rothschild, B.J., Eds., Vol. 12. John Wiley and Sons, New York, 51–112.

Yamazaki, H., Mitchell, J.G., Seuront, L., Wolk, F. and Li, H. (2006) Phytoplankton microstructure in fully developed oceanic turbulence. *Geophys. Res. Lett.*, 33, L01603, doi:10.1029/2005GL024103.

Yamazaki, H. and Squires, K. (1996) Comparison of oceanic turbulence and copepod swimming. *Mar. Ecol. Prog. Ser.*, 144, 299–301.

Yonekawa, S., Sakai, N. and Kitani, O. (1996) Identification of idealized leaf types using simple dimensionless shape factors by image analysis. *Trans. ASAE*, 39, 1525–1533.

Young, I.M. and Crawford, J.W. (1991) The fractal structure of soil aggregates: Its measurement and interpretation. *J. Soil Sci.*, 42, 187–192.

Zamir, M. (2001) Fractal dimensions and multifractality in vascular branching. *J. Theor. Biol.*, 212, 183–190.

Zar, J.H. (1996) *Biostatistical analysis*. Prentice-Hall, Englewood Cliffs, NJ.

Żebrowski, J.J. and Baranowski, R. (2004) Type I intermittency in nonstationary systems-models and human heart rate variability. *Physica a*, 336, 74–83.

Zbilut, J.P. (1988) Dimensional analysis rate variability in heart transplant recipients. *Math. Biosci.*, 90, 49–70.

Zhu, X., Cai, Y. and Yang, X. (2004) On fractal dimensions of China's coastlines. *Math. Geol.*, 36, 447–461.

Zhu, X., Yang, X., Xie, W. and Wang, J. (2000) On spatial fractal character of coastline: A case study of Jiangsu Province, China. *China Ocean Eng.*, 14, 533–540.

Zeide, B. (1998) Fractal analysis of foliage distribution in loblolly pine crowns. *Can. J. For. Res.*, 28, 106–114.

Zeide, B. and Gresham, C.A. (1991) Fractal dimensions of tree crowns in three loblolly pine plantations of coastal South Carolina. *Can. J. For. Res.*, 21, 1208–1212.

Zipf, G.K. (1949) *Human Behavior and the Principle of Least Effort*. Hafner, New York.

Index

Printed and bound by CPI Group (UK) Ltd, Croydon, CR0 4YY

21/10/2024

01777046-0007